Ronny Ramlau, Otmar Scherzer (Eds.)
The Radon Transform

Radon Series on Computational and Applied Mathematics

Volume 22

The Radon Transform

The First 100 Years and Beyond

Edited by
Ronny Ramlau and Otmar Scherzer

DE GRUYTER

Editors
Prof. Dr. Ronny Ramlau
University of Linz
Industrial Mathematics Institute
Altenbergerstr. 69
4040 Linz
Austria
ronny.ramlau@ricam.oeaw.ac.at

Prof. Dr. Otmar Scherzer
University of Vienna
Computational Science Center
Oskar-Morgenstern-Platz 1
1090 Vienna
Austria
otmar.scherzer@univie.ac.at

ISBN 978-3-11-055941-5
e-ISBN (PDF) 978-3-11-056085-5
e-ISBN (EPUB) 978-3-11-055951-4
ISSN 1865-3707

Library of Congress Control Number: 2019935656

Bibliographic information published by the Deutsche Nationalbibliothek
The Deutsche Nationalbibliothek lists this publication in the Deutsche Nationalbibliografie; detailed bibliographic data are available on the Internet at http://dnb.dnb.de.

Typesetting: VTeX UAB, Lithuania
Printing and binding: CPI books GmbH, Leck

www.degruyter.com

Contents

Ronny Ramlau and Otmar Scherzer

100 years of Mathematical Tomography

Abstract: This volume is honoring the 100th anniversary of the publication of the famous paper of Johann Radon: “Über die Bestimmung von Funktionen durch ihre Integralwerte längs gewisser Mannigfaltigkeiten,” which appeared in *Berichte über die Verhandlungen der Königlich-Sächsischen Gesellschaft der Wissenschaften zu Leipzig, Mathematisch-Physische Klasse 69* (1917), pp. 262–277 [2].

Exactly 100 years later, in 2017, the Johann Radon Institute for Computational and Applied Mathematics (RICAM) jointly with the Johannes Kepler University Linz (JKU) held a conference in honor of the publication of Johann Radon's paper. The conference took place from March 27th to 31st in Linz, Austria https://www.ricam.oeaw.ac.at/events/conferences/radon100/ and was attended by about 170 participants who reported on the status of the still growing field of mathematical tomography.

About this volume

In 1917, Johann Radon published his fundamental paper [2], wherein he introduced what is nowadays called the Radon transform. Today, this paper is considered to be the foundation of the area of *Mathematical Tomography*, which is a booming area in applied sciences.

For a long time, this paper did not get the credit which it deserves.

Johann Radon himself appears to have not have commented on these results, beyond their initial publication. In an obituary of Radon [1], written by his colleague Paul Funk (himself a distinguished mathematician at the University of Vienna), many mathematical achievements of Johann Radon were outlined, but Paul Funk left out the Radon transform.

Only much later, driven by applications, was the paper of Johann Radon rediscovered, and today it gets the broad credit it deserves.

This volume has been collected exactly 100 years after the publication of Radon's fundamental paper [2]. It collects surveys and original papers on Mathematical Tomography and shows the remarkable advancement of the field. These new achievements are also driven by fundamental developments in imaging and industry, where computational methods based on mathematical inversion techniques allow us to vi-

Ronny Ramlau, Institute for Industrial Mathematics, Johannes Kepler University, Altenbergerstraße 69, A-4040 Linz, Austria; and Johann Radon Institute for Computational and Applied Mathematics (RICAM), Altenbergerstraße 69, A-4040 Linz, Austria, e-mail: ronny.ramlau@ricam.oeaw.ac.at
Otmar Scherzer, Johann Radon Institute for Computational and Applied Mathematics (RICAM), Altenbergerstraße 69, A-4040 Linz, Austria; and Faculty of Mathematics, University of Vienna, Oskar-Morgenstern Platz 1, A-1090 Vienna, Austria, e-mail: otmar.scherzer@univie.ac.at

https://doi.org/10.1515/9783110560855-201

sualize underlying information which is not directly accessable [3]. The interplay between mathematics and applications was not always as close, demonstrated by the well-known fact that Allan M. Cormack and Godfrey Hounsfield, who received the Nobel-prize in Physiology or Medicine for the development of the first medical CT-scanner, developed their algorithms for image reconstruction not only independently from each other but also without knowledge of Radon's work.

About this book

This book provides an overview on mathematical tomography and the Radon transform, with thirteen papers by well-known experts in the field. All articles were peer reviewed.

Karl Sigmund gives an intriguing insight in the life and work of Johann Radon. He outlines his groundbreaking mathematical achievements and his life in the most turbulent time of the last century.

Michel Defrise and Christine De Mol present applications for PET imaging and blind deconvolution. This paper is devoted to Mario Bertero, a pioneer in Inverse Problems.

Simon Gindikin studies deep results for the Radon transform in a very general setting in a historic context from Leray to Radon.

Joonas Ilmavirta and François Monard consider the problem of inversion of integral transforms on manifolds.

Roman Novikov presents recent and new results on the non-Abelian Radon transform and its inversion.

Victor Palamodov discusses the developments of the Funk–Radon theory over the last centuries.

Gaik Ambartsoumian shows recent results on V-line and Conical Radon transforms and presents applications in imaging.

Alfred Louis discusses and explains ghost artefacts in Radon problems.

Thomas Schuster discusses applications of the Radon transform in vector field tomography. His paper is devoted to Armin Lechleiter, a leading researcher in Inverse Problems, who passed away much too young, at the age of 36.

Gabor Herman applies different algorithms for x-ray inversion and compares the results for medical applications.

Tanja Tarvainen reports about active research fields in photoacoustical imaging and Bayesian inversion.

Shari Moskow and John C. Schotland report on the inverse Born series for solving imaging problems.

Andreas Alpers and Peter Gritzmann represent the field of discrete tomography in this volume.

The editors would like to dedicate the volume to the memory of Joyce McLaughlin, †2018 who was a life long supporter of RICAM, and an expert in Inverse Problems and, in particular, elastography.

Bibliography

[1] P. Funk. Nachruf auf Prof. Johann Radon. In: *Mathematische Nachrichten*, volume 62, pages 191–199, 1958. issn: 0025-584X.

[2] J. Radon. Über die Bestimmung von Funktionen durch ihre Integralwerte längs gewisser Mannigfaltigkeiten. In: *Berichte über die Verhandlungen der KöniglichSächsischen Gesellschaft der Wissenschaften zu Leipzig, Mathematisch-Physische Klasse*, volume 69, pages 262–277, 1917.

[3] R. West. In industry, seeing is believing. In: *Physics world*, pages 27–30, 2003.

Karl Sigmund

1 Johann Radon 1887–1956

Abstract: This paper provides a short biography of the famous Austrian mathematician, Johann Radon, and it outlines the fundamental mathematical achievements.

Keywords: Radon transform, Radon measure, Radon numbers, Radon–Helly theorem, Radon–Riesz theorem

MSC 2010: 01A-60, 01A-70

His name is a household word in mathematics: next to the Radon transform, students learn about the theorems of Lebesgue–Radon–Nikodym, of Helly–Radon, and of Radon–Riesz; they are familiar with the concepts of Radon measure and Radon integral, and may have heard of the Radon curve and the Radon sequence. Johann Radon has certainly left an enduring mark and significantly contributed to the mathematics of the first half of the twentieth century.

By all accounts, Radon was a modest, quiet, and unassuming person, who did not make much ado about his mathematical fame. It is quite probable that the tremendous growth of the field of inverse problems would have surprised him. His colleague and near-contemporary, Paul Funk (1886–1969), a former student of Hilbert and a first-rate mathematician on his own right, did not even mention the Radon transform when he wrote Radon's obituary for the Austrian Academy of Science.

Johann Radon was born on December 16, 1987, in Tetschen on the Elbe River. The small provincial town in Bohemia lay close to Saxony. Today, it is Decin, a border town of the Czech republic. His father, a Sudeten German, was head accountant at a local bank. The mother came from Thuringia. The parents sent their only child to the gymnasium of Leitmeritz (today Litomerice). Actually, they also moved there, after Johann's father had retired. The school-boy's health was frail, but he showed great promise. He did well in mathematics and science, as well as in Latin and Greek. Moreover, he played several instruments, and had a beautiful singing voice. The intention of the gifted young man, at that time, was to become a philosopher (Figure 1.1).

After finishing secondary school with brilliant marks in 1905, Johann Radon enrolled in mathematics and physics at the University of Vienna, and also took up music

Acknowledgement: The author thanks Dr. Sienell from the Austrian Academy of Science and Dr. Maisel from the Archive of the University of Vienna. Special thanks go to Mag. Brigitte Bukovics for her advice and her kind permission to use the photographs.

Karl Sigmund, Faculty of Mathematics, University of Vienna, Oskar-Morgenstern-Platz 1, A-1090 Vienna, Austria, e-mail: Karl.Sigmund@univie.ac.at

https://doi.org/10.1515/9783110560855-001

Figure 1.1: Johann Radon as a schoolboy.

classes. Again, his parents followed him. Among Radon's teachers were Franz Mertens (1840–1927), an eminent number theorist; the brilliant Wilhelm Wirtinger (1865–1945), who contributed to many fields of analysis; and the associate professor of geometry, Gustav Kohn (1859–1921). Among the lecturers, or docents, were Josef Plemelj (1873–1967), who made essential contributions to the Riemann–Hilbert problem and introduced Radon to potential theory, and Alfred Tauber (1866–1942). By then, the field of what soon became known as Tauberian theorems was taking roots. Among the younger post-docs, Hans Hahn (1879–1934) and Heinrich Tietze (1880–1964) stood out. Their future paths would often cross with Radon's. At the time, Hahn mentored Radon in the calculus of variations, and introduced him to the foundations of arithmetic and geometry. There were some brilliant other students, too, senior to Radon: for instance, Roland Weitzenböck (1885–1955), Wilhelm Blaschke (1885–1962), and Eduard Helly (1884–1943).

Obviously, mathematics was in a good shape in Vienna at that time. The main responsible for this development was Gustav von Escherich (1849–1935). Escherich's principal contributions were in the calculus of variations. His scientific work alone would not have guaranteed him an outstanding place. But he was instrumental in putting Vienna on the map, in what concerned mathematics, after centuries of near-insignificance. Escherich had studied under Weierstrass in Berlin. He introduced first-rate standards of precision and clarity into the curriculum at the University of Vienna and managed to link up with international contemporary research.

Among Escherich's many brilliant students, Johann Radon stood out. He obtained his PhD on February 18, 1910, with a thesis on the calculus of variations. In that same year, he also obtained the teacher's diploma. But it was already becoming clear that he was cut out for an academic career. The next step was almost preordained. Like so many other mathematical "rookies" in his generation, Radon traveled with a stipend to study at Hilbert's institute in Göttingen, at that time undoubtedly the mecca of mathematics.

During these years, analysis experienced a profound metamorphosis. The tools developed by Henri Lebesgue's theory of integration and the problems raised by Fredholm's theory of functional operators came together in various, often unexpected ways. This had lasting influence on Radon's development.

After his post-doc semester in Göttingen, Johann Radon was appointed at the University of Brünn (today Brno), as an assistant to Emmanuel Czuber, an insurance mathematician and statistician well known for his work on probability theory. Soon afterward, in 1912, Professor Czuber was appointed at the Technische Hochschule (today: Technical University) in Vienna. Radon followed him there as his assistant, and thus returned to the buzzing capital of the Austro-Hungarian empire.

In 1913, Radon published in the *Sitzungsberichte der Akademie der Wissenschaften zu Wien* a truly seminal work of 144 pages, entitled "Theorie und Anwendungen der absolut additive Mengenfunktionen." It was a masterpiece, combining the measure theoretical ideas of the French school centering around Lebesgue and Borel with the functional analytic work pioneered by Hilbert's famous *Mitteilungen* from 1906 and the work of Fredholm and Hellinger. In Radon's own words, the aim of his paper was "to encompass in a general theory, on the one hand the theory of linear integral equations and on the other hand the theory of linear and bi-linear forms of infinitely many variables." This development of a general measure theory as the basis for functional analysis proved enormously successful. It must have been in the air: A closely parallel undertaking was due to the Hungarian Friedrich (or Frigyes) Riesz, in particular, with the latter's contribution "Über Systeme integrierbarer Funktionen."

Radon's seminal paper was strikingly well rounded and complete. It contains the decomposition theorems for signed measures, and relates measures (now called Radon-measures) with linear forms on spaces of continuous functions. It extends to the n-dimensional case a fundamental theorem which Lebesgue had derived for absolutely continuous measures on the real line. (Otton Nikodym (1887–1974) would generalize it further, and define what is now known as the Radon–Nikodym property for locally convex spaces). In addition, Radon's opus magnum also contains a highly polished duality theory of L^p-spaces, and a theorem on weak versus strong convergence which today is known as Radon–Riesz theorem. (It yields the basic motivation for the study of Radon–Riesz spaces). A final step was taken in 1915 by Maurice Fréchet, who developed measure theory on abstract spaces, rather than on topological spaces.

In 1914, Radon applied at the University of Vienna for the degree of lecturer, and submitted his 1913 paper as a habilitation thesis. The commission was duly impressed. Escherich called it "one of the most important treatises that appeared in the last years," Wirtinger viewed it as "a most important achievement" and on Furtwängler (who had succeeded Mertens after the latter's retirement), it made "a very good impression." After the probationary lecture in June on "The theory of quadratic forms

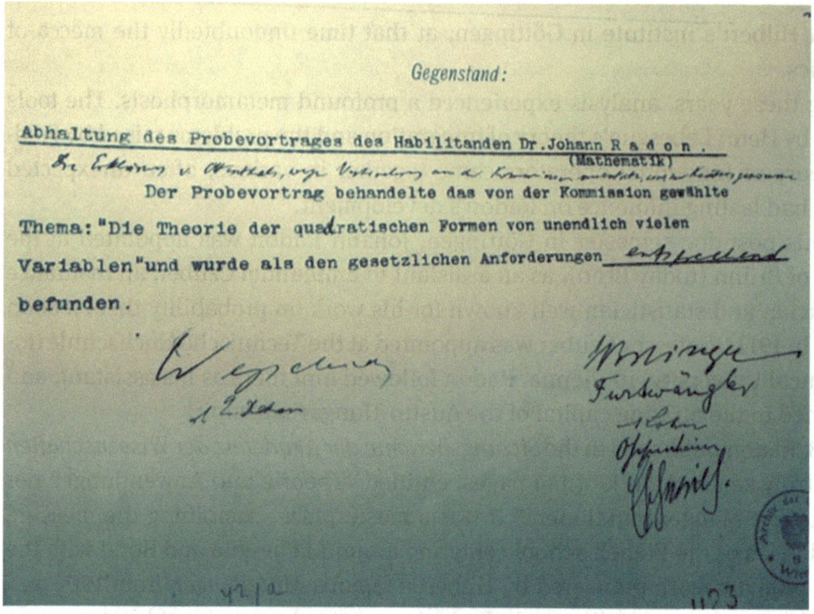
Gegenstand:

Abhaltung des Probevortrages des Habilitanden Dr. Johann R a d o n . (Mathematik)

Der Probevortrag behandelte das von der Kommission gewählte Thema: "Die Theorie der quadratischen Formen von unendlich vielen Variablen" und wurde als den gesetzlichen Anforderungen entsprechend befunden.

Figure 1.2: Radon's probationary lectures satisfies the requirements for his habilitation: he can give lectures at the university.

with infinitely many variables," Radon obtained the title of Privatdozent on August 26, 1914 (Figure 1.2).

By then, the First World War was already in full swing. Johann Radon, however, was exempted from military service, thanks to his extreme myopia. During the war years, he stayed as assistant at the Technische Hochschule, and also lectured at the Universität für Bodenkultur (today, the University for Natural Resources and Life Sciences). In August 1916, he married Maria Rigele, named Mizzi, a cousin of his friend, Weitzenböck. During the next three years, three sons were born: in 1917, Wolfgang, who died very young; in 1918, Hermann; and in 1919, Ludwig. A daughter, Brigitte, followed in 1924 (Figure 1.3).

For Radon, these years were extremely productive from the scientific view-point, too. In particular, in 1917, he published "Über die Bestimmung von Funktionen durch ihre Integralwerte längs gewisser Mannigfaltigkeiten," little dreaming what stellar fate this note would encounter half a century later. In addition to working on boundary value problems for logarithmic potentials, Radon did important work on convex geometry and also, influenced by the research program of Wilhelm Blaschke, on affine geometry.

This last would soon have decisive consequences for Johann Radon's career. After the defeat of 1918, the town-fathers of Hamburg courageously decided to establish a university in their town. The creation of a mathematical center for this university was

Figure 1.3: Conjugal bliss: Johan Radon and his wife Maria, née Rigele.

entrusted to Wilhelm Blaschke, then in his early thirties. Blaschke combined mathematical talents with inspired leadership: within a few years, he built up a vibrant and dynamic mathematical hot spot (Figure 1.4). Among the earliest appointees were Erich Hecke, Kurt Rademacher, Kurt Reidemeister, and a string of Austrians including Johann Radon, Emil Artin, Otto Schreier, and Wolfgang Pauli. (This migration was not all one-sided: in 1923, Kurt Reidemeister moved from Hamburg to Vienna, where he became associate professor in geometry).

When Johann Radon left Vienna with his young family, heading for an associate professorship in Hamburg, he may have hoped to return very soon from the waterkant to the shores of the Danube. Indeed, in 1919, his former professor, Gustav Escherich, had retired. The competition for his succession was intense. In the end, it was Hans Hahn who was appointed, while Johann Radon shared second place with Heinrich Tietze. It was a brilliant list, made up of three former students of Escherich who, by now, all were professors in Germany. Hahn's return from Bonn to Vienna would in due time lead to the creation of the Vienna Circle.

As for Radon, he accepted in 1922 a position as full professor in Greifswald, as the successor of Felix Hausdorff, who had moved to Frankfurt. In 1925, Radon moved to Erlangen, this time as the successor of Tietze; and in 1928, Radon moved to Breslau as the successor of Kneser. In the 1920s, such frequent displacements from one university to another were by no means uncommon. Actually, in 1929, Radon was offered a chair in Leipzig, but this time he declined it.

His scientific work continued at the highest level, and was remarkable for its diversity. Thus in 1920, Radon offered a beautiful proof of a result nowadays known as the theorem of Helly–Radon: If a family of convex bodies in n-dimensional space has the property that each $n+1$ of them intersect, then the whole family has nonempty intersection. This result had already been obtained by Eduard Helly shortly before WWI. At the outbreak of the war, Helly had volunteered for the Austro-Hungarian army. Soon

Figure 1.4: Wilhelm Blaschke as a friend of the young Radon family.

after, he was wounded on the Eastern front and had to spend, not only the rest of the war but several years afterward, in Siberian camps. Since by 1920 it was still unclear whether he would ever manage to return, Radon published his own proof (with full acknowledgement, of course, of Helly's claim to priority). Radon's proof was based on a reformulation of Helly's result: Any subset of R_n with more than $n + 2$ points carries two measures with disjoint support but same centers of gravity.

Radon kept his active interest in the calculus of variations though all his life. In 1927, in "Über die Oszillationstheoreme der konjugierten Punkte beim Problem von Lagrange," he published a solution to a problem which had dogged his former teacher Escherich throughout his career, and later returned to it in a series of lectures presented in Hamburg (Figure 1.5).

Of particular interest are Radon's contributions to the composition of quadratic forms, inspired by seminal work of Hurwitz which had shed deep insights into the study of division algebras. Every natural number n can be written in the form $2^{4k+l}m$, where m is an odd number and $0 \leq l \leq 3$. The Radon number is $\rho(n)$ defined as $2^l + 8k$. The sequence of Radon numbers

$$1, 2, 1, 4, 1, 8, 1, 9, 1, 10, 1, 12, 1, 16, 1, 17, 1, 18, 1, 20\dots$$

has the remarkable property that there exist $\rho(n) - 1$ linearly independent vector fields on the sphere S^{n-1}. In fact, Frank Adams proved in 1963 that this is the precise upper bound.

On several occasions, in the 1930s, it seemed that Radon would be offered to return to the University of Vienna, due to a string of vacancies. Thus in 1934, Hahn had unexpectedly died. But the ministry abolished his chair, apparently as a gesture directed against the Vienna Circle. Next, in 1936, Wirtinger retired. But again, the job was neither offered to Johann Radon, nor to Karl Menger, nor to Emil Artin, eminent candidates all, but to Wirtinger's assistant, Karl Mayrhofer, a mathematician of much lesser caliber who later turned out to have been an illegal National Socialist all along.

Figure 1.5: Radon as professor in Breslau (today, Wroclaw).

Finally, in 1938, when Furtwängler retired, the mathematical nonentity, Anton Huber, another long-standing party member, was appointed in his place. By then, the "Anschluss" of Austria to the Third Reich had already taken place.

In 1939, the Radon family lost their second son, Hermann, who had been grievously ill for many years. Radon learned the tragic news while he himself was hospitalized and recovering from surgery.

World War II broke out, and in 1943, Radon's third son, Ludwig, was fatally wounded on the Russian front. In 1945, the Red Army fought its way into Breslau. Johann Radon, his wife and his daughter had to flee, and found refuge in far-off Innsbruck. Breslau became Wroclav, in Poland. It soon became clear that Radon could never hope to return to his home or to his position; but Vienna became an option again.

In the chaotic aftermath of the war, the two professors of mathematics of the University of Vienna, Karl Mayrhofer and Anton Huber, had to step down from office, due to their National Socialist past. A committee was instructed to look for two candidates. When it was found out that Radon lived in Innsbruck, and actually lectured at the University there, the problem seemed half-solved. But due to intentional or unintentional misinformation, rumors spread that Radon was unwilling to move to Vienna. Fortunately, Hans Thirring, who had returned in 1945 to his chair in theoretical physics, was able to dispel these doubts, and after an exchange of hectic telegrams, Johann Radon was finally appointed to a chair at the University of Vienna (Figure 1.6).

In 1946, Radon had also been offered chairs in Greifswald, Jena, and Leipzig, thanks not only to his scientific reputation, but also to the fact that he was among the few scientists in Germany whose political past was without blemish. However, these universities were all beyond the line that was soon to mark the Iron Curtain. Not that life promised to be easy in bombed-out Vienna, divided as it was into four occupation zones.

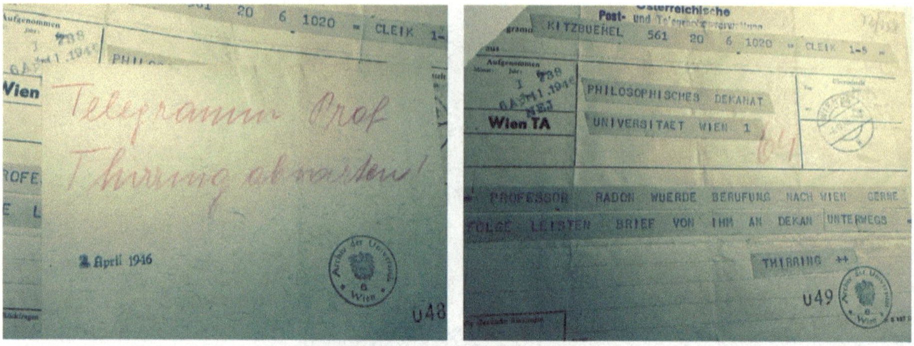

Figure 1.6: A telegram confirms that Radon is ready to move to Vienna.

On October 1, 1946, Johann Radon took up his duties at the University of Vienna. At first, he had to run the mathematical department almost single-handed. The situation improved only in 1948 with the appointment of the young Edmund Hlawka to the second chair of mathematics, and with the able assistance of Nikolaus Hofreiter and Leopold Schmetterer. Thus Radon was able to concentrate again on his favorite fields: real analysis, calculus of variations, and differential geometry. His lectures were widely praised as mathematical gems. More and more, he embodied the ideal of the world-famous scientist, a kind grandfatherly figure. He enjoyed house concerts and family life. His daughter, Brigitte, had acquired the teachers' diploma in mathematics. In 1950, she married Erich Bukovics, an up-and-coming young mathematician who would later become professor at the Technical University. The young couple had two sons.

However, it proved impossible to avoid the administrative burdens. From 1947 onward, Radon was editor-in-chief of the "Monatshefte für Mathematik," which had been founded, generations ago, by his teacher Gustav von Escherich. In 1947, he also became full member of the Austrian Academy of Science. In 1951/52, Radon held office as dean of the huge philosophical faculty. From 1950 to 1952, he headed the Austrian Mathematical Society. In 1953, he became secretary general of the Austrian Academy of Science, and in 1954, he was elected rector of the University (Figure 1.7). The various inaugural lectures gave him some opportunity to return to his early philosophical interests. But he had to write more and more obituaries.

On May 26, 1956, Johann Radon died, aged sixty-nine. He had long suffered from a weak heart and a probably congenital lung-ailment. In that same year, 1956, Alan Cormack discovered tomography, in complete ignorance of the work that Radon had published some 40 years ago in an obscure journal. In 1971, Houndsfield constructed the first CT scanner. In 1979, Houndsfield and McCormack were rewarded with the Nobel Prize. By then, the mathematical roots of medical imaging had been rediscovered, and inverse problems were well on their way of becoming a leading branch of mathematics.

Figure 1.7: Radon as rector of the University of Vienna.

Bibliography

[1] Über das Minimum des Integrals $\int F(x, y, \rho, k)ds$. *Sitzungsber. Kaiserl. Akad. Wiss. Wien, Math.-Nat. Kl.*, 1257–1326, 1910.

[2] Zur Theorie der Mayerschen Felder beim Lagrangeschen Variationsproblem. *Sitzungsber. Kaiserl. Akad. Wiss. Wien, Math.-Nat. Kl.*, 1337–1360, 1911.

[3] Über einige Fragen betreffend die Theorie der Maxima und Minima mehrfacher Integrale. *Montash. Math. Phys.*, 53–63, 1911.

[4] Lineare Scharen orthogonaler Matrizen. *Abh. Math. Sem. Univ. Hamburg*, 1–14, 1912.

[5] Theorie und Anwendungen der absolut additiven Mengenfunktionen. *Kaiserl. Akad. Wiss. Wien, Math.-Nat. Kl.*, 1295–1438, 1913.

[6] Berichtigung, Die Kettenlinie bei allgemeinster Massenverteilung. *Sitzungsber. Kaiserl. Akad. Wiss. Wien, Math.-Nat. Kl.*, 339, 1916.

[7] Die Kettenlinie bei allgemeinster Massenverteilung. *Sitzungsber. Kaiserl. Akad. Wiss. Wien, Math.-Nat. Kl.*, 221–240, 1916.

[8] Über eine besondere Art ebener konvexer Kurven. *Ber. Math.-Phys. Kl. Sächs. Akad. Wiss. Leipzig*, 123–128, 1916.

[9] Über eine Erweiterung des Begriffs der konvexen Funktionen mit einer Anwendung auf die Theorie der konvexen Körper. *Sitzungsber. Kaiserl. Akad. Wiss. Wien, Math.-Nat. Kl.*, 241–258, 1916.

[10] Über die Bestimmung von Funktionen durch ihre Integralwerte längs gewisser Mannigfaltigkeiten. *Ber. Math.-Phys. Kl. Sächs. Akad. Wiss. Leipzig*, 262–277, 1917.

[11] Über affine Geometrie XVI: Die Grundgleichungen der affinen Flächentheorie. *Ber. Math.-Phys. Kl. Sächs. Akad. Wiss. Leipzig*, 91–107, 1918.

[12] Über affine Geometrie XVII: Zur Affingeometrie der Regelflächen. *Ber. Math.-Phys. Kl. Sächs. Akad. Wiss. Leipzig*, 147–155, 1918.

[13] Über die Randwertaufgaben beim logarithmischen Potential. *Sitzungsber. Akad. Wiss. Wien, Math.-Nat. Kl.*, 1123–1167, 1919.

[14] Kurzfassung, Über die Randwertaufgaben beim logarithmischen Potential. *Anz. Akad. Wiss. Math.-Nat. Kl.*, 190, 1919.

[15] Über lineare Funktionaltransformationen und Funktionalgleichungen. *Sitzungsber. Akad. Wiss. Wien, Math.-Nat. Kl.*, 1083–1121, 1919.

[16] Kurzfassung, Über lineare Funktionaltransformationen und Funktionalgleichungen. *Anz. Akad. Wiss. Math.-Nat. Kl.*, 189, 1919.

[17] Kurzbericht, Bestimmung einer Riemannschen Metrik durch Krümmungseigenschaften. *Jber. Deutsch. Math. Ver.*, 76, 1921.
[18] Mengen konvexer Körper, die einen gemeinsamen Punkt enthalten. *Math. Ann.*, 113–115, 1921.
[19] Über statische Gravitationsfelder. *Abh. Math. Sem. Univ. Hamburg*, 268–280, 1922.
[20] Geometrie und Raumvorstellung. *Naturwissenschaften*, 510–511, 1924.
[21] Über konforme Geometrie VI: Kurvennetze auf Flächen und im Raume von Riemann. *Abh. Math. Sem. Univ. Hamburg*, 45–53, 1924.
[22] Berichtigung, Zur Behandlung geschlossener Extremalen in der Variationsrechnung. *Abh. Math. Sem. Univ. Hamburg*, 13–14, 1926.
[23] Hermann Rothe. *Jber. Deutsche Math. Ver.*, 172–175, 1926.
[24] Mechanische Herleitung des Parallelismus von T. Levi-Civita. In F. Klein and W. Blaschke, editors, *Vorlesungen über höhere Geometrie*, pages 331–346. Springer, Berlin, 1926.
[25] Zur Behandlung geschlossener Extremalen in der Variationsrechnung. *Abh. Math. Sem. Univ. Hamburg*, 195–205, 1926.
[26] Über die Differentialgleichungen von Monge. Ihre Beziehungen zur Theorie der partiellen Differentialgleichungen erster Ordnung und zur Variationsrechung. In F. Klein and W. Blaschke, editors, *Vorlesungen über höhere Geometrie*, pages 367–379. Springer, Berlin, 1926.
[27] Über konforme Geometrie V: Neue Kennzeichnung der zyklischen Kurvennetze. *Abh. Math. Sem. Univ. Hamburg*, 313–320, 1926.
[28] Mathematik und Wirklichkeit. *Sitzungsber. Phys.-Med. Soz. Erlangen.*, 181–190, 1926/27.
[29] Über die Oszillationstheoreme der konjugierten Punkte beim Problem von Lagrange. *Sitzungsber. Bayer. Akad. Wiss. Math.-Nat. Abt.*, 243–257, 1927.
[30] Bestimmung einer Riemannschen Metrik durch Krümmungseigenschaften. *Monatsh. Math. Phys.*, 9–24, 1928.
[31] Zum Problem von Lagrange. *Abh. Math. Sem. Univ. Hamburg*, 273–299, 1928.
[32] Lösung einer Aufgabe, gestellt im Jber. Deutsch. Math.-Verein 43, 20-20 (1933). *Jber. Deutsch. Math. Ver.*, 20–22, 1933.
[33] Restausdrücke bei Interpolations- und Quadraturformeln durch bestimmte Integrale. *Monatsh. Math. Phys.*, 389–396, 1935.
[34] Annäherung konvexer Körper durch analytisch begrenzte. *Monatsh. Math. Phys.*, 340–344, 1936.
[35] Bewegungsinvariante Variationsprobleme betreffend Kurvenscharen. *Abh. Math. Sem. Univ. Hamburg*, 70–82, 1937.
[36] Singuläre Variationsprobleme. *Jber. Deutsch. Math. Ver.*, 220–232, 1937.
[37] Ein Satz der Matrizenrechnung und seine Bedeutung für die Analysis. *Monatsh. Math. Phys.*, 198–204, 1939.
[38] Über Tschebyscheff-Netze auf Drehflächen und eine Aufgabe aus der Variationsrechnung. *Mitt. Math. Ges.*, 147–151, 1940.
[39] Ein einfacher Beweis für die Halbstetigkeit der Integrale der Variationsrechnung. *Math. Ann.*, 205–209, 1944.
[40] Wilhelm Wirtinger. Nachruf. *Alm. Akad. Wiss. Wien*, 336–346, 1947.
[41] Zur mechanischen Kubatur. *Monatsh. Math.*, 286–300, 1948.
[42] Godfrey Harold Hardy. Nachruf. *Alm. Öster. Akad. Wiss.*, 313–316, 1949.
[43] Über geschlossene Extremalen und eine einfache Herleitung der isoperimetrischen Ungleichung. *Ann. Mat. Pura Appl.*, 315–320, 1949.
[44] Bemerkungen zu, Über geschlossene Extremalen und eine einfache Herleitung der isoperimetrischen Ungleichung. *Ann. Mat. Pura Appl.*, 309, 1950.
[45] Constantin Caratheodory. *Nachruf. Alm. Öster. Akad. Wiss.*, 323–328, 1950.
[46] Nachruf für R. Suppantschitsch. *Nachr. Österr. Math. Ges.*, 7–8, 1950.

[47] Zur Polynomentwicklung analytischer Funktionen. *Math. Nachr.*, 156–157, 1950/51.
[48] Emanuel Czuber zum Gedächtnis. *Nachr. Österr. Math. Ges.*, 11–13, 1951.
[49] Gleichgewicht und Stabilität gespannter Netze. *Arch. Math.*, 309–316, 1954.
[50] Heinrich Tietze. *Nachr. Österr. Math. Ges.*, 74–78, 1954.
[51] Mathematik und Naturerkenntnis. Inaugurationsvortrag Universität Wien, 1954.
[52] P. M. Gruber, E. Hlawka, W. Nöbauer, and L. Schmetterer, editors. *Collected Works (2 vols)*. Verlag der Österr. Akad. Wiss. Wien, 1987.

Michel Defrise and Christine De Mol

2 On blind imaging, NMF and PET

Abstract: After reviewing the multiplicative iterative algorithms EMML and ISRA which allow to naturally enforce, at each iteration, a positivity constraint on the solution of linear inverse problems, we discuss their extension to blind imaging with positivity constraints and to the related problem of Nonnegative Matrix Factorization (NMF). We then establish a connection between this framework and two problems of interest in medical imaging, namely dynamic Positron Emission Tomography (PET) and the joint estimation of activity and attenuation in time-of-flight PET.

Keywords: Multiplicative algorithms, blind imaging, positivity, nonnegative matrix factorization, dynamic positron emission tomography, time-of-flight positron emission tomography

MSC 2010: 35R30, 44A12

2.1 The multiplicative algorithms EMML and ISRA

In many applications, positivity has proved a very useful property to be enforced on the solution of linear inverse problems, which consist in solving the equation

$$\boldsymbol{K}\boldsymbol{x} = \boldsymbol{y} \tag{1}$$

formulated here in a discrete setting, i. e., where $\boldsymbol{x} \in \mathbb{R}^M$ is a vector of coefficients describing the unknown object, $\boldsymbol{y} \in \mathbb{R}^N$ the vector of (noisy) data and $\boldsymbol{K}$ the linear operator (here a $N \times M$ matrix) modeling the link between the two.

In the presence of noise, instead of trying to solve exactly the above equation, the standard strategy is to minimize some contrast or cost function such as a least-squares distance between $\boldsymbol{Kx}$ and $\boldsymbol{y}$ in the case of Gaussian noise or a Kullback–Leibler discrepancy in the case of Poisson noise. When the problem is ill-conditioned, as typical for discretized ill-posed problems, one can regularize the problem by adding to the contrast function one or more penalties on the solution, such as the classical quadratic squared L^2-norm $\|\boldsymbol{x}\|_2^2 = \sum_m |\boldsymbol{x}_m|^2$ or the sparsity-enforcing L^1-norm $\|\boldsymbol{x}\|_1 = \sum_m |\boldsymbol{x}_m|$.

Note: Dedicated to Professor Mario Bertero on the occasion of his 80th birthday.

Acknowledgement: Michel Defrise acknowledges support by the FWO Project G027514N and the Strategic Research Program SRP10 of the VUB, and Christine De Mol by the network IAP P7/06 StUDys.

***Corresponding author: Christine De Mol,** Department of Mathematics and ECARES, Université libre de Bruxelles, Brussels, Belgium, e-mail: demol@ulb.ac.be

Michel Defrise, Department of Nuclear Medicine, Vrije Universiteit Brussel, Brussels, Belgium, e-mail: Michel.Defrise@vub.be

https://doi.org/10.1515/9783110560855-002

Positivity naturally arises in the case of photon-counting measurements, i. e., in the case of Poisson noise, where the cost function to be minimized is the following generalized Kullback–Leibler divergence (which turns out to be the negative log-likelihood for this case)

$$F(\boldsymbol{x}) = \mathrm{KL}(\boldsymbol{y}, \boldsymbol{K}\boldsymbol{x}) \equiv \sum_{n=0}^{N-1}\left[y_n \ln\left(\frac{y_n}{(\boldsymbol{K}\boldsymbol{x})_n}\right) - y_n + (\boldsymbol{K}\boldsymbol{x})_n\right] \tag{2}$$

subject to the constraint $\boldsymbol{x} \geq \boldsymbol{0}$. It is assumed that $\boldsymbol{K} \geq \boldsymbol{0}$ and $\boldsymbol{y} \geq \boldsymbol{0}$. This means that all elements of the vectors $\boldsymbol{x}$ and $\boldsymbol{y}$ and of the matrix $\boldsymbol{K}$ are nonnegative. Notice that it is such entrywise notion of nonnegativity (or else of positivity) which will be used throughout the paper. To find the minimum of this convex cost function, a celebrated iterative algorithm has been proposed and goes under the name Expectation Maximization Maximum Likelihood (EMML) in medical imaging [42] and Richardson [41]—Lucy [30] in astronomy. The successive iterates are given by

$$\boldsymbol{x}^{(k+1)} = \frac{\boldsymbol{x}^{(k)}}{\boldsymbol{K}^T\boldsymbol{1}_N} \circ \boldsymbol{K}^T\frac{\boldsymbol{y}}{\boldsymbol{K}\boldsymbol{x}^{(k)}} \quad (k = 0, 1, \ldots) \tag{3}$$

using the Hadamard (entrywise) product $\circ$ and division ($\boldsymbol{K}^T$ denotes the transpose of $\boldsymbol{K}$ and $\boldsymbol{1}_N$ is a vector of ones in $\mathbb{R}^N$). When initialized by a strictly positive vector, $\boldsymbol{x}^{(0)} > \boldsymbol{0}$, the algorithm automatically preserves the positivity of the iterates $\boldsymbol{x}^{(k)}$. Another nice feature is that it ensures a monotonic decrease of the cost function, a property that is most easily established by viewing it as a Majorization–Minimization (MM) algorithm [25]. Indeed, as noticed by De Pierro [15], it can be rewritten as

$$\boldsymbol{x}^{(k+1)} = \arg\min_{\boldsymbol{x}} G(\boldsymbol{x}; \boldsymbol{x}^{(k)}) \tag{4}$$

where $G(\boldsymbol{x}; \boldsymbol{a})$ is the following "*surrogate*" cost function for $F(\boldsymbol{x})$

$$G(\boldsymbol{x}; \boldsymbol{a}) = \sum_{n=0}^{N-1}\Bigg[y_n \ln y_n - y_n + (\boldsymbol{K}\boldsymbol{x})_n - \frac{y_n}{(\boldsymbol{K}\boldsymbol{a})_n}\sum_{m=0}^{M-1} K_{n,m}a_m \ln\left(\frac{x_m}{a_m}(\boldsymbol{K}\boldsymbol{a})_n\right)\Bigg] \tag{5}$$

satisfying the surrogating properties

$$G(\boldsymbol{x}; \boldsymbol{a}) \geq F(\boldsymbol{x}) \quad \text{and} \quad G(\boldsymbol{a}; \boldsymbol{a}) = F(\boldsymbol{a}) \tag{6}$$

for all $\boldsymbol{x}$ and $\boldsymbol{a}$, with the restriction in this case that $\boldsymbol{x} \geq \boldsymbol{0}$ and $\boldsymbol{a} \geq \boldsymbol{0}$. Since this surrogate is separable (i. e., it can be written as a sum of terms, where each term depends only on a single unknown component x_m), it can be easily minimized explicitly, yielding the EMML algorithm. The surrogating properties ensure that $F(\boldsymbol{x}^{(k+1)}) \leq F(\boldsymbol{x}^{(k)})$, which is often considered as a desirable property for a good numerical behavior of the algorithm. A possible drawback, however, is slow convergence to a minimizer of the cost function. The convergence proof for the EMML algorithm can be found in the literature (see, e. g., [33] and the references therein).

An analogue of the EMML iterative algorithm for the case of Gaussian noise, i. e., of the minimization of the least-squares (negative log-likelihood) cost function

$$F(\boldsymbol{x}) = \frac{1}{2}\|\boldsymbol{K}\boldsymbol{x} - \boldsymbol{y}\|_2^2 \tag{7}$$

subject to $\boldsymbol{x} \geq \boldsymbol{0}$, assuming $\boldsymbol{K} \geq \boldsymbol{0}$ and $\boldsymbol{y} \geq \boldsymbol{0}$, is called Image Space Reconstruction Algorithm (ISRA) and has been proposed in [12] and [14].

The successive multiplicative updates are given by

$$\boldsymbol{x}^{(k+1)} = \boldsymbol{x}^{(k)} \circ \frac{\boldsymbol{K}^T\boldsymbol{y}}{\boldsymbol{K}^T\boldsymbol{K}\boldsymbol{x}^{(k)}} \tag{8}$$

and, again, positivity is automatically preserved if $\boldsymbol{x}^{(0)} > \boldsymbol{0}$. The algorithm is easily derived through the separable surrogates, for $\boldsymbol{x} \geq \boldsymbol{0}$, $\boldsymbol{a} \geq \boldsymbol{0}$,

$$G(\boldsymbol{x}; \boldsymbol{a}) = \frac{1}{2}\sum_{n=0}^{N-1} \frac{1}{(\boldsymbol{K}\boldsymbol{a})_n} \sum_{m=0}^{M-1} K_{n,m} a_m \left[y_n - \frac{x_m}{a_m}(\boldsymbol{K}\boldsymbol{a})_n\right]^2. \tag{9}$$

Hence the cost function decreases monotonically throughout the iteration process and can be shown to converge to its minimal value. Moreover, the iterates themselves have been shown to converge to the minimizer of (7) when it is unique [14] and to a minimizer of (7) when $\boldsymbol{K}$ is singular [19].

In the case of an ill-conditioned problem, however, the minimizers are unstable with respect to the noise on the data. The algorithms, being unregularized, typically exhibit a semi-convergence behavior. On numerical simulations involving a "true" solution $\boldsymbol{x}$, one can observe that the reconstruction error first decreases with the number of iterations, then passes through a minimum and increases due to noise amplification. In practice, regularization is usually achieved by early stopping of the iterative process, before instabilities occur (see [5] for a thorough discussion of this semi-convergence phenomenon).

2.2 Blind imaging with positivity and NMF

In the previous section, the operator or matrix $\boldsymbol{K}$ modeling the imaging process was supposed to be known. In many instances, however, it is unknown, in which case the inverse problem is said to be "*blind*." The blind problem can be formulated as the minimization of (2) or (7) for both unknowns $\boldsymbol{x}$ and $\boldsymbol{K}$. A major difficulty then arises from the fact that, although the resulting functional is convex with respect to $\boldsymbol{x}$ or $\boldsymbol{K}$ separately, it is not jointly convex, leading to all drawbacks of a nonconvex optimization setting (local minima, saddle points, etc.). Due to such bi-convexity property, however, a natural strategy appears to be an alternate minimization on $\boldsymbol{x}$ (with $\boldsymbol{K}$ fixed) and $\boldsymbol{K}$ (with $\boldsymbol{x}$ fixed). Such alternating minimization approaches date back to [3, 10, 11, 43] for alternating least squares or to [20] for alternating Richardson–Lucy. Notice that when the imaging operator $\boldsymbol{K}$ is translation-invariant, the problem is also referred to as "*Blind Deconvolution.*" There is a vast literature on the subject, which we will not review here. Let us just single out the paper [8] for its convergence result on regularized alternating least squares, as well as the noniterative and nonlinear inversion method of [23], with an interesting uniqueness result, but with some drawbacks analyzed in [9].

In this chapter, we will focus on the case where positivity constraints apply, and in particular on generalizations of the EMML and ISRA algorithms to the blind case.

Let us first remark that the formulation of the problem can be easily extended to include multiple inputs/unknowns ($\boldsymbol{x}$ becomes a $M \times P$ matrix $\boldsymbol{X}$) and multiple outputs/measurements ($\boldsymbol{y}$ becomes a $N \times P$ matrix $\boldsymbol{Y}$). The blind inverse problem is then equivalent to solving the equation $\boldsymbol{KX} = \boldsymbol{Y}$ for both $\boldsymbol{K}$ and $\boldsymbol{X}$ or else to "*Nonnegative Matrix Factorization*" (NMF) when all elements of these three matrices are assumed to be nonnegative. NMF is used as a data dimension reduction method analogous to Singular Value Decomposition (SVD) or Principal Component Analysis (PCA) [18, 28]. In the case of noisy data, exact factorization should be replaced by the minimization of a discrepancy cost function. The inclusion of multiple outputs can contribute to somehow compensate for the difficulty of the blind imaging problem, by making use of the information contained in several images produced by the same instrument (e. g., a telescope). Such formulation is also typically met in so-called "*Hyperspectral Imaging*" problems where the same object is observed with different wavelengths (see, e. g., the review papers [31] and [21]).

Another point of attention is the ill-conditioning of the linear inverse problem which induces to include some regularizing penalties in the cost function. To fix the ideas, we introduce penalties on the Frobenius norm of $\|\boldsymbol{K}\|_F^2 = \sum_{n,m} \boldsymbol{K}_{n,m}^2$ of $\boldsymbol{K}$, as well as on the Frobenius and on the L^1-norm $\|\boldsymbol{X}\|_1 = \sum_{m,p} |\boldsymbol{X}_{m,p}|$ of $\boldsymbol{X}$ and tune them with the (positive) regularization parameters μ, ν, and λ, respectively. The following algorithm, however, can be easily generalized to include other separable penalties. Using at each iteration step, the surrogate cost function (5) for the Kullback–Leibler discrepancy, the alternating minimization, for $\boldsymbol{K}, \boldsymbol{X}$ nonnegative (assuming $\boldsymbol{Y}$ nonnegative, too) of the resulting (bi-convex) cost function

$$F(\boldsymbol{K},\boldsymbol{X}) = \mathrm{KL}(\boldsymbol{Y},\boldsymbol{KX}) + \frac{\mu}{2}\|\boldsymbol{K}\|_F^2 + \lambda\|\boldsymbol{X}\|_1 + \frac{\nu}{2}\|\boldsymbol{X}\|_F^2 \tag{10}$$

with

$$\mathrm{KL}(\boldsymbol{Y},\boldsymbol{KX}) = \sum_{n=0}^{N-1}\sum_{p=0}^{P-1}\left[(\boldsymbol{Y})_{n,p}\ln\left(\frac{(\boldsymbol{Y})_{n,p}}{(\boldsymbol{KX})_{n,p}}\right) - (\boldsymbol{Y})_{n,p} + (\boldsymbol{KX})_{n,p}\right] \tag{11}$$

can be done column by column for $\boldsymbol{X}$ and row by row for $\boldsymbol{K}$. The successive multiplicative updates for $\boldsymbol{K}$ and $\boldsymbol{X}$ are then explicitly given by

$$\boldsymbol{K}^{(k+1)} = \frac{2\boldsymbol{A}^{(k)}}{\boldsymbol{B}^{(k)} + \sqrt{\boldsymbol{B}^{(k)} \circ \boldsymbol{B}^{(k)} + 4\mu \boldsymbol{A}^{(k)}}} \tag{12}$$

where

$$\boldsymbol{A}^{(k)} = \boldsymbol{K}^{(k)} \circ \frac{\boldsymbol{Y}}{\boldsymbol{K}^{(k)}\boldsymbol{X}^{(k)}} (\boldsymbol{X}^{(k)})^T \tag{13}$$

$$\boldsymbol{B}^{(k)} = \mathbf{1}_{N\times P}\, (\boldsymbol{X}^{(k)})^T \tag{14}$$

($\mathbf{1}_{N\times P}$ denotes the $N \times P$ matrix of ones) and by

$$\boldsymbol{X}^{(k+1)} = \frac{2\boldsymbol{C}^{(k+1)}}{\boldsymbol{D}^{(k+1)} + \sqrt{\boldsymbol{D}^{(k+1)} \circ \boldsymbol{D}^{(k+1)} + 4\nu \boldsymbol{C}^{(k+1)}}} \tag{15}$$

where

$$\boldsymbol{C}^{(k+1)} = \boldsymbol{X}^{(k)} \circ (\boldsymbol{K}^{(k+1)})^T \frac{\boldsymbol{Y}}{\boldsymbol{K}^{(k+1)}\boldsymbol{X}^{(k)}} \tag{16}$$

$$\boldsymbol{D}^{(k+1)} = \lambda \mathbf{1}_{M\times P} + (\boldsymbol{K}^{(k+1)})^T \mathbf{1}_{N\times P}. \tag{17}$$

The algorithm is to be initialized with strictly positive but otherwise arbitrary $\boldsymbol{K}^{(0)}$ and $\boldsymbol{X}^{(0)}$.

As a special case for $\lambda = \mu = \nu = 0$, we recover the blind algorithm proposed by Lee and Seung [28], which in turn reduces to the EMML/Richardson–Lucy algorithm for $\boldsymbol{K}$ fixed.

Similarly, for the case of Gaussian noise, we can minimize the following bi-convex cost function alternately row by row for $\boldsymbol{K}$, and column by column for $\boldsymbol{X}$ (assumed to be nonnegative, as well as $\boldsymbol{Y}$),

$$F(\boldsymbol{K},\boldsymbol{X}) = \frac{1}{2}\|\boldsymbol{Y} - \boldsymbol{K}\boldsymbol{X}\|_F^2 + \frac{\mu}{2}\|\boldsymbol{K}\|_F^2 + \lambda\|\boldsymbol{X}\|_1 + \frac{\nu}{2}\|\boldsymbol{X}\|_F^2 \tag{18}$$

using the surrogate (9), and derive the successive multiplicative updates

$$\boldsymbol{K}^{(k+1)} = \boldsymbol{K}^{(k)} \circ \frac{\boldsymbol{Y}(\boldsymbol{X}^{(k)})^T}{\boldsymbol{K}^{(k)}\boldsymbol{X}^{(k)}(\boldsymbol{X}^{(k)})^T + \mu\boldsymbol{K}^{(k)}} \tag{19}$$

$$\boldsymbol{X}^{(k+1)} = \boldsymbol{X}^{(k)} \circ \frac{(\boldsymbol{K}^{(k+1)})^T\boldsymbol{Y}}{(\boldsymbol{K}^{(k+1)})^T\boldsymbol{K}^{(k+1)}\boldsymbol{X}^{(k)} + \nu\boldsymbol{X}^{(k)} + \lambda\mathbf{1}_{M\times P}} \tag{20}$$

again to be initialized with strictly positive but otherwise arbitrary $\boldsymbol{K}^{(0)}$ and $\boldsymbol{X}^{(0)}$. We recover as special cases blind algorithms proposed in [22] for $\mu = 0$, $\nu = 0$, in [29] for $\lambda = 0$, $\mu = 0$, $\nu = 0$, and ISRA for $\boldsymbol{K}$ fixed and $\lambda = \mu = \nu = 0$.

As concerns the convergence of the iterates in both the Poisson and Gaussian cases, the following results can be established [26, 27] under the assumption that μ and either ν or λ are strictly positive and that Y has at least one strictly positive element in each row and each column:

1. The cost function decreases monotonically due the surrogate properties and monotonicity is strict iff $(\boldsymbol{K}^{(k+1)}, \boldsymbol{X}^{(k+1)}) \neq (\boldsymbol{K}^{(k)}, \boldsymbol{X}^{(k)})$.
2. The sequence of the values of the cost function $F(\boldsymbol{K}^{(k)}, \boldsymbol{X}^{(k)})$ converges.
3. An asymptotic regularity property holds for the sequence of iterates: $\forall n, m, p$, $\lim_{k\to+\infty}(\boldsymbol{K}^{(k+1)}_{n,m} - \boldsymbol{K}^{(k)}_{n,m}) = 0; \lim_{k\to+\infty}(\boldsymbol{X}^{(k+1)}_{m,p} - \boldsymbol{X}^{(k)}_{m,p}) = 0$.
4. As a consequence, by Ostrowski's theorem ([35], Theorem 26.1), the set of accumulation points of the sequence of iterates $(\boldsymbol{K}^{(k)}, \boldsymbol{X}^{(k)})$ is compact and connected.
5. Hence, if this set is finite, the iterates $(\boldsymbol{K}^{(k)}, \boldsymbol{X}^{(k)})$ converge. Moreover, it can be shown that they converge to a stationary point $(\boldsymbol{K}^*, \boldsymbol{X}^*)$, i. e., a point satisfying the first-order Karush–Kuhn–Tucker conditions.

Notice that the algorithms above, as well as these convergence results, can be generalized to the case where an additional normalization constraint is implemented at each iteration [26], namely $\sum_m \boldsymbol{K}_{n,m} = 1$, a natural constraint met in several applications such as NMF for hyperspectral imaging.

Some applications lead to "*semi-blind*" or "*myopic*" problems, where partial knowledge is available on the structure of $\boldsymbol{K}$. Examples include deconvolution problems, where $\boldsymbol{K}$ might be a Toeplitz matrix, and the joint estimation of activity and attenuation in PET discussed in Section 2.4. Alternate minimization is still applicable to these semi-blind problems provided the known structure of $\boldsymbol{K}$ is enforced at each iteration.

Unfortunately, for the time being, a complete convergence proof of the iterates to a stationary point is not available and does not seem so easy to obtain despite several attempts in that direction. This does not preclude other convergence results obtained with different types of algorithms such as proximal alternating minimization and projection methods for nonconvex problems [2, 6, 7] or constrained gradient methods based on the Scaled Gradient Projection (SGP) algorithm [36].

2.3 Nonnegative matrix factorization in dynamic PET

We briefly describe an application of the previous algorithms to medical imaging. Positron emission tomography allows estimating the spatial distribution in the body of a radioactive tracer, which decays by emitting positrons. An emitted positron annihilates with an electron to produce two 511 keV photons, which are emitted back to back. Pairs of annihilation photons are detected in a ring of detectors around the patient, and the number of photon pairs (*coincident events*) detected by a pair of detectors is related to the integral of the tracer concentration along the line of response (LOR) linking the two detectors [32]. These coincidence data are histogrammed in L lines of response (LOR), each identified by a pair of detectors in the scanner. Using the EMML algorithm, these data are reconstructed to yield an estimate of the tracer concentration in each voxel of a discretized image matrix.

Most clinical protocols in PET are static: they involve a single measurement taken typically one hour after injecting the radiolabeled tracer, when its biodistribution can be considered as stabilized. Additional diagnostic information is potentially provided by the time evolution of the tracer concentration. In dynamic PET, data are measured for a set of T so-called *time frames* starting directly after injecting the tracer. Image reconstruction—typically via the EMML algorithm—from these data leads to a set of T activity images, which can be organized as a nonnegative $N \times T$ matrix $\boldsymbol{Y}$, with $\boldsymbol{Y}_{n,t}$ the estimated activity in voxel n at time t. Note that in this section we consider the reconstructed activity images as the data, hence the use of the symbol $\boldsymbol{Y}$.

A large variety of methods has been developed to reconstruct and analyze dynamic PET data (see a review in [37]), and in a number of cases it has been shown that the time evolution of the tracer activity allows for an improved discrimination between tissues, leading to an improved diagnostic or prognostic accuracy compared to static imaging.

A first category of methods describes the time evolution of the activity by means of a system of differential equations, which model the metabolism of the specific tracer under study. These equations depend on a small set of kinetic parameters (typically 3 or 4), which can be estimated from the time activity curve $\boldsymbol{Y}_{n,t}$ of each voxel n, and convey meaningful information such as metabolic rates for a biochemical reaction or a transport rate. These methods often require the knowledge of the arterial concentration of the tracer as a function of time.

A second category includes generic methods, which model the time evolution in each voxel as a linear combination of $M < T$ basis time-activity vectors. These vectors can be organized as a $M \times T$ matrix $\boldsymbol{X}$, with $\boldsymbol{X}_{m,t}$ the basis time-activity vector m at time t. The dynamic image data are then modeled as

$$\boldsymbol{Y} = \boldsymbol{K}\boldsymbol{X} \tag{21}$$

with a $N \times M$ matrix $\boldsymbol{K}$, with elements $\boldsymbol{K}_{n,m}$ equal to the weight of the time basis function m in voxel n. The problem consists in jointly estimating $\boldsymbol{K}$ and $\boldsymbol{X}$ from the reconstructed dynamic images $\boldsymbol{Y}$. In contrast with the methods based on tracer-specific kinetic models, the estimated time-activity vectors $\boldsymbol{X}$ cannot in general be interpreted in terms of tissue types or metabolic processes but the achieved dimensionality reduction ($M < T$) is useful as a preprocessing step for an automated clustering of the image voxels.

The matrices $\boldsymbol{X}$ and $\boldsymbol{K}$ are determined up to a nonsingular $M \times M$ matrix. Several penalties are well adapted to dynamic PET and improve identifiability and stability. Viewing $\boldsymbol{X}$ as the time evolution of the tracer in M compartments (associated to tissue types or biochemical processes), and $\boldsymbol{K}$ as weighing the fraction of each compartment in a voxel, naturally suggests imposing nonnegativity of all elements of $\boldsymbol{X}$ and $\boldsymbol{K}$. Moreover, penalties can enforce sparsity of $\boldsymbol{K}$ (assuming that most voxels contain few tissue types), smoothness of the time evolution in each compartment, and spatial smoothness of $\boldsymbol{K}$. Assuming Gaussian noise in the reconstructed images $\boldsymbol{Y}$, and a unit covariance matrix, one is led to the same nonnegative matrix factorization problem as in Section 2.2.

Note that similar factorization methods can be applied before rather than after image reconstruction, in which case $\boldsymbol{Y}$ in (21) represents the raw dynamic data measured by the scanner, and a Poisson noise model is appropriate (see [24, 37]). Also, PCA data reduction methods have been used instead of NMF [1].

2.4 Joint estimation of activity and attenuation in time-of-flight PET

This section describes a second problem in positron emission tomography, which also leads to a bi-convex minimization problem similar to those described in Section 2.2. We consider the static case with a single measurement but for a time-of-flight PET scanner. A time-of-flight (TOF) PET scanner collects data in the same way as described in Section 2.3, but in addition the detectors measure the arrival time of the two detected photons emitted by the annihilation of a positron. The difference between the arrival times of the two photons localizes the origin of the annihilation along the line of response (LOR), with an accuracy determined by the timing resolution of the detectors (typically 300 ps). The time difference is histogrammed into T time-of-flight bins. As a result, TOF-PET data are stored in a vector $\boldsymbol{y} = \{\boldsymbol{y}_n, n = 0, \dots, N-1\}$, where $N = LT$ and $\boldsymbol{y}_{n=\ell T+t}$ is the number of events detected for LOR $\ell = 0, \dots, L-1$ and time bin $t = 0, \dots, T-1$. The activity image is parameterized as a vector $\boldsymbol{x} = \{\boldsymbol{x}_m \geq 0, m = 0, \dots, M-1\}$, $\boldsymbol{x}_m$ being the tracer concentration in voxel m. The goal of PET is to estimate $\boldsymbol{x}$ from the data $\boldsymbol{y}$.

The noise-free data is related to the activity image by

$$\boldsymbol{y} = \boldsymbol{K}\boldsymbol{x} + \boldsymbol{b} \tag{22}$$

with $\boldsymbol{b} \in \mathbb{R}^N$ a known nonnegative background, and a $N \times M$ matrix $\boldsymbol{K}$. The matrix element $\boldsymbol{K}_{n,m}$ is the probability that a pair of photons emitted in voxel m is detected in the LOR $\ell = n/T$ with time-of-flight difference bin $t = n\%T$ (n modulo T). This matrix is modeled as a product $\boldsymbol{K} = \boldsymbol{Q}\boldsymbol{P}$. The $N \times M$ matrix $\boldsymbol{P}$ is the detection probability calculated assuming that the photons only interact in the detectors. The $N \times N$ diagonal matrix $\boldsymbol{Q}$ models the data attenuation caused by photon interactions within the object. The attenuation is independent of the time difference and, therefore,

$$\boldsymbol{Q}_{n,n'} = \boldsymbol{\delta}_{n,n'} \boldsymbol{q}_{n/T}, \tag{23}$$

with $1 \geq \boldsymbol{q}_\ell > 0$, $\ell = 0, \ldots, L-1$, equal to the probability that a photon pair emitted along LOR ℓ escapes the body without interacting. The matrix $\boldsymbol{P}$ only depends on the scanner and is known, but the attenuation $\boldsymbol{Q}$ depends on the object.

In practice, PET scans are performed with a hybrid PET/CT scanner and the attenuation is estimated by the Beer–Lambert law,

$$\boldsymbol{q}_\ell = \exp\left\{ - \sum_{m=0}^{M-1} \boldsymbol{s}_{\ell,m} \boldsymbol{\mu}_m \right\}, \quad \ell = 0, \ldots, L-1 \tag{24}$$

where $\boldsymbol{s}_{\ell,m} \geq 0$ is the intersection length of LOR ℓ with voxel m, and $\boldsymbol{\mu}_m$, the linear attenuation coefficient in voxel m, is measured using a CT scan of the patient. This section deals with more challenging situations where the CT scan is not available or is unreliable due to patient motion between the PET and CT scans. The matrix $\boldsymbol{Q}$ (or equivalently the vector $\boldsymbol{q} \in \mathbb{R}^L$) is then unknown and must be estimated jointly with the activity $\boldsymbol{x}$.

The data $\boldsymbol{y}$ in PET are modeled as independent Poisson variables with expectation given by (22). Given $\boldsymbol{y}$ and the background $\boldsymbol{b}$, maximum likelihood estimation then leads to the minimization of the generalized Kullback–Leibler divergence

$$F(\boldsymbol{x}, \boldsymbol{q}) = \mathrm{KL}(\boldsymbol{y}, \boldsymbol{QPx} + \boldsymbol{b}), \tag{25}$$

with the diagonal matrix $\boldsymbol{Q}$ defined by (23). We consider here a nonpenalized likelihood, assuming that the data parameterization (voxel size) and noise level (number of detected coincident events $\sum_n \boldsymbol{y}_n$) are sufficient to guarantee a good stability. Note that $F(\boldsymbol{x}, \boldsymbol{q}) = F(a\boldsymbol{x}, \frac{\boldsymbol{q}}{a})$ for any $a > 0$; therefore, the solution is determined at best up to a global scale factor. Up to this scale factor, the solution to the equation $\boldsymbol{y} = \boldsymbol{QPx} + \boldsymbol{b}$ has been shown to be unique for a continuous-continuous model of the joint estimation problem [16] with noise-free data. Unfortunately, no general result on the identifiability of $\boldsymbol{x}$, $\boldsymbol{q}$ is known in the discrete setting considered here. See [4] for a general overview on this joint estimation problem in TOF-PET.

Following [39], we minimize (25) by alternate minimization as in the previous sections, noting that fixing either $\boldsymbol{Q}$ or $\boldsymbol{x}$ reduces the bi-convex problem to KL minimization for a linear inverse problem. The activity and the attenuation updates are therefore given by the Richardson–Lucy formula:

$$\boldsymbol{x}^{(k+1)} = \frac{\boldsymbol{x}^{(k)}}{\boldsymbol{P}^T \boldsymbol{Q}^{(k)} \boldsymbol{1}_N} \circ \boldsymbol{P}^T \boldsymbol{Q}^{(k)} \frac{\boldsymbol{y}}{\boldsymbol{Q}^{(k)} \boldsymbol{P} \boldsymbol{x}^{(k)} + \boldsymbol{b}} \tag{26}$$

$$q^{(k+1)} = \frac{q^{(k)}}{VPx^{(k+1)}} \circ V \frac{(Px^{(k+1)}) \circ y}{Q^{(k)}Px^{(k+1)} + b} \tag{27}$$

where the $L \times N$ matrix V sums over all time bins for a given LOR: $V_{\ell,n} = \delta_{\ell,n/T}$.

Several remarks are in order:

1. In the absence of background ($b = 0$) the attenuation update (27) simplifies and yields a closed-form expression for the minimizer of the convex cost function at fixed activity x:

$$q^*(x) = \arg\min_q F(x,q) = \frac{Vy}{VPx}. \tag{28}$$

The parameter q can be eliminated by inserting (28) into (25) (with $b = 0$). The reduced nonconvex cost function $F(x, q^*(x))$ is then minimized using a majorization–minimization algorithm:

$$x^{(k+1)} = \frac{x^{(k)}}{P^T V^T \frac{Vy}{VPx^{(k)}}} \circ P^T \frac{y}{Px^{(k)}}. \tag{29}$$

Properties of this algorithm are analyzed in [17]. In particular, the KL divergence $F(x^{(k)}, q^{(k)})$ is nonincreasing and converges, the activity estimates $x^{(k)}$ are nonnegative, and the rescaled activity estimates $\tilde{x}^{(k)} = x^{(k)}/\|x^{(k)}\|$ are asymptotically regular. It can also be shown that the bi-convex cost function $F(x, q)$ has no local minimum if the data are consistent, that is if there is some nonnegative x_0, q_0 such that $y = Q_0Px_0$. This property has little impact because measured data are inconsistent and one easily finds small-scale toy examples where several local minima can be identified.

2. The algorithm (26), (27) can be easily adapted to include a quadratic penalty on x and q.
3. An alternative approach jointly estimates the attenuation coefficient vector $\mu \in \mathbb{R}^M$ instead of the attenuation factors $q \in \mathbb{R}^L$, using (24). This approach facilitates regularization because prior information on the attenuation coefficient μ is easily derived knowing a typical material composition of the object; for example, $\mu \sim 0.1\,\text{cm}^{-1}$ for water at 511 keV. In contrast the values of the attenuation factors q are more difficult to define a priori. Joint estimation of μ and x by alternate minimization has been explored previously for non-TOF PET [13, 34], but in that case the data do not provide sufficient information for a reliable estimation. Thanks to the additional information provided by the time-of-flight measurement, the recent application to TOF-PET [38] leads to a reliable estimation and opens the way to practical implementations.
4. When a CT measurement of the attenuation coefficient μ is available but corresponds to a different position of the patient than the TOF-PET data, joint estimation has also been proposed to estimate the activity x and a nonrigid geometric warping that maps the CT map μ onto the deformed frame required for PET [40].

Bibliography

[1] Y. Anzai, S. Minoshima, G. T. Wolf, and R. L. Wahl. Head and neck cancer: detection of recurrence with three-dimensional principal components analysis at dynamic FDG PET. *Radiology*, 212(1):285–290, 1999.

[2] H. Attouch, J. Bolte, P. Redont, and A. Soubeyran. Proximal alternating minimization and projection methods for nonconvex problems: an approach based on the Kurdyka–Lojasiewicz inequality. *Math. Oper. Res.*, 35(2):438–457, 2010.

[3] G. R. Ayers and J. C. Dainty. Iterative blind deconvolution method and its applications. *Opt. Lett.*, 13(7):547–549, 1988.

[4] Y. Berker and Y. Li. Attenuation correction in emission tomography using the emission data—a review. *Med. Phys.*, 43(2):807–832, 2016.

[5] M. Bertero and P. Boccacci. *Introduction to Inverse Problems in Imaging*. Institute of Physics Publishing, 1998.

[6] J. Bolte, P. L. Combettes, and J. C. Pesquet. Alternating proximal algorithm for blind image recovery. In *2010 IEEE International Conference on Image Processing*, pages 1673–1676, 2010.

[7] J. Bolte, S. Sabach, and M. Teboulle. Proximal alternating linearized minimization for nonconvex and nonsmooth problems. *Math. Program.*, 146(1):459–494, 2014.

[8] M. Burger and O. Scherzer. Regularization methods for blind deconvolution and blind source separation problems. *Math. Control Signals Syst.*, 14(4):358–383, 2001.

[9] A. S. Carasso. False characteristic functions and other pathologies in variational blind deconvolution. A method of recovery. *SIAM J. Appl. Math.*, 70(4):1097–1119, 2009.

[10] T. F. Chan and C. K. Wong. Total variation blind deconvolution. *IEEE Trans. Image Process.*, 7(3):370–375, 1998.

[11] T. F. Chan and C. K. Wong. Convergence of the alternating minimization algorithm for blind deconvolution. *Linear Algebra Appl.*, 316(1–3):259–285, 2000.

[12] M. E. Daube-Witherspoon and G. Muehllehner. An iterative image space reconstruction algorithm suitable for volume ECT. *IEEE Trans. Med. Imaging*, 5(2):61–66, 1986.

[13] A. De Pierro and F. Crepaldi. Activity and attenuation recovery from activity data only in emission computed tomography. *Comput. Appl. Math.*, 25(2-3):205–227, 2006.

[14] A. R. De Pierro. On the convergence of the iterative image space reconstruction algorithm for volume ECT. *IEEE Trans. Med. Imaging*, 6(2):174–175, 1987.

[15] A. R. De Pierro. On the relation between the ISRA and the EM algorithm for positron emission tomography. *IEEE Trans. Med. Imaging*, 12(2):328–333, 1993.

[16] M. Defrise, A. Rezaei, and J. Nuyts. Time-of-flight PET data determine the attenuation sinogram up to a constant. *Phys. Med. Biol.*, 57(4):885–899, 2012.

[17] M. Defrise, A. Rezaei, and J. Nuyts. Transmission-less attenuation correction in time-of-flight PET: analysis of a discrete iterative algorithm. *Phys. Med. Biol.*, 59(4):1073–1096, 2014.

[18] D. Donoho and V. Stodden. When does non-negative matrix factorization give a correct decomposition into parts? In S. Thrun, L. K. Saul, and B. Schölkopf, editors, *Advances in Neural Information Processing Systems 16*, pages 1141–1148. MIT Press, 2004.

[19] P. Eggermont. Multiplicative iterative algorithms for convex programming. *Linear Algebra Appl.*, 130:25–42, 1990.

[20] D. A. Fish, A. M. Brinicombe, E. R. Pike, and J. G. Walker. Blind deconvolution by means of the Richardson–Lucy algorithm. *J. Opt. Soc. Am. A*, 12(1):58–65, 1995.

[21] M. Hong, M. Razaviyayn, Z. Q. Luo, and J. S. Pang. A unified algorithmic framework for block-structured optimization involving big data: with applications in machine learning and signal processing. *IEEE Signal Process. Mag.*, 33(1):57–77, 2016.

[22] P. O. Hoyer. Non-negative sparse coding. In *Proceedings of the 12th IEEE Workshop on Neural Networks for Signal Processing*, volume 1, pages 557–565, 2002.

[23] L. Justen and R. Ramlau. A non-iterative regularization approach to blind deconvolution. *Inverse Probl.*, 22(3):771–800, 2006.

[24] E. Krestyannikov, J. Tohka, and U. Ruotsalainen. Joint penalized-likelihood reconstruction of time-activity curves and regions-of-interest from projection data in brain PET. *Phys. Med. Biol.*, 53(11):2877–2896, 2008.

[25] K. Lange, D. R. Hunter, and I. Yang. Optimization transfer using surrogate objective functions. *J. Comput. Graph. Stat.*, 9(1):1–20, 2000.

[26] L. Lecharlier. *Blind inverse imaging with positivity constraints*. PhD Thesis, Université libre de Bruxelles, Faculté des Sciences, 2014.

[27] L. Lecharlier and C. De Mol. Regularized blind deconvolution with Poisson data. *J. Phys. Conf. Ser.*, 464(1):012003, 2013.

[28] D. D. Lee and H. S. Seung. Learning the parts of objects by non-negative matrix factorization. *Nature*, 401:788–791, 1999.

[29] D. D. Lee and H. S. Seung. Algorithms for non-negative matrix factorization. In T. K. Leen, T. G. Dietterich, and V. Tresp, editors, *Advances in Neural Information Processing Systems 13*, pages 556–562. MIT Press, 2001.

[30] L. B. Lucy. An iterative technique for the rectification of observed distributions. *Astron. J.*, 79:745–754, 1974.

[31] W.-K. Ma, J. Bioucas-Dias, T.-H. Chan, N. Gillis, P. Gader, A. Plaza, A. Ambikapathi, and C.-Y. Chi. A signal processing perspective on hyperspectral unmixing: insights from remote sensing. *IEEE Signal Process. Mag.*, 31(1):67–81, 2014.

[32] S. R. Meikle and R. D. Badawi. Quantitative techniques in PET. In D. L. Bailey, D. W. Townsend, P. E. Valk, and M. N. Maisey, editors, *Positron Emission Tomography, Basic Sciences*, pages 93–126. Springer Verlag, 2005.

[33] F. Natterer and F. Wübbeling. *Mathematical Methods in Image Reconstruction*. Society for Industrial and Applied Mathematics, 2001.

[34] J. Nuyts, P. Dupont, S. Stroobants, R. Benninck, L. Mortelmans, and P. Suetens. Simultaneous maximum a posteriori reconstruction of attenuation and activity distributions from emission sinograms. *IEEE Trans. Med. Imaging*, 18(5):393–403, 1999.

[35] A. Ostrowski. *Solution of Equations in Euclidean and Banach Spaces*. Academic Press, 1973.

[36] M. Prato, A. L. Camera, S. Bonettini, and M. Bertero. A convergent blind deconvolution method for post-adaptive-optics astronomical imaging. *Inverse Probl.*, 29(6):065017, 2013.

[37] A. Reader and J. Verhaeghe. 4D image reconstruction for emission tomography. *Phys. Med. Biol.*, 59(22):R371–R418, 2014.

[38] A. Rezaei, M. Defrise, G. Bal, C. Michel, M. Conti, C. Watson, and J. Nuyts. Simultaneous reconstruction of activity and attenuation in time-of-flight PET. *IEEE Trans. Med. Imaging*, 31(12):2224–2233, 2012.

[39] A. Rezaei, M. Defrise, and J. Nuyts. ML-reconstruction for TOF-PET with simultaneous estimation of the attenuation factors. *IEEE Trans. Med. Imaging*, 33(7):1563–1572, 2014.

[40] A. Rezaei, C. Michel, M. E. Casey, and J. Nuyts. Simultaneous reconstruction of the activity image and registration of the CT image in TOF-PET. *Phys. Med. Biol.*, 61(4):1852–1874, 2016.

[41] W. H. Richardson. Bayesian-based iterative method of image restoration. *J. Opt. Soc. Am.*, 62(1):55–59, 1972.

[42] L. A. Shepp and Y. Vardi. Maximum likelihood reconstruction for emission tomography. *IEEE Trans. Med. Imaging*, 1(2):113–122, 1982.

[43] Y.-L. You and M. Kaveh. A regularization approach to joint blur identification and image restoration. *IEEE Trans. Image Process.*, 5(3):416–428, 1996.

Simon Gindikin

3 From Radon to Leray

Abstract: We discuss the influence of the Radon transform on multidimensional complex analysis. In the first turn, it is Leray's construction of an universal multidimensional analogue of the Cauchy formula–*the Cauchy–Fantappie formula*. It was extended by fundamental conceptions of analytic functionals, analytic duality, $\bar{\partial}$-cohomology, and different complex versions of the Radon transform. In our exposition, the conception of *the Cauchy–Radon transform* plays an essential role. It is the result of the replacement of δ-function in the integrand of the Radon transform by a Cauchy kernel. It appears as a modification of the Radon transform for which the inversion formula is independent of the evenness of the dimension. This then gives new possibilities to consider analogs of the Radon transform in non-Euclidean situation.

Keywords: Radon transform, Radon–Cauchy transform, inversion formula, hyperfunctions, Cauchy–Fantappie integral formula, Martineau's duality

MSC 2010: 44A12, 46F15, 32A26

Without doubts, tomography is the most spectacular application of the Radon transform. However, there also were a few absolutely fundamental results in pure mathematics, which appeared under the strong influence of the Radon transform. I would emphasize applications to differential equations of F. John, multidimensional Cauchy integral formulas of Leray and the horospherical transform on symmetric homogeneous spaces of Gelfand. In this exposition, we will talk about Leray's results and their development. Our exposition has a narrow focus on complex parallels to the Radon transform. We do not consider more broad possibilities, corresponding to the Radon–John transform for integration along planes of codimension higher than one, such as the Penrose transform.

3.1 The Radon–Cauchy transform

We will make a small, but essential, modification of the usual Radon transform at $\mathbb{R}^n$. Let $\mathbb{R}^n_x$, $\mathbb{R}^n_\xi$ be dual copies with the duality form

$$\langle \xi, x \rangle = \xi_1 x_1 + \cdots + \xi_n x_n$$

Simon Gindikin, Departm. of Math., Hill Center, Rutgers University, 110 Frelinghysen Road, Piscataway, NJ 08854, USA, e-mail: gindikin@math.rutgers.edu

https://doi.org/10.1515/9783110560855-003

and $L(\xi,p)$ are hyperplanes

$$\langle \xi, x\rangle = p.$$

Let $f(x) \in C_0^\infty(\mathbb{R}_x^n)$. Then its Radon transform is

$$\mathcal{R}f(\xi,p) = \int_{\mathbb{R}^n} f(x)\delta(\langle \xi, x\rangle - p)\,dx = \int_{L(\xi,p)} f(x)(\langle \xi, dx\rangle \rfloor dx).$$

where $\langle \xi, dx\rangle \rfloor dx$ is such a form α that $(\xi, dx)\wedge\alpha = dx$ (the interior product; its restriction on $L(\xi,p)$ is unique). In the second formula, we take the (real) residue on $L(\xi,p)$.

The inversion formulas have different structures for odd and even n. We will use the canonical form in $\mathbb{R}_\xi^n$:

$$\omega(\xi, d\xi) = \sum_{j\le n}(-1)^j \xi_j \bigwedge_{l\ne j} d\xi_j.$$

Then for odd $n = 2k+1$ we have

$$f(x) = c\int_\Gamma \mathcal{R}f_p^{(n-1)}(\xi, \langle \xi, x\rangle)\omega(\xi, d\xi), \quad c = (-1)^k/2(2\pi)^{2k}.$$

We take the derivation of the order $n-1$ on p and integrate it along any cycle Γ around 0 at $\mathbb{R}_\xi^n$ which intersects almost all rays from 0 once (it could be a sphere). In other words, $(n-1)$-form in the integrand can be pushed down on the projective space $\mathbb{R}P_\xi^{n-1}$ with the homogeneous coordinates ξ

For even $n = 2l$, we have

$$f(x) = c\int_0^\infty \frac{dp}{p}\int_\Gamma \mathcal{R}f_p^{(n-1)}(\xi, \langle \xi, x\rangle - p)\omega(\xi, d\xi), \quad c = (-1)^l/(2\pi)^{n-1}.$$

In the first case, we have a local inversion formula: $f(x)$ is reconstructed through integrals along hyperplanes (infinitesimally) close to x. For even n, the inversion formula is nonlocal. This difference for odd and even dimensions had a deep connection with the Huygens principle. However, for some aims it is convenient to have an universal formula which works for all dimensions.

For this reason, we suggest a modification of this definition in which we replace the δ-function in the definition of the Radon transform by a Cauchy kernel. Let us consider

$$\mathcal{C}f(\xi, p - i\varepsilon) = \int_{\mathbb{R}^n} \frac{f(x)}{\langle \xi, x\rangle - p + i\varepsilon}\,dx, \quad \varepsilon > 0.$$

This transform is well-defined since for $\varepsilon > 0$ there are no singularities in the integrand. Of course, we can take $\varepsilon < 0$ as well. We can connect it with the fact that (complex) hyperplanes $\langle \zeta, z\rangle = p - i\varepsilon$ in $\mathbb{C}_z^n$ have no real points.

Using the elementary theory of distributions [4], we can define the boundary values $\mathcal{C}f(\xi,p-i0) = \lim_{\varepsilon\to 0}\mathcal{C}f(\xi,p-i\varepsilon)$ in the sense of distributions. Let us call this boundary function $\mathcal{C}f(\xi,p-i0)$ by *the Cauchy–Radon transform* of f.

We have

$$\lim_{\varepsilon\to 0}\frac{1}{p-i\varepsilon} = (p-i0)^{-1} = p^{-1} + i\pi\delta(p).$$

As a result, there is the following connection between the Radon and Cauchy–Radon transforms. We have

$$2\mathcal{R}f(\xi,p) = \mathcal{C}f(\xi,p-i0) + \mathcal{C}f(\xi,p+i0).$$

In the opposite direction, we have

$$\mathcal{C}f(\xi,p) = \mathcal{R}F(\xi,p) * (p-i0)^{-1};$$

so we take here the convolution on p.

The remarkable fact is that, if one were to operate with the Radon–Cauchy transform, the difference between even and odd dimensions disappears. Namely, *we have the universal inversion formula which works in both cases*:

$$f(x) = \left(\frac{i}{2\pi}\right)^n \int_\Gamma \mathcal{C}f_p^{(n-1)}(\xi,\langle\xi,x\rangle - i0)\omega(\xi,d\xi).$$

To understand the nature of this phenomenon, let us remark that if we were to applying the inversion formula for odd dimensions in the case of even dimensions the result will be zero; the same thing will happen applying the formula for even dimensions in the case of odd dimensions. This observation together with the connection between the Radon and Cauchy–Radon transforms gives directly the inversion of the Cauchy–Radon transform. The appearance of these zeros is the consequence of the symmetry of $\mathcal{R}$ on p and the aim of the transition to the Cauchy–Radon transform is "to destroy" this symmetry.

It is convenient to illustrate using the connection of the Radon and Fourier transforms. Let us remind that the Fourier and Radon transforms are connected by the one-dimensional Fourier transform

$$\mathcal{F}f(\lambda\xi) = \mathcal{F}_{p\to\lambda}(\mathcal{R}(\xi,p))$$

for fixed ξ. Here, in the first position $\mathcal{F}$ is the n-dimensional Fourier transform and in the 2nd one it is one-dimensional. We have

$$(\mathcal{F}(p-i0)^{-1}) = \chi(\lambda)-$$

the characteristic function of the positive half-line. Then

$$\mathcal{F}_{p\to\lambda}(\mathcal{C}f(\xi,p)) = \mathcal{F}f(\lambda\xi)\chi(\lambda).$$

So we destroy the symmetry by taking only the restriction $\mathcal{F}f$ on the positive half-line on λ for each ξ. Now, it is easy to see another way to prove the inversion formula for the Cauchy–Radon transform.

We can rewrite the inversion formula in the following way:

$$f(x) = \left(\frac{i}{2\pi}\right)^n \int_{\Gamma}\int_{\mathbb{R}^n} f((y)(\langle\xi,(y-x)\rangle - i0)^{-n}\omega(\xi,d\xi)\wedge dy.$$

The second part of this integral identity can be considered as the composition of the Cauchy–Radon transform and its inversion. This real form which participates in this integral coincides with the form in Leray's Cauchy–Fantappie formula (see below). This formal coincidence of Radon and Leray forms was probably not remarked upon by Leray.

3.2 Connections with hyperfunctions and concave tubes

The Cauchy–Radon transform is an example of using complex language in real analysis. It is a very deep phenomenon, about which there is an aphorism (sometimes attributed to Weierstrass): "Real analysis lives on the boundary of complex one." One of most impressive constructions of such type, is hyperfunctions of Sato. In the simplest case, they interpret most generalized functions on $\mathbb{R}^n$ as $\bar{\partial}$-cohomology from $H^{(n-1)}(\mathbb{C}^n \setminus \mathbb{R}^n; \mathcal{O})$. We will connect this cohomology with the Cauchy–Radon transform.

Let $\mathbb{C}(\xi)$ be the subspace $\{\mathrm{Im}\langle\xi,z\rangle < 0\}$. These subspaces give the covering of the supplementary $\mathbb{R}^n$ to $\mathbb{C}^n$ by Stein manifolds. It would be possible to take just a finite covering but we prefer to work with complete covering by $\mathbb{C}(\xi)$, $\xi \in \mathbb{R}^n$, and use for the representation of the cohomology our language of smoothly parameterized Cech cohomology [1]. Namely, we consider the space Z of $(n-1)$-forms $\Omega(z|\xi,d\xi)$ on $\xi \in \mathbb{R}P^{n-1}$ which holomorphically depend on parameters $z \in \mathbb{C}(\xi)$. These forms have domains of parameters z depending on ξ but the action of the differential d_ξ on these forms is well-defined and all these forms are closed as forms of maximal degree. If B is the subspace of d_ξ-exact forms, then $H^{(n-1)} = Z/B$.

Let $\Omega_f(z|\xi,d\xi)$ be the result the substitution $p = \langle\xi,z\rangle$, $z \in \mathbb{C}(\xi)$, in the integrand of the inversion formula for the Cauchy–Radon transform. In such a way, we have an operator from $C_0^\infty(\mathbb{R}^n)$ in Z. The characteristic property of these forms is that they are constant on z for fixed ξ along planes $\langle\xi,z\rangle = p$. It turns out that such forms cannot be exact. So we have the operator in cohomology—*hyperfunctions*. So the Cauchy–Radon transform can be interpreted as an operator in hyperfunctions. It can be extended on different spaces of functions and distributions. If a hyperfunction can be represented by a form $\Omega \in Z$ which is constant along the planes $\langle\xi, z\rangle = p$, then we can consider this form as the Cauchy–Radon transform of the hyperfunction. I believe that such a version of the Radon transform on hyperfunctions deserves careful consideration.

Let us mention a few applications. Let $V \subset \mathbb{R}^n$ be an open convex cone without lines and let $\mathcal{C}f(\xi, p) = 0$ for ξ outside of the closure $\mathrm{cl}(V)$ of V. Then

$$\bigcap_V \mathbb{C}(\xi) = T(-V^*),$$

the tube domain $\mathbb{R}^n - iV^*$ for the dual cone V^*. If we extend the Cauchy–Radon transform in the subspaces $\mathbb{C}(\xi)$, then the inversion formula represents the holomorphic extension of $f(x)$ in the tube $T(-V^*)$.

This construction can be extended on general convex tube domains. For simplicity, we consider strictly convex tube domains $D = T(U) = \mathbb{R}^n + iU \subset \mathbb{C}_z^n$, $z = x + iy$. Here, $U \subset \mathbb{R}_y^n$ is strictly convex with a smoothed boundary bU. We can present U as the intersection of supporting real half-spaces

$$Y(\xi, p(\xi)) = \{\langle\xi, y\rangle < p(\xi)\}$$

where p is homogeneous of degree 1. Let $\mathbb{C}(\xi, p(\xi)) = \mathbb{R}_x^n + iY(\xi, p(\xi))$ be the corresponding complex half-plane. Then

$$D = \bigcap_\xi \mathbb{C}(\xi, p(\xi)).$$

Above we considered such construction for the $U = -V^*$.

Let $f(z)$ be a holomorphic function at the closure of D which fast decreases on x uniformly on y (in the space $\mathcal{S}$). Let $y(\xi)$ is defined by the condition $\langle\xi, y(\xi)\rangle = p(\xi)$. So it is the tangent point of the boundary bY with such a tangent hyperplane. Then we can define a generalized Cauchy–Radon transform

$$\mathcal{C}f(\xi, p) = \int\limits_{[\mathbb{R}_x^n + iy(\xi)]} \frac{f(z)\,dx}{\langle\xi, z\rangle - p}, \quad \operatorname{Im} p < p(\xi).$$

We already have the familiar inversion formula for the Cauchy–Radon transform:

$$f(z) = \left(\frac{i}{2\pi}\right)^n \int\limits_\Gamma \mathcal{C}f_p^{(n-1)}(\xi, \langle\xi, z\rangle)\omega(\xi, d\xi), \quad z \in D = T(U),$$

but this time it is for holomorphic functions in the tube $D \subset \mathbb{C}^n$.

Returning to tubes $T(V)$ over cones, let us consider the situation when $\mathcal{C}f(\xi, p) = 0$ for $\xi \in \mathrm{cl}(V)$. Then $\mathcal{C}f(\xi, p)$ gives a cohomology class in $H^{(n-1)}(\mathbb{C}^n \setminus T(V^*), \mathcal{O})$, so that is in a concave tube. We can interpret it as the cohomological Laplace transform for functions with supports in concave cones. For more details about the cohomological Laplace transform for nonconvex tubes, see [6]. These general considerations use a version of Radon–John transform—the generalization where hyperplanes are replaced by planes of higher codimension.

3.3 Horospherical Cauchy–Radon transform on the hyperbolic space

We will discuss how to modify the conception of the Cauchy–Radon transform for non-Euclidean version of the Radon transform. Let us start from the case of the hyperbolic space. Radon in his original paper considered the analogue of his transform for hyperbolic geodesic hyperplanes. However, following Gelfand, from the point of hyperbolic harmonic analysis, it is more informative to consider the horospherical version of the hyperbolic Radon transform.

Let us consider the model of the hyperbolic space H^n in the one sheet of the two-sheeted hyperboloid in $\mathbb{R}^n$:

$$\square(x) = (x_1)^2 - (x_2)^2 - \cdots - (x_{n+1})^2 = 1, \quad x_1 > 0.$$

Let Ξ be the cone $\{\square(\xi) = 0, \xi \neq 0\}$ and the dot-product $\langle \xi, x\rangle$ correspond to the quadratic form $\square(x)$. Then the horospheres $E(\xi, p)$ are isotropic sections (paraboloids) by the planes; $\langle \xi, x\rangle = p$. We define integrals over horospheres as the horospheric transform but we are interesting here in the horospherical Cauchy transform

$$\mathcal{C}f(\xi, p - i\varepsilon) = \int_{H^n} \frac{f(x)\mu_H}{\langle \xi, x\rangle - p - i\varepsilon}, \quad \varepsilon < 0$$

where $d\mu_H = d(\square(x)) \rfloor dx$. We consider the boundary values $\mathcal{C}f(\xi, p - i0)$. Let $S(x)$ for $x \in H^n$ is the intersection of the cone Ξ by the plane $\langle \xi, x\rangle = 1$. Then

$$f(x) = c \int_{S(x)} \mathcal{C}f_p^{(n-1)}(\xi, 1 - i0)\omega(\xi, d\xi).$$

So we see that the inversion formula in the hyperbolic case looks the same as in Euclidean case. This is not specific for the Cauchy version of the horospherical transform, but is also true for the usual hyperbolic horospherical transform in an appropriate normalization. I believe this coincidence is a very deep circumstance. It is generalized on arbitrary symmetric spaces.

There is a hyperbolic version of the Fourier transform—spectral decomposition of the Laplace–Beltrami operator on H^n. To obtain it, we need to compose the Mellin transform for dilations of the cone Ξ ($\xi \to \lambda\xi$) with the horospherical transform. The pair of “twins” (hyperbolic horospherical and Fourier transforms) is similar to the Euclidean pair (the Radon and Fourier transforms).

3.4 The horospherical Cauchy transform on the sphere

Let $S^n \subset \mathbb{R}^{n+1}$ be the n-dimensional sphere:

$$\square(x) = x_1^2 + \cdots + x_{n+1}^2 = 1.$$

Its (geodesic) hyperplanes are sections by subspaces $\langle \xi, x \rangle = 0$—big hyperspheres. The Minkowski–Funk transform—the integration along big hyperspheres—was known before the Radon transform and discussed in Radon's original paper. It turns out that the conception of the Cauchy–Radon transform gives a way for a horospherical Radon transform on the sphere [2].

At first glance, it looks strange since there are no horospheres on the sphere. However, there are no real horospheres but we can consider complex ones. Using Cauchy-type kernels, we can define the Cauchy transform for them if *they do not contain real points*: the intersection with the complexification $\mathbb{C}S^n$ has no points from S^n. So we remark that the horospherical Cauchy transform makes sense for a more broad class of horospheres than just complex translations of real horospheres.

More exactly, let $\mathbb{C}S^n \subset \mathbb{C}_z^n$:

$$\square(z) = 1$$

be the complexification of S^n and $\Xi \subset \mathbb{C}_\zeta^n$, $\zeta = \xi + i\eta$, is defined by

$$\square(\zeta) = 0, \quad \zeta \neq 0.$$

(Complex) horospheres $L(\zeta)$ are isotropic sections of $\mathbb{C}S^n$:

$$\langle \zeta, z \rangle = 1, \quad \zeta \in \Xi.$$

These sections are paraboloids. We have on Ξ:

$$\square(\xi) = \square(\eta), \quad \langle \xi, \eta \rangle = 0.$$

We are interested by horospheres without real points (points of S^n), since we can use them for the definition of the horospherical Cauchy transform. Let Ξ_0 be the subset of such $\zeta \in \Xi$ that horospheres $L(\zeta)$ do not intersect S^n. It turns out that Ξ_0 is given by the condition

$$\square(\xi) = \square(\eta) < 1.$$

The complex orthogonal group $\mathrm{SO}(n+1;\mathbb{C})$ acts on the Ξ and the conditions on Ξ_0 are invariant relative to its maximal compact subgroup $\mathrm{SO}(n+1)$. By the action of this subgroup on Ξ, we can transform any ζ to the form $(a, ia, 0, \ldots, 0)$. It is evident that such point lies in Ξ_0 if and only if $|a| < 1$. SO$(n+1)$-invariant form of this condition is just $\square(\xi) < 1$ and we have the above description of Ξ_0.

Its boundary $\partial\Xi_0$ corresponds to complex horospheres $L(\zeta)$ with

$$\square(\xi) = \square(\eta) = 1, \quad \langle \xi, \eta \rangle = 0.$$

Such horosphere intersects S^n at the unique point

$$x = \xi.$$

So we have a fibering

$$\partial\Xi \to S^n$$

with $(n-1)$-dimensional spheres which can be identified with spheres at tangent planes to S^n. Let us through $S(x)$, $x \in S^n$, denote the fiber over x:

$$\zeta = x + i\eta, \quad \langle \eta, x \rangle = 0.$$

The horospheres $L(\zeta)$, $\zeta \in S(x)$, will participate in the reconstruction $f(x)$.

The group $\mathbb{C}^\times$ acts on Ξ, defining the cone structure on Ξ. Its part, $\Lambda = \{\lambda \in \mathbb{R}; |\lambda| < 1\}$ acts on the domain Ξ_0 as contractions:

$$\zeta \mapsto \lambda\zeta.$$

It contracts the boundary $\partial\Xi_0$ inside Ξ_0: $\square(\lambda\xi) = \lambda^2 < 1$.

We constructed the geometrical background for the horospherical Cauchy transform on the sphere. Now we are ready to give this definition. Let us consider $f \in C^\infty(S^n)$ and

$$\mathcal{C}f(\zeta) = \int_{S^n} \frac{f(x)}{\langle \zeta, x \rangle - 1} \omega(x; dx), \quad \zeta \in \Xi_0;$$

ω is here the invariant form on S^n. This integral is well-defined and represents a holomorphic function of $\zeta \in \Xi_0$. So we have an operator

$$\mathcal{C} : C^\infty(S^n) \to \mathcal{O}(\Xi_0).$$

This transform can be extended also on boundary points $\zeta = \xi + i\eta \in \partial\Xi_0$ since then the integrand for $\mathcal{C}f(\zeta)$ will be singular only at the point $x = \xi$ and it is easy to regularize the integral. We can interpret it as boundary values of $\mathcal{C}f(\zeta)$, $\zeta \in \Xi_0$, using the constructed contractions:

$$\mathcal{C}f(\zeta) = \lim_{\varepsilon \to 0} \mathcal{C}f((1-\varepsilon)\zeta), \quad \zeta \in \partial\Xi_0.$$

Let us define the operator D_p considering horospheres $\langle\zeta,x\rangle = 1-i\varepsilon$; it is equivalent to differentiation along the contractions of Ξ_0. Then the inversion formula has the familiar form:

$$f(x) = c \int\limits_{[\square(\eta)=1.\langle\eta,x\rangle=0]} (D_p)^{n-1}Cf(x+i\eta)\omega(\eta;d\eta), \quad c = (i/2\pi)^n, x \in S^n.$$

So we take the average of boundary values along $S(x)$. As in considered above cases, the horospherical Cauchy transform has some associated Fourier transform. As we remarked on Ξ_0 acts the circle $\zeta \mapsto \exp(i\theta)\zeta$, commutating with the action of $SO(n+1)$. Composition of the horospherical Cauchy transform and the decomposition in the corresponding Fourier series gives the spherical Fourier transform.

Let us emphasize once more the universal structure of considered inversion formulas. The difference in harmonic analysis on homogeneous spaces appears when we add the action of an Abelian group and in this way the difference lies in Abelian harmonic analysis.

3.5 Integral formula Cauchy–Fantappie

The end of nineteenth and the beginning of twentieth century were the golden age of multidimensional complex analysis when Weierstrass, Poincare, Cousin, Hartogs, E. Levi discovered specifically multidimensional phenomena (holomorphy domains, pseudoconvexity, etc.) and stated basic problems. However, as later remarked Oka, for a long time yet problems were stated, rather than solved. Results basically were about the whole space or direct products of one-dimensional domains. The essential restriction was that unknown was a multidimensional analogue of the classical Cauchy formula. For a long time, just in the case of products of one-dimensional domains, could one apply the combination of one-dimensional formulas on different variables. Only in 1930th, A. Weil suggested a formula for analytic polyhedrons. This case is quite special but sufficient for the approximation of arbitrary holomorphy domains. It gave a possibility to Oka to solve basic problems of several complex variables. In 1940th, there was found another remarkable integral formula of Bochner–Martinelli, which clarified several important situations. Its important property was universality: it works for arbitrary domains as the classical one-dimensional Cauchy formula. However, it has a weak point: its kernel is not holomorphic on exterior variables. It seriously restricts its possibility for applications.

New direction of the problem of multidimensional Cauchy formulas started in 1956 when Leray [7] discovered the Cauchy–Fantappie integral formula. May be it was one of the simplest results of Leray but with very essential consequences. On other side, to this day, in several complex variables there were appeared new methods for solving basic problems, alternative to integral formulas. Let us describe this construction.

We consider $\mathbb{C}^n_z \times \mathbb{C}^n_\zeta$ with n-form dz and $(n-1)$-form $\omega(\zeta, d\zeta)$. Let $f(z)$ be a holomorphic function in a domain $D \subset \mathbb{C}^n_z$. We call *the Cauchy–Fantappie form* the form

$$\Omega_{\mathrm{CF}} = \frac{(n-1)!}{(2\pi)^n} \frac{f(z)\, dz \wedge \omega(\zeta, d\zeta)}{\langle \zeta, z - w\rangle^n}.$$

On ζ, this form can be pushed down on the projective space $\mathbb{C}P^{(n-1)}_\zeta$. So for fixed w, it will be closed outside the singularities as holomorphic form of maximal degree. This is a crucial observation. Leray's idea was that specific Cauchy formulas can be obtained by the integration of this form on different cycles.

More precisely, let D be a simply connected domain with a smooth boundary bD and $w \in D$ and let $\zeta(w, z)$ will be a vector-function on bD that

$$\langle \zeta(w,z), z\rangle \neq \langle \zeta(w,z), w\rangle.$$

Then we have a cycle Γ over bD and

$$\int_\Gamma \Omega_{\mathrm{CF}} = f(w).$$

To prove this formula, it is enough to verify it for one cycle. Fortunately, it was already known for the case when D is the complex ball.

So for Leray, in the case of many variables the universality of Cauchy formula is already not the uniqueness, as for one variable, but is some formula with extra parameters (ζ) which generates different formulas. This adding of some parameters, different from variables, dramatically simplifies the picture and works in some other examples. For specific domains, the choice of vector-functions $\zeta(w, z)$ on the boundary bD, satisfying the requirement, can be a difficult assignment. A few examples:

(1) Of course,

$$\zeta(w, z) = \bar{z} - \bar{w}$$

works and we have the above mentioned Martinelli–Bochner formula

$$f(w) = \frac{(n-1)!}{(2\pi)^n} \int_{bD} \frac{f(z)\, dz \wedge \omega(\zeta, d\zeta)}{\langle \bar{z} - \bar{w}, z - w\rangle^n}.$$

As we mentioned, this formula is true for arbitrary domains but the kernel in it is not holomorphic on the exterior variables w.

(2) Let D be a convex domain with a smooth boundary, defined by a condition

$$\rho(z, \bar{z}) < 0$$

and $\partial\rho(z)$, $z \in bD$ be the holomorphic gradient. So

$$\langle \partial\rho(z), z - w \rangle = 0$$

is holomorphic tangent (supporting) hyperplane at the point $z \in bD$. So this hyperplane does not intersect D and $\zeta(w,z)$ gives the integral formula for the convex domains

$$f(w) = \frac{(n-1)!}{(2\pi)^n} \int_{bD} \frac{f(z)\, dz \wedge \omega(\zeta, d\zeta)}{\langle \partial\rho(z), z - w \rangle^n}.$$

The remarkable property of this integral formula for convex domains is that $\zeta(w,z)$ in it, is independent of w. It was a source of several interesting constructions of Martineau [8] (see below).

(3) As a the difference from one variable integral formulas for several variables essentially depend from boundaries of domains, including their smoothness. As a result, the cycle Γ in the Cauchy–Fantappie formula sometimes cannot be chosen as one-valued function on the boundary bD. Let us illustrate it on the simplest example of the bidisk: $D \subset \mathbb{C}^2 = \{|z_1| < 1, |z_2| < 1\}$. Let us separate 3 parts of its boundary bD:

$$\begin{aligned} &(A) \quad |z_1| < 1, \quad |z_2| = 1; \\ &(B) \quad |z_1| = 1, \quad |z_2| < 1; \\ &(C) \quad |z_1| = 1, \quad |z_2| = 1. \end{aligned}$$

For the cycle Γ, we take $\zeta \equiv (0,1)$ on (A); $\zeta \equiv (0,1)$ on (B) and $\zeta = (t-1, t_2)$, $t_1 + t_2 = 1$, $t_1, t_2 \geq 0$. So in Γ over the points of (C) there are segments. The restriction of Ω_{CF} on Γ we have zero on (A), (B), and on C we have

$$\int_0^1 \frac{dt_1}{t_1(z_1 - w_1) + (1 - t_1)(z_2 - w_2)} = \frac{1}{(z_1 - w_1)(z_2 - w_2)}$$

and we have the integral formula for the bidisk:

$$f(w) = \frac{1}{(2\pi i)^2} \int_{[|z_1|=1, |z_2|=1]} \frac{f(z)\, dz_1 \wedge dz_2}{(z_1 - w_1)(z_2 - w_2)}.$$

Most nontrivial specification of the Cauchy–Fantappie integral formula is the formula of Henkin–Ramires for strictly convex domains which gave to Henkin the possibility to construct multidimensional constructive complex analysis just on the basis of integral formulas.

Without a doubt, the fact that the Cauchy–Fantappie form Ω_{CF} algebraically coincides with the Radon inversion formula written as an integral identity of the Cauchy–Radon transform (see the end of Section 3.1), deserves careful deliberation. This coincidence was not remarked on by Leray. This form in these two situations is used quite differently. In the one case, we operate with real functions and in the beginning take the integrals along hyperplanes. In the second one, we work with holomorphic functions and pick up a special cycle Γ over the boundary for the integration.

Some convergence appears for tube domains in $\mathbb{C}^n$ where we can include real hyperplanes on the boundary in the cycle Γ and integrate along them producing the Cauchy–Radon transform (cf. Section 3.2 here). May be this algebraic form is a construction of a higher level. The famous Feynman formula (which we just applied for $n = 2$) has also a clear association with this form.

We already mentioned that one of the first applications of the Cauchy–Fantappie formula was the theory of analytic functionals of Martineau [8] in convex domains which were connected with the integral formula for convex domains (see example 2 above). It is more convenient to make these considerations in the projective space. Let D be a convex domain in the projective space $\mathbb{C}P^n$ with the homogeneous coordinates z and $\mathbb{C}P^n_\zeta$ the dual space of hyperplanes $\langle\zeta, z\rangle = 0$; so ζ are homogeneous coordinates in the dual projective space. Correspondingly, $D' \subset \mathbb{C}P^n_\zeta$ is the dual domain of hyperplanes which do not intersect D. It is convex as well. Instead of holomorphic functions, we consider sections of line bundles $f \in \mathcal{O}^{-k}(D)$–homogeneous functions of z of degree $-k$. For any $\zeta \in D'$ (such that the hyperplane $\langle\zeta, z\rangle = 0$ lies outside D), we have

$$\frac{1}{\langle\zeta, z\rangle}$$

is an element of $\mathcal{O}^{-1}(D)$. As a result, any analytic functional F on this space induces an element $F(\zeta)$ in $\mathcal{O}^{-1}(D')$—the evaluations of these special functions. It turns out that any $F(\zeta) \in \mathcal{O}^{-1}(D')$ uniquely extends up to an analytic functional using the integral formula for convex domains. Of course, it associates with the Cauchy–Radon transform. For $k \neq 1$, we can use degrees of the equation of the hyperplanes.

Let D^* be the concave domain which is the supplementary to the closure of D'. There are no holomorphic functions (or sections of line bundles) in this domain but there are $(n-1)$-dimensional $\bar{\partial}$-cohomology. It is possible to connect these cohomology with analytic functionals on $\mathcal{O}^{-1}(D)$ On other side, it is possible to define an operator of integration of this cohomology along projective hyperplanes inside D^*. These hyperplanes are parameterized by points of D and we have the operator from functionals on $\mathcal{O}^{-1}(D)$ to this space which can be treated as an analogue of the Radon transform. Its inversion can be done using the same remarkable differential form which appeared in the inversion of the Cauchy–Radon transform and in the Cauchy–Fantappie formula, but this time it applies again in a different way [5, 3]. This transform is an example of the Penrose transform—a complex analogue of the Radon transform.

Bibliography

[1] M. Eastwood, T. Bailey, and S. G. Gindikin. Smoothly parameterized Čech cohomology of complex manifolds. *J. Geom. Anal.*, 15:9–23, 2005.

[2] S. Gindikin. Complex horospherical transform on real sphere. In *Geometric Analysis of PDE and Several Complex Variables*. Contemporary Mathematics, volume 368, pages 227–232. Amer. Math. Soc., 2005.

[3] S. Gindikin. Inversion of the Penrose transform and the Cauchy–Fantappie formula. *J. Geom. Phys.*, 78:127–131, 2014.

[4] S. Gindikin, I. Gelfand, and M. Graev. *Selected Topics in Integral Geometry*. AMS, 2003.

[5] S. Gindikin and G. Khenkin. Integral geometry for $\bar{\partial}$-cohomology in q-linearly concave domains in $\mathbb{C}P^n$. *Funct. Anal. Appl.*, 12(4):6–23, 1978.

[6] H. Ishi and S. Gindikin. Cohomological Laplace transform on non-convex cones and Hardy spaces of $\bar{\partial}$-cohomology on non-convex tube domains. *J. Lie Theory*, 28(1):245–263, 2018.

[7] J. Leray. Function de variables complexe: sa représentation comme somme de puissances négatives de functions linéaires. *Atti Accad. Naz. Lincei, Rend. Cl. Sci. Fis. Mat. Nat.*, 20(8):589–590, 1956.

[8] A. Martineau. Indicatrices des fonctionellea analitiques at inversion de la transformation de Roiurier–Borel pat la transformation de la Laplace. *C. R. Acad. Sci.*, 255(16):2888–2890, 1962.

Joonas Ilmavirta and François Monard

4 Integral geometry on manifolds with boundary and applications

Abstract: We survey recent results on inverse problems for geodesic x-ray transforms and other linear and nonlinear geometric inverse problems for Riemannian metrics, connections, and Higgs fields defined on manifolds with boundary.

Keywords: Integral geometry, x-ray transforms, Riemannian manifolds, inverse problems

MSC 2010: 44A12, 53C65, 58J32, 58J40, 58J90, 35R30, 86A22

4.1 Introduction

Johann Radon and his contemporaries formulated several integral geometric problems, not only in linear but also in nonlinear settings [42, 123]. Such problems, namely *travel-time tomography* and *boundary rigidity* as later formulated in [62, 72], are concerned with recovering a Riemannian metric from the shortest length between any two boundary points. Such problems and their cousins (described below), now make the field of *integral geometry*, or how to reconstruct geometric features of a manifold from integral functionals defined over that manifold.

Nowadays, this field forms the basis of several noninvasive approaches to imaging internal properties of materials: seismology [42, 123], or how to reconstruct the density inside the Earth from first arrival times of seismic wavefronts; medical imaging since the development of x-ray computerized tomography [25, 75, 119]; single-photon emission computerized tomography using the attenuated x-ray transform [74, 76, 78]; vector tomography in helio-seismology [51, 52, 94]; ocean imaging [73]; x-ray diffraction strain tomography [19, 59], and tomography in elastic media [103, Chapter 7], [108];

Acknowledgement: J. I. was supported by the Academy of Finland (decision 295853). F. M. was partially supported by NSF grants DMS-1712790 and DMS-1814104. We are grateful for discussions and feedback with Colin Guillarmou, Sean Holman, Thibault Lefeuvre, Jere Lehtonen, Gabriel Paternain, Mikko Salo, Vladimir Sharafutdinov, Plamen Stefanov, Gunther Uhlmann, András Vasy, and Hanming Zhou.

***Corresponding author: François Monard,** Department of Mathematics, University of California Santa Cruz, 1156 High St., Santa Cruz, CA 95064, USA, e-mail: fmonard@ucsc.edu
Joonas Ilmavirta, Department of Mathematics and Statistics, University of Jyväskylä, P.O. Box 35, FI-40014 University of Jyväskylä, Finland, e-mail: joonas.ilmavirta@jyu.fi

https://doi.org/10.1515/9783110560855-004

neutron imaging, as applied to the imaging of vertebrate remains [102] and shales [11]. Nonlinear integral geometric problems also continue to find new applications: recently, neutron spin tomography [99] as a means to measure magnetic fields in materials, has arisen as a novel method which can be of use in electrical engineering, superconductivity, etc. The transform to invert in this case is a nonlinear operator, the so-called "non-Abelian x-ray transform" of the magnetic field; see Problem 3 below.

Recent breakthroughs have fueled the field, exploiting a combination of old and new methods. Examples of such methods are: the systematic use of analysis on the unit sphere bundle combining energy methods (also coined "Pestov identities"), initiated by Mukhometov [71] and generalized in [91, 103], and harmonic analysis on the tangent fibers [15, 82, 85]; in dimensions three and higher, the discovery in [120] that the existence of a foliation of the domain by strictly convex hypersurfaces, local or global, yields a powerful and robust approach to integral geometric inversions [116, 118, 125, 127], via a successful use of Melrose's scattering calculus [60]; the systematic use of analytic microlocal analysis to produce "generic" results, implying the unique identifiability of unknown parameters in an open and dense subset of all cases [45, 112, 128]; finally, recent results in the context of Anosov flows, leading to positive results for certain geometries with trapped sets [33, 34, 38].

This review article aims at giving an overview of the arsenal of these methods, and to describe to what extent they help coping with various geometric settings, whose complexity is mainly governed by two features of the flow considered: the presence of conjugate points and/or infinite-length trajectories.

Scope of the article. The article will be devoted to manifolds with variable curvature, with less emphasis on homogeneous spaces for which the methods employed in, e. g., [41], exploit homogeneity to a large extent and may not generalize. The emphasis will be put on manifolds with boundary, though many results enjoy counterparts in the realm of closed manifolds. The focus will be on mostly analytic methods, rather than topological or purely geometrical. The integration will be done over rays (no integration over higher-dimensional manifolds; see, however, the recent preprint involving an integral transform over two-dimensional leaves [100]). Recent topical reviews have been published on some of the topics covered in what follows [80, 84, 121], and we have attempted to minimize overlap.

It is our hope that this review article does justice to the field and its community, and we apologize in advance for any missing reference which would deserve to be included here. Let us mention that although the following topics are directly related to the current article, lack of time has prevented us to discuss range characterization issues, as provided, e. g., in [3, 7, 69, 86, 92, 98] and cases where the boundary is nonconvex, for which recent results appear in [36].

Notation:
- (M,g), ∂M, TM, T^*M, SM, $\partial_{+/-}SM$: a typical Riemannian manifold, its boundary, its tangent, cotangent, unit tangent bundles, and incoming/outgoing boundaries.
- $\varphi_t(x,v) = (\gamma_{x,v}(t), \dot\gamma_{x,v}(t))$: geodesic flow on SM.
- $C^\infty(A;B)$: space of smooth sections of a bundle $B \overset{\pi}{\to} A$, e. g., a smooth map $f : A \to B$ such that $f(x) \in \pi^{-1}(x)$ for every $x \in A$.
- d_g: boundary distance function of a metric g, defined on $\partial M \times \partial M$.
- $\tau : SM \to \mathbb{R}$: first exit time of the geodesic $\gamma_{x,v}(\cdot)$ out of M.
- $\mathcal{S}_g$: scattering relation of a metric g.
- X: geodesic vector field on SM.
- $\mathcal{I}$: ray transform over functions on SM.
- I_0: restriction of $\mathcal{I}$ to functions on M.
- $I_\perp h := \mathcal{I}[X_\perp h]$: restriction of $\mathcal{I}$ to solenoidal one-forms in two dimensions.
- $\mathcal{I}_{A,\Phi}$ or $\mathcal{I}_{\nabla,\Phi}$: transform with connection ∇ (associated with connection one-form A) and Higgs field Φ over sections of a bundle $E \to SM$.
- $I_{A,\Phi,0}$: restriction of $\mathcal{I}_{A,\Phi}$ to $C^\infty(M;E)$.
- $C_{A,\Phi}$: scattering data of the pair (A,Φ).

4.1.1 Main problems

We fix $(M,\partial M,g)$ a Riemannian manifold with boundary and $\mathcal{G}$ the set of all geodesics through M and ∇ the Levi–Civita connection. The manifold (M,g) has a unit tangent bundle

$$SM = \{(x,v) \in TM,\ |v|^2_{g(x)} = 1\}$$

with inward (+) and outward (–) boundaries

$$\partial_\pm SM = \{(x,v) \in SM,\ x \in \partial M,\ \pm g(v,\nu_x) > 0\},$$

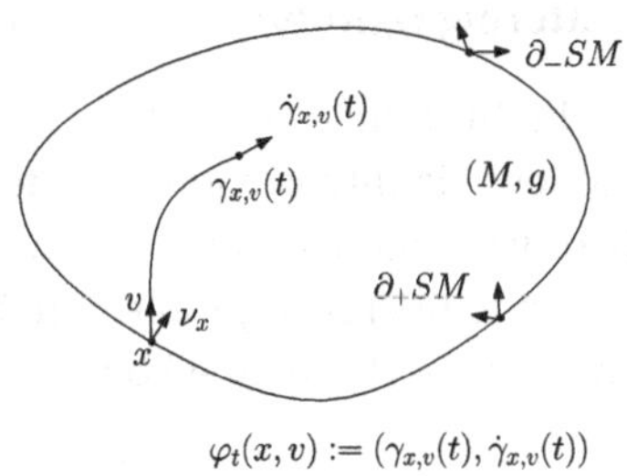

and where the geodesic flow $\varphi_t : SM \to SM$ is well-defined, with infinitesimal generator the geodesic vector field $X = \frac{d}{dt}\varphi_t(x,v)|_{t=0}$.

Given $(x,v) \in SM$, we denote $\tau(x,v)$ the first time $t \ge 0$ for which $\gamma_{x,v}(t) \in \partial M$, and we call M *nontrapping* if $\sup_{SM} \tau$ is finite. We say that ∂M is *strictly convex* if the second fundamental form is positive definite.

In what follows, symmetric (covariant) tensors of degree $m \ge 0$ will be denoted $S^m(T^*M)$. We will restrict our attention to smooth metrics, unless otherwise explicitly stated.

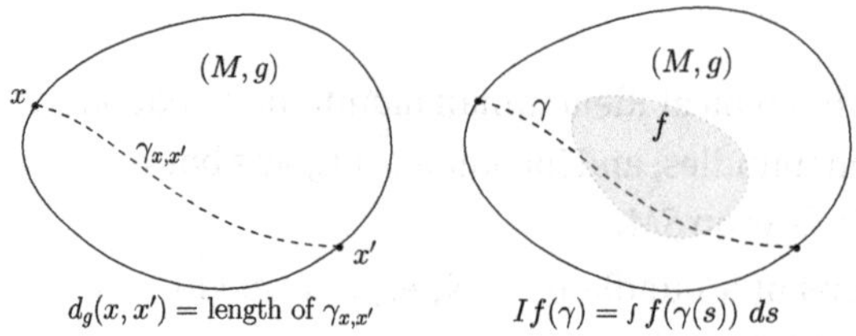

Figure 4.1: Settings for Problems 1 (left) and 2 (right).

4.1.1.1 Reconstruction of functions, metrics, and tensor fields

Given two boundary points $(x, x') \in \partial M \times \partial M$, we define the boundary distance

$$d_g(x, x') := \inf_\gamma \int |\dot\gamma(t)|_{g(\gamma(t))}\, dt,$$

where the infinimum is taken over all absolutely continuous curves in M with end-points x, x'. This defines a *boundary distance function* $d_g\colon \partial M \times \partial M \to [0, \infty)$, see Figure 4.1 (left). We also define the *scattering relation* $\mathcal{S}_g\colon \partial_+ SM \to \partial_- SM$, given by $\mathcal{S}_g(x, v) = \varphi_{\tau(x,v)}(x, v)$.

Both maps above have a natural invariance: if $\psi\colon M \to M$ is a diffeomorphism fixing every boundary point of M, then $d_{\psi^* g} = d_g$ and $\mathcal{S}_{\psi^* g} = \mathcal{S}_g$. This invariance is written as an equivalence relation: $g \sim_1 g'$ iff there exists $\psi\colon M \to M$ diffeomorphism fixing ∂M such that $g' = \psi^* g$. We can now formulate three nonlinear inverse problems:

Problem 1 (Boundary, lens and scattering rigidity). *Given (M, g) a Riemannian manifold with boundary:*
Boundary rigidity: *Does d_g determine g modulo $\sim_1$?*
Lens rigidity: *Does $(\tau|_{\partial_+ SM}, \mathcal{S}_g)$ determine g modulo $\sim_1$?*
Scattering rigidity: *Does $\mathcal{S}_g$ determine g modulo $\sim_1$?*

In this article, we will not discuss scattering rigidity. In addition, it is well-known that lens rigidity is equivalent to boundary rigidity for simple manifolds, while lens rigidity is a more natural setting in general.

On to the linear problem, fixing f a symmetric m-tensor, the *geodesic x-ray transform* $If\colon \mathcal{G} \to \mathbb{R}$ is defined by

$$If(\gamma) = \int f_{\gamma(t)}(\dot\gamma(t)^{\otimes m})\, dt, \quad \gamma \in \mathcal{G}, \tag{1}$$

see Figure 4.1 (right). Such a linear transform has a natural kernel for $m \ge 1$, namely: if h is an $m-1$-tensor vanishing at ∂M, and σ denotes symmetrization, then $I(\sigma\nabla h) \equiv 0$. This kernel is therefore made of so-called *potential tensors*, and we write $f \sim_2 f'$ iff they differ by a potential tensor field. The relation $\sim_2$ is an equivalence relation. In general, the x-ray transform of a function $f\colon SM \to \mathbb{R}$ can be defined as

$$\mathcal{I}f(\gamma) = \int f(\gamma(t), \dot\gamma(t))\, dt. \tag{2}$$

This can be seen as a generalization of (1); see Section 4.2.1.3.

Problem 2 (Tensor tomography (TT(m))). *Does If determine $f \in S^m(T^*M)$ modulo $\sim_2$? If $m = 0$, does If determine f?*

Problem 2 for $m = 0$ and $m = 2$ arises as a linearization of Problem 1. When TT(m) is true for $m \geq 1$, we also say that I is *solenoidal-injective* (or in short, *s-injective*) or injective over solenoidal tensors. This is because by virtue of Sharafudtinov's decomposition, every m-tensor f with L^2 components is $\sim_2$-equivalent to a unique solenoidal tensor field f^s (i. e., satisfying $\delta f^s = 0$ with δ the formal adjoint of $-\sigma\nabla$), satisfying a continuity estimate of the form $\|f^s\|_{L^2} \leq C\|f\|_{L^2}$ for some constant $C(M, m)$.

4.1.1.2 Reconstruction of connections, Higgs fields, and sections of bundles

Now fix an n-dimensional vector bundle $E \xrightarrow{\pi} M$ and (A, Φ) a (connection, Higgs field) pair on this bundle; see also Section 4.2.2 below. In a local trivialization, A is a $n \times n$ matrix of one-forms and Φ is a $n \times n$ matrix of functions, and such quantities allow to lift any path $c(t)$ on M into a path $\tilde{c}(t)$ on E (in the sense that $\pi(\tilde{c}(t)) = c(t)$ for every t) by solving the ODE

$$\frac{d\tilde{c}}{dt} + (A_{c(t)}(\dot{c}(t)) + \Phi(c(t)))\tilde{c}(t) = 0.$$

Then a curve $\varphi_t(x, v)$ If $(x, v) \in \partial_+ SM$, and let S belong to the fiber above x. Assuming that the geodesic $\gamma_{x,v}$ exits M for the first time at $\tau > 0$ with $x' = \gamma_{x,v}(\tau) \in \partial M$, then the solution $\tilde{\gamma}_{x,v}$ of the ODE above with curve $c = \gamma_{x,v}$, augmented with the initial condition $\tilde{\gamma}_{x,v}(0) = S$ allows to uniquely "parallel-transport" the state S to the state $\tilde{\gamma}_{x,v}(\tau)$ above x', which we denote $C_{A,\Phi}(x, v)S$, see Figure 4.2 (left). A natural question is to ask whether the *scattering data* (or *non-Abelian ray transform*) $C_{A,\Phi}(x, v)S$, known for all $(x, v, S) \in \partial_+ SM \times \mathbb{C}^n$, determines the pair (A, Φ). To formulate this problem, we first rule out a natural obstruction.

We write $(A, \Phi) \sim_3 (B, \Psi)$ if there exists $Q \in C^\infty(M, GL(n, \mathbb{C}))$ with $Q|_{\partial M} = Id$, such that $B = Q^{-1}dQ + Q^{-1}AQ$ and $\Phi = Q^{-1}\Psi Q$. When this is true, it is easy to see that $C_{A,\Phi}(x, v)S = C_{B,\Psi}(x, v)S$, since in this case, if $\tilde{\gamma}_{x,v}$ is the (A, Φ)-lift of $\gamma_{x,v}$, then $Q\tilde{\gamma}_{x,v}$ is the (B, Ψ)-lift of $\gamma_{x,v}$, and both lifts agree at both endpoints.

Problem 3 (Non-Abelian x-ray transform). *Does $C_{A,\Phi}$ determine (A, Φ) modulo $\sim_3$?*

In the case $n = 1$, it is easy to see that $C_{A,\Phi}(x, v)1 = \exp(\mathcal{I}[A + \Phi](x, v))$ so that the problem is a usual x-ray transform. In the case $n = 3$, Problem 3 also applies to neutron spin tomography [19], a case where $A = 0$ and where Φ, valued in the Lie algebra $\mathfrak{so}(3)$, models the unknown magnetic field.

The linear counterpart of Problem 3 is as follows: let E, A, Φ as above, and fix m a tensor order. If f a section of $E \otimes S^m(T^*M)$ (an E-valued symmetric m-tensor), for $\gamma \in \mathcal{G}$,

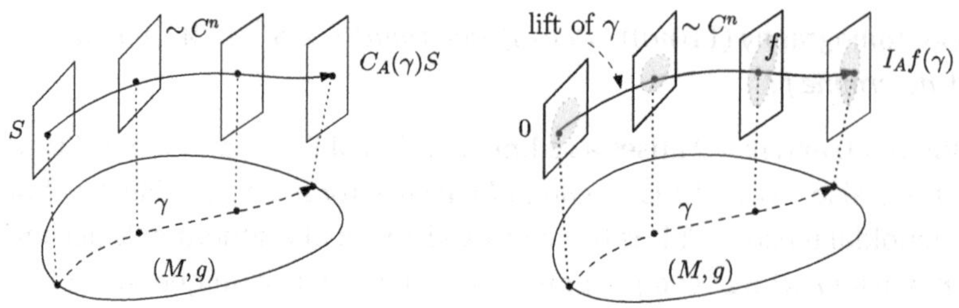

Figure 4.2: Settings for Problems 3 (left) and 4 (right).

the *attenuated x-ray transform*[1] $I_{A,\Phi}f(\gamma)$ is the integral over γ of all values of f above each point of γ (paired m times with $\dot\gamma$), parallel-transported to a common point via (A,Φ), see Figure 4.2 (right). If $m \geq 1$, this problem has a natural obstruction and to point it out, it is natural to view the transform as defined over sums of m-tensor/$m-1$-tensor denoted by $f_m + f_{m-1}$: we say that $f_m + f_{m-1} \sim_4 f'_m + f'_{m-1}$ if there exist an $m-1$ tensor p vanishing at ∂M such that $f_m = f'_m + \sigma\nabla p + Ap$ and $f_{m-1} = f'_{m-1} + \Phi p$. Whenever $f_m + f_{m-1} \sim_4 f'_m + f'_{m-1}$, we have that $I_{A,\Phi}(f_m + f_{m-1}) = I_{A,\Phi}(f'_m + f'_{m-1})$. The natural question is therefore the following.

Problem 4 (Attenuated x-ray transform). *Does $I_{A,\Phi}(f_m + f_{m-1})$ determine (f,h) modulo $\sim_4$? If $m = 0$ (f is a section of E), does $I_{A,\Phi}f$ determine f?*

4.1.2 The inverse problems agenda in a geometric context

For each one of Problems 1–4, one may ask the typical inverse problems questions:
(i) Is the operator injective, modulo the natural obstructions?
(ii) If yes, in what topology is the inverse continuous?
(iii) How to explicitly and efficiently invert the operator?
(iv) How to characterize the range of the operator?
(v) In the presence of noisy data, what is a proper regularization approach and how is it statistically optimal?

The answers to questions (i)–(v) strongly depend on the underlying geometric features of the manifold, the geometry and topology of M (namely, the presence or absence of conjugate points[2] and/or trapped geodesics[3]), the structure of the connection and

1 Or x-ray transform with connection and Higgs field.

2 Two points x, y are *conjugate* along a geodesic γ if there exists a Jacobi field over γ vanishing at both x and y.

3 The *trapped set* of (M,g) is the set of points $(x,v) \in SM$ such that the length of $\gamma_{x,v} \cap M$ is infinite, where $\gamma_{x,v}$ is the unique maximal geodesic satisfying $\gamma(0) = x$ and $\dot\gamma(0) = v$.

Higgs field (rank, structure group, etc.), the dimension of the manifold (including significant differences in the landscapes of results between dimension two, and higher dimensions), and the presence of weights in the transforms.

Many answers are positive in the case of homogeneous spaces [29, 41, 96] and in the case of *simple*[4] *geometries*: in the case of simple surfaces, it is known that such surfaces are boundary distance rigid [62, 93], and that ray transforms are injective over functions [72] and over solenoidal tensors of any order [82], also when one includes many types of connections and Higgs fields [20, 27, 69, 77, 81, 106, 122]; for higher-dimensional simple manifolds, generic injectivity and stability results are known [112, 128], injectivity of x-ray transforms is known over functions and vector fields, and for higher-order tensor fields, the result is true under stronger assumptions on the geometry [4, 5, 71, 91, 97, 103, 104, 107].

In geometries with *conjugate points*, another separation between two- and higher-than-three dimensions occurs: in two dimensions, conjugate points on surfaces unconditionally destroy stability of x-ray transforms [70, 114] while the question needs to be refined in higher dimensions and exhibits a tradeoff between the order of conjugate points considered and the dimension of the manifold [44]. In fact, there is more at play in higher dimensions: the mere existence of a foliation by strictly convex hypersurfaces allows to prove global injectivity and stability [87, 116, 120]. Such a criterion allows for conjugate points and some form of trapped geodesics as well, and as such shifts the focus to the following question: which manifolds admit strictly convex foliations? Injectivity questions remain open on surfaces with conjugate points, except for the case of circularly symmetric ones, where injectivity over solenoidal tensor fields is known to hold [104], and injectivity over piecewise constant functions holds [48].

In geometries with *trapped geodesics*, one may easily construct counterexamples to injectivity, and thus one must assume some thing about the trapped set. Under the crucial assumption that the trapped set be *hyperbolic* for the geodesic flow (a condition which is always true on manifolds with negative sectional curvatures), injectivity, and stability can be restored in many cases [33, 34, 38].

See also the recent topical reviews [121] on Problems 1–2, and [80] on Problems 3–4.

4.1.3 Outline

The remainder of the article is organized as follows.

We devote Section 4.2 to introductory material and notation, describing the geometrical framework needed to discuss integral geometry. This includes basic geometry

4 A Riemannian manifold (M, g) is *simple* if it is nontrapping, ∂M is strictly convex and M contains no conjugate points.

of the sphere bundle (on which the geodesic flow lives) and natural operators on it, transport equations, tensor fields, conjugate points, and two-dimensional structure. We also briefly discuss trapping (Section 4.2.3) and connections (Section 4.2.2).

Sections 4.3–4.8 then present results, arranged by methods.

- In Section 4.3, we discuss energy estimates known as Pestov identities. We give the fundamental commutators in Section 4.3.1 before deriving a Pestov identity on simple manifolds (Section 4.3.2). We then extend the methods to other geometrical settings (Section 4.3.4), connections and Higgs' fields (Section 4.3.5) and generalized geodesic flows (Section 4.3.6).
- In Section 4.4 we discuss explicit approaches to injectivity and inversion in two dimensions.
- Section 4.5 covers invariant distributions and their relation to tensor tomography, including their construction via iterated Beurling series (Section 4.5.3).
- In Section 4.6, we discuss applications of microlocal analysis to integral geometry. This includes analysis of cases with and without conjugate points, geometry of Fourier integral operators, and general families of curves.
- In Section 4.7, we turn to layer stripping arguments and methods based on considerations of convexity. These rely on a combination of local support theorems and a global foliation of the manifold. We discuss different methods to obtain local support theorems.
- While the results up to this point have been mainly linear, we discuss some relations between linear and nonlinear problems in Section 4.8.

Section 4.9 concludes with a small collection of open questions.

4.2 Geometric setting and tools

A natural reformulation of integral geometric problems involving the integration of objects along curves, is by viewing the integrand as a source term for a "geometric" transport equation posed on the tangent bundle, and apply various PDE methods (energy identities, etc.) to that equation. Such ideas are not new and form the basis of V. A. Sharafutdinov's pioneering monograph [103]. The main difference of our presentation (which largely follows [85]) is in how to represent integrands of tensor field type as natural objects to be integrated over a flow in phase space: in [103], a section of $S^m(T^*M)$ is identified with a so-called semibasic tensor field on TM (i. e., covariant in horizontal directions and contravariant in vertical ones in a certain sense). Here, tensor fields are regarded as scalar functions on the sphere bundle SM, whose tensorial nature is encoded in the finite expansions in spherical harmonics on the unit tangent fibers. This latter identification somewhat allows to bypass the proliferation of indices as one increases the tensor order.

4.2.1 The geometry of the unit sphere bundle

4.2.1.1 Vertical and horizontal vectors on the sphere bundle

Given (M, g) a Riemannian manifold, local charts on the tangent bundle TM may be written as (x, y), where the tangent vector looks locally like $y = y^i \frac{\partial}{\partial x_i}$ (here and below, repeated indices are summed over). The double tangent bundle $T(TM)$ admits a horizontal-vertical splitting which embodies whether one is differentiating vertically (along a fiber), or horizontally (along the base, while keeping a tangent vector "fixed"). Horizontal directions depend on the Riemannian metric, while vertical ones only on the smooth structure.

Specifically, the *vertical subbundle* of $T(TM)$ is defined so that the fiber $V(x, y)$ at $(x, y) \in TM$ is $V(x, y) = \ker(d_{(x,y)}\pi)$, where $\pi: TM \to M$ is the canonical projection. To define the *horizontal subbundle*, we define a connection map $K: T(TM) \to TM$ fiber by fiber. Take any $\xi \in T_{(x,y)}TM$ and $\sigma: (-\epsilon, \epsilon) \to TM$ a curve with $\sigma(0) = (x, y)$ and $\dot{\sigma}(0) = \xi$. We may write this curve as $\sigma(t) = (y(t), Z(t))$, where y is a curve on M and Z a vector field along it. Upon defining $K_{(x,y)}\xi := (\nabla_y Z)(0)$, the fiber of the horizontal bundle at (x, y) is then $H(x, y) := \ker(K_{(x,y)})$. Each fiber of the $T(TM)$ then decomposes as

$$T_{(x,y)}TM = H(x,y) \oplus V(x,y), \quad (x,y) \in TM.$$

In local coordinates $V(x, y) = \langle \frac{\partial}{\partial y^i}, 1 \le i \le n \rangle$ while $H(x, y) = \langle \delta_{x_i} := \frac{\partial}{\partial x_i} - \Gamma_{ij}^k y^j \frac{\partial}{\partial y^k}, 1 \le i \le n \rangle$. With the splitting above, the maps

$$d_{(x,y)}\pi|_{H(x,y)}: H(x,y) \to T_x M, \quad K_{(x,y)}|_{V(x,y)}: V(x,y) \to T_x M$$

are linear isomorphisms, allowing us to freely identify horizontal and vertical vectors on $T_{(x,y)}TM$ with vectors on $T_x M$. These isomorphisms become isometries (and the splitting, orthogonal) upon introducing the *Sasaki metric* at $(x, y) \in TM$ defined for $\xi, \eta \in T_{(x,y)}TM$ by

$$\langle \xi, \eta \rangle_{x,y} := g_x(K_{(x,y)}(\xi), K_{(x,y)}(\eta)) + g_x(d_{(x,y)}\pi(\xi), d_{(x,y)}\pi(\eta)),$$

or equivalently in coordinates, with $\xi = X^i \delta_{x_i} + Y^i \frac{\partial}{\partial y^i}$ and $\eta = \widetilde{X}^i \delta_{x_i} + \widetilde{Y}^i \frac{\partial}{\partial y^i}$,

$$\langle \xi, \eta \rangle_{x,y} = g_{ij} X^i \widetilde{X}^j + g_{ij} Y^i \widetilde{Y}^j.$$

The unit sphere bundle SM of a Riemannian manifold M is the subbundle of TM consisting of unit tangent vectors of unit length:

$$SM = \{(x,v);\ x \in M,\ v \in T_x M,\ |v| = 1\}.$$

There the horizontal-vertical splitting becomes

$$T_{x,v}SM = \mathbb{R}X(x,v) \oplus \mathcal{H}(x,v) \oplus \mathcal{V}(x,v), \quad (x,v) \in SM, \tag{3}$$

where $\mathbb{R}X \oplus \mathcal{H}(x,v) = H(x,v)$ and $\mathcal{V}(x,v) = \ker d_{x,v}(\pi|_{SM})$. Elements of $\mathcal{H}(x,v)$ and $\mathcal{V}(x,v)$, when identified as vectors of T_xM, are both orthogonal to v, so smooth sections of $\mathcal{H}$ and $\mathcal{V}$ can be isomorphically identified with smooth sections in $\mathcal{Z} := C^\infty(SM,N)$, where we define the bundle $N \to SM$ by

$$N := \bigcup_{(x,v)\in SM} \{v\}^\perp, \quad \{v\}^\perp := \{w \in T_xM,\ g_x(w,v) = 0\}. \tag{4}$$

According to the decomposition (3), the total gradient of a scalar function u on the sphere bundle SM consists of three Sasaki-orthogonal components: the geodesic derivative Xu (scalar-valued), and the vertical and horizontal gradients $\overset{\mathrm{v}}{\nabla} u$ and $\overset{\mathrm{h}}{\nabla} u$ (each identified with elements of $\mathcal{Z}$). In particular, we have two differential operators

$$\overset{\mathrm{v}}{\nabla}\colon C^\infty(SM) \to \mathcal{Z}, \quad \overset{\mathrm{h}}{\nabla}\colon C^\infty(SM) \to \mathcal{Z}.$$

Roughly speaking, the vertical gradient of $u(x,v)$ is the gradient of with respect to v and the horizontal gradient is the component of the gradient with respect to x orthogonal to v. If $\dim(M) = 2$, then these two gradients can be regarded as scalars as done in Section 4.2.1.5 below.

The adjoints are the vertical and horizontal divergences which we denote $-\overset{\mathrm{v}}{\mathrm{div}}$ and $-\overset{\mathrm{h}}{\mathrm{div}}$ with the following mapping properties:

$$\overset{\mathrm{v}}{\mathrm{div}}\colon \mathcal{Z} \to C^\infty(SM), \quad \overset{\mathrm{h}}{\mathrm{div}}\colon \mathcal{Z} \to C^\infty(SM).$$

The geodesic vector field also acts on $\mathcal{Z}$ by covariant differentiation along the geodesic flow.

4.2.1.2 The x-ray transform and transport equations on *SM*

In the framework just described, given $F \in L^2(SM)$, the x-ray transform of F defined on (1) can be viewed as the inward restriction $u|_{\partial_+ SM}$ of the solution u to a transport problem

$$Xu = -F \quad (SM), \quad u|_{\partial_- SM} = 0. \tag{5}$$

With this setting in mind, injectivity questions and inversion formulas can be tackled by classical PDE methods on manifolds: for instance, injectivity over functions means: if $Xu = -F(x)$ throughout SM and $u|_{\partial SM} = 0$, does this imply $F = 0$?

Similarly, to address tensor tomography, there is a natural way to identify a symmetric m-tensor field f on M with a scalar field $\ell_m f$ on SM, given by a mapping

$$\begin{aligned} &\ell_m: C^\infty(S^m(T^*M)) \to C^\infty(SM), \quad \ell_m: L^2(S^m(T^*M)) \to L^2(SM) \\ &\ell_m f(x,v) = f_x(v,\dots,v), \quad (x,v) \in SM. \end{aligned} \tag{6}$$

Via this identification, the x-ray transform of f is again given by $If := u|_{\partial_+ SM}$, where u solves the transport problem (5) with right-hand side $\ell_m f$. Whenever the context allows, we will not distinguish f and $\ell_m f$.

4.2.1.3 Tensor fields and spherical harmonics

The L^2 space of every fiber of the sphere bundle can be decomposed into eigenspaces of the vertical Laplacian

$$-\overset{\vee}{\operatorname{div}}\,\overset{\vee}{\nabla}: C^\infty(SM) \to C^\infty(SM).$$

Namely, on each spherical fiber, the vertical Laplacian coincides with the Laplacian of the function $v \mapsto u(x,v)$ on the manifold (S_xM, g_x), whose spectrum is the same as that of the spherical Laplacian $\Delta_{\mathbb{S}^{n-1}}$, given by $\lambda_m = m(n+m-2)$ for $m = 0,1,2,\dots$, with eigenfunctions the spherical harmonics. The corresponding eigenspaces induce an $L^2(SM)$-orthogonal decomposition

$$L^2(SM) = \bigoplus_{m=0}^{\infty} H_m(M), \quad H_m := \ker(-\overset{\vee}{\operatorname{div}}\,\overset{\vee}{\nabla} - \lambda_m Id) \cap L^2(SM), \tag{7}$$

which on each fiber over M is just the spherical harmonic decomposition in S^{n-1}. Let us also set $\Omega_m := H_m(M) \cap C^\infty(SM)$. Then any function $f \in L^2(SM)$ splits as $f = \sum_{m=0}^\infty f_m$ so that for almost every $x \in M$ the function $v \mapsto f_m(x,v)$ is a spherical harmonic of order m. The zeroth component f_0 of a function on the sphere bundle is the fiberwise average.

Tensor fields and finite harmonic content

In the decomposition above, an mth order tensor field f, via its identification (6) with $\ell_m f$, can be regarded as a function on SM which only contains spherical harmonics up to order m and of the same parity as m. Conversely, if a scalar function u on SM contains spherical harmonics up to a finite order m and they all have the same parity, then there is a tensor field f so that $\ell_m f = u$.

Since $\ell_{m+1}(\sigma\nabla h) = X(\ell_m h)$ and $If = \mathcal{I}(\ell_m f)$ (see equations (1), (2), (6)), the tensor tomography Problem 2 can be recast as follows: If $f: SM \to \mathbb{R}$ only contains spherical harmonics up to order m and integrates to zero over all (lifted) geodesics of M, is there a function h with spherical harmonics up to order $m-1$ so that $f = Xh$? In terms of the transport equation (5), the question is whether the spherical harmonic expansion of u ends at order $m-1$.

Decomposition of X

The geodesic vector field behaves nicely with respect to the decomposition (7): it maps Ω_m into $\Omega_{m-1} + \Omega_{m+1}$ [40, Proposition 3.2]. Hence on Ω_m we can write

$$X = X_+ + X_-, \quad \text{where } X_\pm : \Omega_m \to \Omega_{m\pm1} \quad (\text{convention: } \Omega_{-1} \equiv 0)$$

and such that, for $u \in \Omega_m$ and $w \in \Omega_{m+1}$ and one of them vanishes on ∂SM, we have

$$(X_+u, v) = -(u, X_-v).$$

In particular, the transport equation (5), upon projecting onto each harmonic subspace Ω_k, can be equivalently viewed as the tridiagonal system of equations

$$X_+u_{m-1} + X_-u_{m+1} = -f_m, \quad m = 0, 1, \dots. \tag{8}$$

4.2.1.4 Jacobi fields and conjugate points

Given a geodesic γ and p, q two distinct points on it, we say that p and q are *conjugate* along γ if there exists a nontrival Jacobi field along γ which vanishes at both p and q. Specifically, if $p = \gamma(t_1)$ and $q = \gamma(t_2)$ for some $t_1 < t_2$, there exists $J(t) \in T_{\gamma(t)}M$ a nontrivial solution of

$$D_t^2 J(t) + R(J(t), \dot\gamma(t))\dot\gamma(t) = 0, \quad J(t_1) = 0, \quad J(t_2) = 0,$$

where D_t denotes Levi–Civita covariant differentiation and R denotes the Riemannian curvature tensor. Since a pair of points can be conjugate along more than one geodesic (e. g., antipodal points on a sphere), it can be useful to keep track along which geodesic a pair of points is conjugate. A way to do this is to keep track of the tangent vectors, and to consider conjugate pairs as a subset of $SM \times SM$; see also Section 4.6.3.3.

An equivalent definition which is more amenable to generalizing this concept to other flows, is to say that, with $\varphi_t(x, v) = (\gamma_{x,v}(t), \dot\gamma_{x,v}(t))$ denoting the geodesic flow on SM, the points (x, v) and $\varphi_t(x, v)$ are conjugate (along the geodesic $\gamma_{x,v}$) if

$$\mathcal{V}(x, v) \cap d\varphi_{-t}|_{\varphi_t(x,v)}\mathcal{V}(\varphi_t(x, v)) \neq \{0\}.$$

In other words, conjugate points occurs when the differential of the flow maps vertical vectors into vertical vectors.

As we will see below, many positive results hold in the absence of conjugate points. In their presence, two-dimensional problems usually become unstable, and higher-dimensional ones require further discussion; see, in particular, Sections 4.6 and 4.7.

4.2.1.5 Additional structure in two dimensions

In two dimensions, the unit circle bundle SM admits a global framing by three global sections of $T(SM)$: a first section is the geodesic vector field $X = \frac{d}{dt}|_{t=0}\varphi_t(x,v)$; a second is the generator of the rotation group on the fibers $V = \frac{d}{dt}|_{t=0}\rho_t(x,v)$ (assuming the surface to be oriented, giving rise to a rotation-by-$\pi/2$ operator $v \mapsto v_\perp$, then $\rho_t(x,v) = (x, (\cos t)v + (\sin t)v_\perp)$); finally, their commutator $X_\perp := [X,V]$ gives the third one. Such vector fields admits the following structure equations encoding the geometry

$$[X,V] = X_\perp, \quad [X_\perp, V] = -X, \quad [X, X_\perp] = -\kappa V, \tag{9}$$

with κ denoting the Gaussian curvature. The *Sasaki metric* on SM makes $(X, X_\perp, V)$ orthonormal, with Liouville measure denoted $d\Sigma^3$. In two dimensions, we identify $\overset{h}{\nabla} u = -(X_\perp u)v^\perp$ and $\overset{v}{\nabla} u = (Vu)v^\perp$ (smooth sections of N defined in (4)) with the functions $X_\perp u$ and Vu (smooth functions on SM), respectively.

Locally (or globally, if M is simply connected), SM can be parameterized in *isothermal coordinates* (x,y,θ), where $g = e^{2\lambda(x,y)}(dx^2 + dy^2)$, θ is the angle between a tangent vector v and ∂_x, namely a tangent vector v sitting above (x,y) has the expression $v = e^{-\lambda(x,y)}\binom{\cos\theta}{\sin\theta}$, the Liouville form reads $d\Sigma^3 = e^{2\lambda}\,dx\,dy\,d\theta$, and the canonical frame reads

$$\begin{aligned} X &= e^{-\lambda}(\cos\theta\partial_x + \sin\theta\partial_y + (-\sin\theta\partial_x\lambda + \cos\theta\partial_y\lambda)\partial_\theta), \quad V = \partial_\theta, \\ X_\perp &= -e^{-\lambda}(-\sin\theta\partial_x + \cos\theta\partial_y - (\cos\theta\partial_x\lambda + \sin\theta\partial_y\lambda)\partial_\theta). \end{aligned}$$

Jacobi fields. The structure equations (9) make it convenient to compute Jacobi fields (or variations of the exponential map). For $\xi \in T_{(x,v)}(SM)$, we may decompose $d\varphi_t(\xi)$ along the frame $\{X(t), X_\perp(t), V(t)\}$ at the basepoint $\varphi_t(x,v)$ as

$$d\varphi_t(\xi) = \zeta_1(x,v,t)X(t) + \zeta_2(x,v,t)X_\perp(t) + \zeta_3(x,v,t)V(t).$$

Equations (9) provide us a differential system in t for the coefficients ζ_j (see, e. g., [61, Section 4.2]):

$$\dot\zeta_1 = 0, \quad \dot\zeta_2 + \zeta_3 = 0, \quad \dot\zeta_3 - \kappa(\gamma_{x,v}(t))\zeta_2 = 0.$$

In particular, we may express the variation fields $d\varphi_t(X_\perp) = aX_\perp(t) - \dot a V(t)$ and $d\varphi_t(V) = -bX_\perp(t) + \dot b V(t)$ in terms of two functions $a(x,v,t)$, $b(x,v,t)$ defined for

$(x, v) \in SM$ and $t \in (-\tau(x, -v), \tau(x, v))$, solving the scalar Jacobi equation

$$\ddot{a} + \kappa(\gamma_{x,v}(t))a = \ddot{b} + \kappa(\gamma_{x,v}(t))b = 0, \quad \begin{bmatrix} a & b \\ \dot{a} & \dot{b} \end{bmatrix}(0) = \begin{bmatrix} 1 & 0 \\ 0 & 1 \end{bmatrix}. \tag{10}$$

Here, the function $b(x, v, t)$ is the one that detects conjugate points on M. Specifically, if $t > 0$ is such that $b(x, v, t) = 0$, then the points x and $x' = \gamma_{x,v}(t)$ are conjugate along $\gamma_{x,v}$ as the Jacobi field $J(t) = b(x, v, t)\dot{\gamma}_{x,v}^{\perp}(t)$ vanishes at both x and x'.

4.2.2 Connections and Higgs fields

To set the stage similar to Section 4.1.1.2, let $E \to M$ a Hermitian vector bundle[5] over M. We assume that the fiber over each point is a copy of $\mathbb{C}^r$, where r is called the *rank* of the bundle. Let ∇^E a connection on E. We say that ∇^E is *Hermitian* (or *unitary*) if the following identity is satisfied:

$$Y(u, u')_E = (\nabla^E_Y u, u')_E + (u, \nabla^E_Y u')_E,$$

for all vector fields Y on M and sections $u, u' \in C^\infty(M; E)$. Via the canonical projection $\pi : SM \to M$, such a bundle and its connection can be pulled back into a bundle

$$\pi^* E := \{(x, v; e), (x, v) \in SM, e \in E_x\}$$

over SM with Hermitian connection $\pi^*\nabla^E$, which is where geodesic transport equations will be naturally written.[6] Following the spherical harmonic decomposition on the tangent spheres, one may still decompose an element $u \in C^\infty(SM; E)$ into a sum $u = \sum_{k=0}^\infty u_k$.

The geodesic vector field can be viewed as acting on sections of E by $\mathbb{X}u := \nabla^E_X u$ for a section $u \in C^\infty(SM; E)$, and for $f \in C^\infty(SM; E)$, this incarnation of the x-ray transform is given by $I_\nabla f := u|_{\partial_+ SM}$, where u solves the transport problem

$$\mathbb{X}u = -f \quad (SM), \quad u|_{\partial_- SM} = 0. \tag{11}$$

Note that in a local trivialization, the connection can be represented as a $r \times r$ matrix of one-forms A, and then $\mathbb{X}$ reads as $\mathbb{X} = X + A$, where X acts componentwise.

One may also add a *Higgs field* Φ, that is to say, a smooth section of the endomorphism bundle End(E) such that at every $x \in M$, Φ_x is a linear operator $\Phi_x : E_x \to E_x$. A Higgs field Φ is called a *skew-Hermitian Higgs field* if the endomorphisms Φ_x on each fiber are skew-Hermitian. The Higgs field is the "matrix" generalization of a position-dependent attenuation coefficient, and given $f \in C^\infty(SM; E)$, we can then define the attenuated transform $\mathcal{I}_{\nabla,\Phi} f = u|_{\partial_+ SM}$, where $u \in C^\infty(SM; E)$ solves the transport problem

$$(\mathbb{X} + \Phi)u = -f \quad (SM), \quad u|_{\partial_- SM} = 0.$$

5 In the sense that each fiber is a vector space endowed with a Hermitian inner product $(\cdot, \cdot)_E$.

6 The notational distinction between (E, ∇^E) and their pullbacks will be omitted as in [38].

4.2.3 Trapped geodesics and the hyperbolicity condition

So far, all metrics considered assumed that all geodesics exit the domain M in finite time. If this is no longer the case, we say that the manifold is *trapping*, and define the incoming (–) and outgoing (+) tails

$$\Gamma_\pm := \{(x,v) \in SM, \tau(x,\mp v) = +\infty\},$$

as well as the *trapped set* $K = \Gamma_- \cap \Gamma_+$, invariant by the flow, and consisting of those points which are trapped both forward and backward in time (in general, $(x,v) \in \Gamma_-$ is trapped forward and $(x,v) \in \Gamma_+$ is trapped backward). Geodesics cast from $\partial_+ SM \cap \Gamma_-$ never exit the domain and as such cannot be detected. Moreover, the data would blow up at such geodesics.

Without specific assumption on K, a trapped set can easily generate an infinite-dimensional kernel for an x-ray transform, as the following example suggests: glue a hemisphere on top of a Euclidean cylinder to make it simply connected. Any function supported on the cylinder, circularly symmetric, integrating to zero along the longitudinal direction is in the kernel of the x-ray transform.

A dynamical condition which allows to produce positive answers on manifolds with nontrivial topology [33, 35, 37, 38] is to assume that the trapped set be *hyperbolic* for the geodesic flow. Namely, one may define the *stable bundle* $E_- \subset T_{\Gamma_-}SM$ and *unstable bundle* $E_+ \subset T_{\Gamma_+}SM$ such that

$$\begin{aligned}
&\forall (x,v) \in \Gamma_-,\ \forall t > 0,\ \forall w \in E_-(x,v), \quad \|d\varphi_t(x,v)w\| \le Ce^{-\gamma t}\|w\|,\\
&\forall (x,v) \in \Gamma_+,\ \forall t < 0,\ \forall w \in E_+(x,v), \quad \|d\varphi_t(x,v)w\| \le Ce^{\gamma t}\|w\|,
\end{aligned}$$

with C, γ some uniform positive constants. Upon defining the same bundles over K by restriction, $E_s := E_-|_{T_K SM}$ and $E_u := E_+|_{T_K SM}$, we say that the set K is hyperbolic if and only if

$$\forall (x,v) \in K, \quad T_{(x,v)}SM = \mathbb{R}X(x,v) \oplus E_s(x,v) \oplus E_u(x,v).$$

The assumption of hyperbolic trapping has the following advantages:

(i) The Liouville volume of $\Gamma_+ \cup \Gamma_-$ is zero, and so is the measure of $\Gamma_+ \cup \Gamma_-$ on the boundary ∂SM. In particular, this gives hope to make the x-ray transform valued in some $L^p(\partial_+ SM)$-spaces.

(ii) Solving transport equations of the form $Xu = -f$ on SM may develop singularities even when f is smooth; however, there is good control over the created singularities. Namely, upon defining the dual bundles $E_\pm^* \subset T_{\Gamma_\pm}^* SM$ by

$$E_-^*(E_- \oplus \mathbb{R}X) = 0, \quad E_+^*(E_+ \oplus \mathbb{R}X) = 0,$$

then for each $f \in C^0(SM)$, it is established in [34, Section 4.2] that the boundary value problem (5) has a unique solution in $L^1(SM) \cap C^0(SM \setminus \Gamma_-)$ given by $u = R_+ f$ with

$$R_+ f(x,v) := \int_0^{\tau(x,v)} f(\varphi_t(x,v))\, dt,$$

and that, if $f \in C_c^\infty(SM)$, then $WF(R_+ f) \subset E_-^*$. Then in the presence of trapping, the x-ray transform $\mathcal{I}$ may be defined as

$$\mathcal{I}f := (R_+ f)|_{\partial_+ SM \setminus \Gamma_-}, \quad \mathcal{I}\colon C^0(SM) \to C^0(\partial_+ SM \setminus \Gamma_-)$$

and extends as a bounded operator

$$\mathcal{I}\colon L^p(SM) \to L^2(\partial_+ SM), \quad \forall p > 2,$$

see [34, Lemma 3.4, Proposition 2.4]. Similarly, the transform over bundles $\mathcal{I}_{\nabla,\Phi}$ also makes sense outside Γ_-, and the results established in [38] are described in that setting.

4.3 Pestov identities

4.3.1 Commutators of derivatives on the sphere bundle

Recall the definition of the natural derivatives X, $\overset{\mathrm{h}}{\nabla}$, $\overset{\mathrm{v}}{\nabla}$, $\overset{\mathrm{h}}{\mathrm{div}}$ and $\overset{\mathrm{v}}{\mathrm{div}}$ on the sphere bundle SM as introduced in Section 4.2.1.1.

The geodesic vector field X acts as a differential operator $X\colon C^\infty(SM) \to C^\infty(SM)$ by

$$Xu(x,v) = \partial_t u(\varphi_t(x,v))|_{t=0},$$

where φ_t is the geodesic flow. The same definition can be used to define also the operator $X\colon \mathcal{Z} \to \mathcal{Z}$, when one uses the covariant derivative along the flow. This gives rise to two incarnations of X given by

$$X\colon C^\infty(SM) \to C^\infty(SM), \quad X\colon \mathcal{Z} \to \mathcal{Z},$$

and we will not distinguish between the two in notation. In addition, we define the curvature operator $R\colon \mathcal{Z} \to \mathcal{Z}$ by $R(x,v)w = R_x(w,v)v$, where the R_x on the right-hand side is the Riemann curvature tensor at $x \in M$.

The starting point of deriving Pestov identities is the following commutator formulas.

Lemma 4.3.1 ([85, Lemma 2.1]). *The following commutator formulas hold on $C^\infty(SM)$ and $\mathcal{Z}$:*

$$[X, \overset{\mathrm{v}}{\nabla}] = -\overset{\mathrm{h}}{\nabla}, \quad [X, \overset{\mathrm{h}}{\operatorname{div}}] = -\overset{\mathrm{v}}{\operatorname{div}} R,$$
$$[X, \overset{\mathrm{h}}{\nabla}] = R \overset{\mathrm{v}}{\nabla}, \quad [X, \overset{\mathrm{v}}{\operatorname{div}}] = -\overset{\mathrm{h}}{\operatorname{div}}, \quad \textit{and}$$
$$\overset{\mathrm{h}}{\operatorname{div}} \overset{\mathrm{v}}{\nabla} - \overset{\mathrm{v}}{\operatorname{div}} \overset{\mathrm{h}}{\nabla} = (n-1)X.$$

To emphasize the commutator nature of the third formula in Lemma 4.3.1, one can write it as

$$\overset{[\mathrm{h},\,\mathrm{v}]}{\operatorname{div} \nabla} = (n-1)X.$$

We place the commutator symbols around the labels "h" and "v" (they are commuted), not around "div" and "∇" (they remain in the same order).

In two dimensions, the horizontal and vertical gradients can be considered as vector fields (globally if the underlying manifold is orientable). The corresponding commutator formulas were given in equation (9).

In addition to commuting operators, we need to integrate by parts. Let us denote the inner product of $u, v \in L^2(SM)$ by (u, v), and similarly for L^2 sections of N. If $u, v \in C^\infty(SM)$ and $Z, W \in \mathcal{Z}$, we have

$$(\overset{\mathrm{v}}{\nabla} u, Z) = -(u, \overset{\mathrm{v}}{\operatorname{div}} Z),$$
$$(Xu, w) = -(u, Xw) + \int_{\partial(SM)} uw\langle v, \nu \rangle, \quad \text{and}$$
$$(XZ, W) = -(Z, XW) + \int_{\partial(SM)} \langle Z, W \rangle \langle v, \nu \rangle.$$

We will not integrate by parts with horizontal derivatives, so these formulas will suffice. For more details on these operators, we refer to [85].

4.3.2 Simple Riemannian manifolds

A smooth and compact Riemannian manifold with boundary is called simple if the manifold is simply connected, the boundary is strictly convex (the second fundamental form is positive definite), and there are no conjugate points.

Lemma 4.3.2. *If M is simple, then any vector field $W \in \mathcal{Z}$ satisfies*

$$\|XW\|^2 - (RW, W) \geq 0$$

and equality holds if and only if $W = 0$.

To prove the lemma, convert the integral over SM to integrals over individual geodesics using the Santaló formula. The resulting integral along a geodesic is precisely the index form, which is positive definite due to the lack of conjugate points.

Lemma 4.3.3. *If $u \in C^\infty(SM)$ vanishes at the boundary, then*

$$\|\overset{\vee}{\nabla} Xu\|^2 = \|X\overset{\vee}{\nabla} u\|^2 - (R\overset{\vee}{\nabla} u, \overset{\vee}{\nabla} u) + (n-1)\|Xu\|^2.$$

This is known as the Pestov identity.

To prove the lemma, convert $\|\overset{\vee}{\nabla} Xu\|^2 - \|X\overset{\vee}{\nabla} u\|^2$ into inner products in $L^2(SM)$, integrate by parts, use commutator formulas to simplify the resulting operator, and simplify the result. The same result for closed manifolds with a full proof can be found in [85, Proposition 2.2].

4.3.2.1 X-ray tomography of scalars and one-forms

Lemmas 4.3.2–4.3.3 lead to an elegant proof of one of the most basic injectivity results on manifolds.

Theorem 4.3.4 ([72]). *If M is a compact and simple manifold with boundary, then the geodesic x-ray transform on M is injective on $C^\infty(M)$.*

We refer to [103, Section 4.9] for a discussion of the history of this method. The idea of the proof is to recast the injectivity as unique solvability of the PDE (5) or

$$\begin{cases} \overset{\vee}{\nabla} Xu = 0 & \text{in } SM, \\ u = 0 & \text{in } \partial(SM) \end{cases}$$

and using the Pestov identity to show that the only solution is indeed $u = 0$.

Proof of Theorem 4.3.4. Let $f \in C^\infty(SM)$ be a function with $If = 0$. Define a function $u\colon SM \to \mathbb{R}$ by

$$u(x,v) = \int_0^{\tau(x,v)} f(\gamma_{x,v}(t))\,dt,$$

where $\gamma_{x,v}\colon [0,\tau(x,v)] \to M$ is the maximal unit speed geodesic starting at x in the direction v. Simplicity ensures that no geodesics are trapped. Since the boundary is strictly convex, all geodesics exit transversally. It is easy to check that therefore u is smooth in the interior of SM.

Because $If = 0$, the function u vanishes at the boundary, since $u|_{\partial_- SM} = 0$ and $u|_{\partial_+ SM} = If$. It is also smooth up to the boundary. One way to see this is to use boundary determination: studying very short geodesics almost tangent to the boundary shows that f and its normal derivatives of all orders must vanish at ∂M. (A similar argument works for broken rays as well, provided that certain weighted ray transforms on the boundary are injective [46]. The same method can be applied in the present case without the need for using transforms on the boundary.)

The function u satisfies the transport equation $Xu = -f$, where we have identified $C^\infty(M) \ni f = \pi^* f \in C^\infty(SM)$. The function f is independent of direction, so $\overset{\mathrm{v}}{\nabla} Xu = 0$. Using Lemma 4.3.3 with $\overset{\mathrm{v}}{\nabla} Xu = 0$ for u and Lemma 4.3.2 for $\overset{\mathrm{v}}{\nabla} u$ gives $\|Xu\| = 0$, implying $f = 0$. □

The same method can also be applied to one-forms.

Theorem 4.3.5 ([5]). *If M is a compact and simple manifold with boundary, then the geodesic x-ray transform on M is solenoidally injective on smooth one-forms. That is, if f is a smooth one-form that integrates to zero over all maximal geodesics, there is $h \in C^\infty(M)$ which vanishes at the boundary and satisfies $f = dh$.*

Proof sketch. The proof is similar to the scalar case presented above, and starts by defining $u \in C^\infty(SM)$ and observing that it vanishes at the boundary. The left-hand side of the Pestov identity no longer vanishes, but it cancels one term on the right precisely, because $\|\overset{\mathrm{v}}{\nabla} f\|^2 = (n-1)\|f\|$, where we have again identified f as a function on SM. This leads to $\|X\overset{\mathrm{v}}{\nabla} u\|^2 - (R\overset{\mathrm{v}}{\nabla} u, \overset{\mathrm{v}}{\nabla} u) = 0$, which by Lemma 4.3.2 implies that $\overset{\mathrm{v}}{\nabla} u = 0$. Therefore, there is a function $h \in C^\infty(M)$ so that $u = -\pi^* h$. The transport equation $Xu = -f$ is then equivalent with $dh = f$, so h is the desired function. □

4.3.2.2 Tensor tomography

If f is a tensor field of order 0, the left-hand side of the Pestov identity of Lemma 4.3.3 vanishes. If f is of order 1, the term precisely cancels the $\|Xu\|^2$ term. If the order is $m \geq 2$, the Pestov identity no longer has this convenient positive definiteness. However, using the Pestov identity not for the whole u but for individual terms u_k in its spherical harmonic decomposition has turned out to be useful.

In two dimensions, the x-ray transform is solenoidally injective on simple manifolds for tensor fields of any order.

Theorem 4.3.6 ([82]). *The geodesic x-ray transform is solenoidally injective on the space of smooth tensor fields of any order $m \geq 0$ on a simple Riemannian surface.*

One can also dispense with simplicity, if certain properties are assumed of the x-ray transform at ranks zero and one.

Theorem 4.3.7 ([82]). *Let M be a compact nontrapping surface with a strictly convex and smooth boundary, so that the x-ray transform is solenoidally injective for tensor* *fields of orders zero and one, and that the adjoint of the x-ray transform on scalars is surjective. Then the geodesic x-ray transform is solenoidally injective on the space of smooth tensor fields of any order $m \geq 0$.*

In higher dimensions, it is not known whether solenoidal injectivity is always true on simple manifolds. However, with a stronger version of not having conjugate points, we can still formulate the result. Namely, given $\alpha \geq 0$, we say that the manifold (M, g) is *α-controlled* if every $W \in \mathcal{Z}$ with zero boundary values satisfies

$$\|XW\|^2 - (RW, W) \geq \alpha \|XW\|^2.$$

In this context, one may show that a simple manifold with strictly convex boundary is 0 controlled. Then the theorem below gives a positive answer to $\mathrm{TT}(m)$ under a stronger condition.

Theorem 4.3.8 ([85, Theorem 11.8]). *The geodesic x-ray transform is solenoidally injective on the space of smooth tensor fields of order $m \geq 1$ on a simple n-dimensional Riemannian manifold which is α-controlled for*

$$\alpha \geq \frac{(m-1)(m+n-2)}{m(m+n-1)}.$$

Earlier results were given by Sharafudtinov in [103, Theorem 4.3.3]. There solenoidal injectivity holds over m-tensors under the curvature bound condition

$$k^+(M, g) := \sup_{(x,v) \in \partial_+ SM} \int_0^{\tau(x,v)} t K^+(\varphi_t(x, v))\, dt < \frac{1}{m+1},$$

where we have defined $K^+(x, v) := \max(0, K(x, v))$ and where $K(x, v)$ is the supremum of sectional curvatures over all two-planes of $T_x M$ containing v. This bound was further improved by the same author from an $1/(m+1)$ bound to an $(m+2n-1)/m(m+n)$ one in his lecture notes [105]. Such conditions on $k^+(M, g)$ allow to relate the criterion of absence of β-conjugate points with the geometry (curvature), via sufficient but not necessary conditions.

There are a number of different ways to use the Pestov identity to obtain tensor tomography results. The basic idea is to show that the integral function u (solution to the transport equation $Xu = -f$) is a tensor field of order $m-1$ (has spherical harmonic content only up to degree $m-1$) if f has order m and $I^m f = 0$. In a certain sense, it is trivial that there is a potential, but the nontrivial part is to show that it is a tensor field.

In two dimensions, one can conveniently use (anti)holomorphic integrating factors and reduce the problem to showing that certain shifted versions of u are (anti)holomorphic [82]. Alternatively, one can use the iterated Beurling transform to bound high order harmonic content of u as outlined in Section 4.5.3. Careful analysis shows that the products of constants of continuity constants for the Beurling transforms are uniformly bounded [58, 85]. Simpler estimates can be used to derive weaker bounds; as long as the products grow sufficiently slowly, one can still conclude that u has finite degree [49]. The Pestov identity may also be combined with other integral identities.

Let us briefly outline the idea with uniformly bounded constants. Iterating the estimate for Beurling transforms, one can conclude that $\|X_+u_k\| \lesssim \|X_+u_{k+2N}\|$ when $k \geq m$ and $N \in \mathbb{N}$. By regularity considerations $\|X_+u_{k+2N}\|_{L^2} \to 0$ as $N \to \infty$, so in fact $X_+u_k = 0$. If u_k satisfies $X_+u_k = 0$, then it corresponds to a trace-free conformal Killing tensor field of order k vanishing at the boundary. Such tensor fields do not exist in simple geometry [16]. For more details on tensor tomography, see [84, 85].

4.3.3 β-conjugate points, the terminator value β_{Ter} and α-control

The notion of α-control just introduced above provides a continuous parameter which allows to encode previous geometric criteria (e. g., simplicity, conditions on curvature, etc.) as threshold conditions on α (or its related so-called terminator value β_{Ter}, as explained below), thereby allowing to refine previous statements. Before mentioning more results, we briefly visit the concepts of α-control and terminator value now.

Let J be a vector field along a geodesic γ. We say that J is a *β-Jacobi field* if it satisfies the β-Jacobi equation

$$D_t^2 J(t) + \beta R(J(t), \dot{\gamma}(t))\dot{\gamma} = 0,$$

where D_t is the covariant derivative and R the Riemann curvature tensor. The constant β describes how sensitive these generalized Jacobi fields are to curvature. These generalizations were introduced to x-ray tomography in [14, 90] and are also extensively used on closed manifolds with Anosov geodesic flow [83, 85]. For $\beta = 1$, we obtain the usual Jacobi fields.

We say that two points on γ are *β-conjugate* if there is a nontrivial β-Jacobi field vanishing at these two points. We then say that M is free of β-conjugate points if no two points are β-conjugate along any geodesic. As one may show that if M is free of β-conjugate points, it is also free of β'-conjugate points for any $\beta' \in [0, \beta)$, this justifies the definition of *terminator value* for the manifold M, given by

$$\beta_{\mathrm{Ter}} := \sup\{\beta \geq 0;\ M \text{ is free of } \beta\text{-conjugate points}\} \in [0, \infty].$$

As seen below, some classical geometric conditions can be reformulated as threshold conditions on β_{Ter}. Namely:
- If the manifold is compact and nontrapping, then $\beta_{\mathrm{Ter}} > 0$.
- There are no conjugate points if and only if $\beta_{\mathrm{Ter}} \geq 1$.
- The manifold has nonpositive curvature if and only if $\beta_{\mathrm{Ter}} = \infty$.

The main use of the terminator value is when relating it to α-control on manifolds, as stated at the end of the previous section. Recall that the manifold (M, g) is *α-controlled* if

$$\|XW\|^2 - (RW, W) \geq \alpha\|XW\|^2, \quad W \in \mathcal{Z}, \quad W|_{\partial M} = 0.$$

Then the terminator value is related to controllability as follows.

Lemma 4.3.9 ([85, Proposition 7.1 and Remark 11.3]). *If a compact manifold (closed or with boundary) satisfies $\beta_{\mathrm{Ter}} \geq \beta$, then the manifold is $(\beta - 1)/\beta$-controlled.*

Tying this lemma with the comments above on β_{Ter}, one may draw the following conclusions (see [85, Lemma 11.2]): a nontrapping manifold with strictly convex boundary is 0-controlled if it has no conjugate points, α-controlled for $\alpha > 0$ if it is simple, and 1-controlled if and only if it has nonpositive sectional curvature.

4.3.4 Other spaces

Pestov identities can also be used on other types of manifolds. We mention some examples.

4.3.4.1 Closed manifolds

Tensor tomography on closed manifolds (compact, without boundary) can be studied in a similar way with Pestov identities. Simple manifolds have boundary, and the corresponding closed manifolds are Anosov manifolds. On an Anosov surface, we have solenoidal injectivity for tensor fields of order zero [13], one [13], two [83, Theorem 1.1], and order $m \geq 3$ in [33, Theorem 1.4]. For the case $m \geq 3$, earlier results were written if the terminator value is at least $\frac{m+1}{2}$ [83, Theorem 1.3], or if the manifold is negatively curved [12]. More details are covered in the two-dimensional survey [84].

On Anosov manifolds of dimension $n \geq 3$, the x-ray transform is solenoidally injective for tensor fields of order $m \geq 2$ if the terminator value satisfies $\beta_{\mathrm{Ter}} > \frac{m(m+n-1)}{2m+n-2}$. For tensor fields of order zero or one, no such condition on the terminator value is needed [13].

4.3.4.2 Pseudo-Riemannian manifolds

On pseudo-Riemannian manifolds one does not usually study all geodesics, but only the light-like ones. The x-ray transform restricted to this set of null geodesics is also known as the light ray transform. Light rays as sets are conformally invariant, unlike Riemannian geodesics.

No Pestov identity is known on general pseudo-Riemannian or Lorentzian manifolds for x-ray or light ray transforms. However, when the spacetime is a product of space and time, a Pestov identity can be used to prove injectivity of the light ray transform for scalars and one-forms [47], under the assumption that both space and time are nonpositively curved Riemannian manifolds with strictly convex boundary and dimension at least two. The dimension assumption rules out Lorentzian manifolds. The Pestov identity is obtained by finding a Pestov-like identity on both space and time manifolds, and then combining them with suitable weights.

If the product $M = M_1 \times M_2$ of two Riemannian manifolds, (M_i, g_i) is equipped with the pseudo-Riemannian product metric $g_1 \oplus -g_2$, then the Pestov identity reads [47, Lemma 3]

$$\begin{aligned}(n_2-1)\|\overset{\vee}{\nabla}_1 Xu\|^2 + (n_1-1)\|\overset{\vee}{\nabla}_2 Xu\|^2\\ = (n_2-1)\|X\overset{\vee}{\nabla}_1 u\|^2 + (n_1-1)\|X\overset{\vee}{\nabla}_2 u\|^2\\ - (n_2-1)(R_1\overset{\vee}{\nabla}_1 u, \overset{\vee}{\nabla}_1 u) - (n_1-1)(R_1\overset{\vee}{\nabla}_2 u, \overset{\vee}{\nabla}_2 u)\\ + (n_1-1)(n_2-1)\|Xu\|^2.\end{aligned}$$

Here, $n_i = \dim(M_i)$ and the various operators on the two underlying Riemannian manifolds are indicated by a subscript. The identity is for smooth functions on the compact light cone bundle $LM := SM_1 \times SM_2$ on which the null geodesic flow lives.

Theorem 4.3.10 ([47, Theorem 1]). *Suppose (M_i, g_i), $i = 1, 2$, are two simple Riemannian manifolds with nonpositive sectional curvature. Equip the product $M = M_1 \times M_2$ with the pseudo-Riemannian metric $c(x)(g_1 \oplus -g_2)$, where $c \in C^\infty(M)$ is a smooth conformal factor. Then a smooth function f supported outside the edges $(\partial M_1 \times \partial M_2)$ integrates to zero over all null geodesics if and only if $f = 0$, and a smooth one-form α supported outside the edges integrates to zero over all null geodesics if and only if $\alpha = dh$, where $h \in C^\infty(M)$ vanishes at the boundary.*

4.3.4.3 Convex obstacles

Consider a compact manifold M with strictly convex boundary with a strictly convex obstacle $O \subset M$. Instead of integrating a function $f: M \to \mathbb{R}$ over all geodesics through M, we integrate a function $f: M \setminus O \to \mathbb{R}$ over all geodesics on $M \setminus O$ which have endpoints on ∂M and reflect specularly on ∂O. In two dimensions, specular reflections can be characterized by saying that the angle of incidence equals the angle of reflection.

One can employ a similar approach to that of Theorem 4.3.4. One defines a function u by integrating the unknown function $f \in \ker I$ along the broken ray until ∂M is met. This function $u(x, v)$ vanishes when $x \in \partial M$ but not a priori when $x \in \partial O$. This produces a boundary term in the Pestov identity, which can be simplified using the symmetry of u under specular reflection.

To be precise, in two dimensions one obtains the Pestov identity [50, Lemma 6]

$$\begin{aligned}\|VXu\|^2 &= \|XVu\|^2 + \|Xu\|^2 \\ &\quad - (KVu, Vu) - (\kappa Vu, Vu)_{L^2(\partial(SN))},\end{aligned}$$

valid for sufficiently regular functions on the manifold $N = M \setminus O$ which vanish at ∂M and satisfy a reflection condition at ∂O. Here, the other terms are in the space $L^2(SN)$ as usual, and κ is the curvature of ∂O. If obstacle is strictly convex, then $\kappa < 0$ and the boundary term has the correct sign.

Regularity is tricky in the presence of reflections; the function u is not a priori smooth even for smooth f and smooth geometry. Singularities occur at tangential reflections at ∂O.

This method was used to prove injectivity of the broken ray transform by Eskin [21] in the Euclidean plane with several reflecting obstacles and by the first author and Salo [50] on nonpositively curved Riemannian surfaces with a single reflecting obstacle.

The method was recently extended to any dimensions and tensors of any rank in [49], still assuming a single reflecting obstacle. The scalar-valued curvature K is replaced with the curvature operator R as described above. In the boundary term, the scalar curvature κ is replaced by the second fundamental form. Assuming nonpositive sectional curvature of the manifold and strict concavity of the reflector (as seen from the interior of N, equivalent with the strict convexity of the obstacle O) gives positivity to all terms.

4.3.4.4 Noncompact manifolds

Pestov identities have recently been successfully used on some noncompact context to prove tensor tomography and boundary rigidity. The identity looks exactly the same, but one needs to be far more careful with integrability and regularity.

A positive answer to tensor tomography (Problem 2) was given in [57, 58] for some cases of *Cartan–Hadamard*[7] manifolds. For such manifolds, $\beta_{\mathrm{Ter}} = \infty$ so we may expect good α-control, which helps to control terms of index form type appearing in Lemma 4.3.3. The results extend to tensor fields with noncompact support, however suitable decay at infinity is needed, a nonartificial requirement since otherwise counterexamples exist even in the Euclidean case. Namely, the following two results are

7 A Cartan–Hadamard manifolds is a complete, simply connected Riemannian manifold with nonpositive sectional curvature.

proved in [58], generalizing earlier results in [57] to dimensions greater than 3 and tensor fields of arbitrary order. Below, for $x \in M$ and $\Pi \subset T_xM$ a two-plane, we denote $K_x(\Pi)$ the sectional curvature of the two-plane Π. Using any distinguished point $o \in M$ as the "origin," we say that a function f (or a tensor field)

- has polynomial decay at infinity of order η if $x \mapsto |f(x)|(1 + d(x, o))^\eta$ is bounded,
- has exponential decay at infinity of order η if $x \mapsto |f(x)|e^{\eta d(x,o)}$ is bounded.

The first theorem is of an asymptotically hyperbolic flavor, while the second is asymptotically Euclidean.

Theorem 4.3.11 ([58, Theorem 1.1]). *Let (M, g) be a Cartan–Hadamard manifold of dimension $n \geq 2$, and assume that there exists $K_0 > 0$ such that*

$$-K_0 \leq K_x(\Pi) \leq 0, \quad x \in M,\ \Pi \subset T_xM.$$

If $f \in C^1(S^m(T^*M))$ and ∇f have exponential decay at infinity of order $\eta > \frac{n+1}{2}\sqrt{K_0}$, and if $I_mf = 0$, then $f = \sigma\nabla h$ for some $h \in C^1(S^m(T^*M))$ with exponential decay at infinity at rate $\eta - \varepsilon$ for any $\varepsilon > 0$ (if $m = 0$ then $f \equiv 0$).

Theorem 4.3.12 ([58, Theorem 1.2]). *Let (M, g) be a Cartan–Hadamard manifold of dimension $n \geq 2$, and assume that the function*

$$\mathcal{K}(x) := \sup_{\Pi \subset T_xM} |K_x(\Pi)|, \quad x \in M$$

decays strictly faster than quadratically at infinity. If $f \in C^1(S^m(T^*M))$ has polynomial decay at infinity of order $\eta > \frac{n+2}{2}$ and ∇f has decay of order $\eta + 1$ and if $I_mf = 0$, then $f = \sigma\nabla h$ for some $h \in C^1(S^m(T^*M))$ with polynomial decay at infinity of order $\eta - 1$.

While the results on Cartan–Hadamard manifolds allow to treat more than one type of geometry at infinity, the recent work [30] focuses on the *asymptotically hyperbolic* context, however, covers the case where a hyperbolic trapped set is present using the tools of Section 4.2.3, and also treats some nonlinear results. It is convenient here to picture (M, g) as a manifold-with-boundary where the metric g has a specific singular behavior at the boundary such that geodesics never reach it in finite time. In addition, geodesics making it to the boundary as $t \to \pm\infty$ always hit the boundary normally, so one must look at second-order information to define the space of geodesics and other related objects like the scattering relation.

Under a no-conjugate points assumption, the authors prove a positive answer to Problem 2 in [30, Theorem 1] for classes of tensors with suitable decay conditions at infinity which agree with those of Theorem 4.3.11. The proof uses Pestov identities in the interior, after using the specific structure of the geodesics in a neighborhood of ∂M to prove that a tensor field whose ray transform vanishes agrees to infinite order with a potential tensor near ∂M.

The authors go on to studying nonlinear inverse problems, determining features of the metric up to gauge from renormalized geodesic lengths; first, at the boundary in [30, Theorem 2], then globally if the manifold is real-analytic and such that $\pi_1(M, \partial M) = 0$ in [30, Theorem 3]. Finally, [30, Theorem 4] establishes a deformation rigidity result (a variant of the lens rigidity problem in Problem 1) in the case of nontrapping asymptotically hyperbolic metrics with nonpositive sectional curvature.

4.3.5 Unitary connections and skew-Hermitian Higgs fields

The method of Pestov identities generalizes to the case of transport with Hermitian connection and skew-Hermitian Higgs field, as treated in the recent works [38, 81]. Let $E \to M$ a bundle as in Section 4.2.2, equipped with a Hermitian connection ∇ and a skew-Hermitian Higgs field Φ.

To write Pestov identities, one must then generalize the Sasaki-related objects (e. g., horizontal/vertical gradients) to sections of the (pullback) bundle $E \to SM$. For $u \in C^\infty(SM; E)$, $\nabla^E u \in C^\infty(SM; T^*(SM) \otimes E)$, and using the Sasaki metric on $T(SM)$ we can identify this with an element of $C^\infty(SM; T(SM) \otimes E)$, splitting according to (3) into

$$\nabla^E u = (\mathbb{X}u, \overset{\mathsf{h}}{\nabla}{}^E u, \overset{\mathsf{v}}{\nabla}{}^E u), \quad \mathbb{X}u := \nabla^E_X u,$$

where $\overset{\mathsf{h}}{\nabla}{}^E u$ and $\overset{\mathsf{v}}{\nabla}{}^E u$ can be viewed as smooth sections of $N \otimes E$.

Then the following Pestov identity can be derived for any $u \in C^\infty(SM; E)$ vanishing at ∂SM (see [38, Proposition 3.3]):

$$\begin{aligned} \|\overset{\mathsf{v}}{\nabla}{}^E \mathbb{X}u\|^2 &= \|\mathbb{X}\overset{\mathsf{v}}{\nabla}{}^E u\|^2 - (R\overset{\mathsf{v}}{\nabla}{}^E u, \overset{\mathsf{v}}{\nabla}{}^E u) \\ &\quad - (F^E u, \overset{\mathsf{v}}{\nabla}{}^E u) + (n-1)\|\mathbb{X}u\|^2, \end{aligned} \tag{12}$$

where R is the Riemann curvature tensor of (M, g), viewed as an operator on the bundles N and $N \otimes E$ over SM by the actions

$$R(x, v)w := R_x(w, v)v, \quad R(x, v)(w \otimes e) := (R_x(w, v)v) \otimes e, \quad e \in E_x,$$

and where $F^E \in C^\infty(SM; N \otimes \mathrm{End}_{\mathrm{sk}}(E))$ is the curvature of the connection ∇^E; see [38, equation (3.5)].

On simple surfaces, where E is trivial and we globally represent $\mathbb{X} = X + A$ for some skew-Hermitian matrix of one-forms A, equation (12) takes the form (see [81, Lemma 6.1])

$$\begin{aligned} \|V(X + A)u\|^2 &= \|(X + A)Vu\|^2 - (\kappa Vu, Vu) - (\star F_A u, Vu) \\ &\quad + \|(X + A)u\|^2, \end{aligned} \tag{13}$$

where $\star F_A \in C^\infty(SM)$ is again related to the curvature of the connection. We now explain two possible ways to exploit the identities (12)–(13) to produce a positive answer to tensor tomography questions.

Theorem 4.3.13 ([81, Theorem 1.3]). *Let (M,g) be a simple surface and E a bundle over M with Hermitian connection A and skew-Hermitian Higgs field Φ. If $f \in C^\infty(SM,\mathbb{C}^n)$ is of the form $f = f_0 + f_1$, and if $I_{A,\Phi}f = 0$, then $f_1 = (d+A)p$ and $f_0 = \Phi p$ for some smooth function $p : M \to \mathbb{C}^n$ vanishing at ∂M.*

Proof. (sketch) *Case* $\Phi = 0$. To prove this, we assume that u and f are as in (11) with $u|_{\partial SM} = 0$, and the result is proved when we show that $Vu = 0$ (so that $p := u_0$ does the trick). We explain how $v := \sum_{k<0} u_k$ is zero, as the proof for $\sum_{k>0} u_k$ is similar. Equation (13) applied to v becomes

$$\|(X+A)Vv\|^2 - (\kappa Vv, Vv) - (\star F_A v, Vv) + \|((X+A)v)_0\|^2 = 0.$$

The sum of the first two terms is nonnegative due to the simplicity of the metric. Moreover, one can establish, using holomorphic integrating factors for scalar connections, that the injectivity of I_A does not depend on perturbing the connection by a scalar one. In particular, A can be perturbed into A' by a scalar term for which $i \star F_{A'} \le 0$ in the sense of Hermitian operators. Assuming this is the case, we obtain that $-(\star F_A v, Vv) \ge 0$. This forces all terms to be zero, and thus $v = 0$.

Case $\Phi \neq 0$. In the setting of the proof above, if a Higgs field is present, one must write a Pestov identity in the form of (13) for the operator $V(X + A + \Phi)$ instead of $V(X + A)$. Some additional terms appear in the identity, which can be controlled by $C_{A,\Phi}\sum_{k=-\infty}^{-1} |k||v_k|^2$, and perturbing the connection with a scalar one so that $-(\star F_A v, Vv) \gg 0$, one can again control these terms in a coercive fashion and enforce $v = 0$. □

The second setting is that of manifolds with negative sectional curvature, where the answer to the tensor tomography problem can be made positive for tensors of arbitrary order.

Theorem 4.3.14 ([38, Theorems 4.1, 4.6]). *Let (M,g) a compact manifold with negative sectional curvature and E a Hermitian bundle with Hermitian connection ∇^E and Φ a skew-Hermitian Higgs field. If $u \in C^\infty(SM;E)$ satisfies $(\mathbb{X}+\Phi)u = -f$ where f has finite degree and if $u|_{\partial SM} = 0$, then u has finite degree.*

Proof. (sketch) In the case $\Phi = 0$, the proof consists in applying (12) to the high-frequency content of u (say, $T_{\ge m}u := \sum_{k\ge m} u_k$ for m large enough), and show that for m large enough, the curvature term $(-R\overset{\vee}{\nabla}{}^E u, \overset{\vee}{\nabla}{}^E u)$ overtakes the contribution of the connection term $(-F^E u, \overset{\vee}{\nabla}{}^E u)$. In particular, [38, Lemma 4.2] shows that for m large enough, $\overset{\vee}{\nabla}{}^E(T_{\ge m}u)$ is controlled by $\overset{\vee}{\nabla}{}^E(T_{\ge m+1}\mathbb{X}u) = -\overset{\vee}{\nabla}{}^E(T_{\ge m+1}f)$, but since the latter vanishes identically for m large enough, so does the former. If a skew-Hermitian Higgs field Φ is added, when controlling the high frequencies of u, the terms involving Φ can still be overtaken thanks to the negative curvature. □

4.3.6 Magnetic and thermostat flows

The method by energy identities has been generalized to other types of non-geodesic flows on manifolds, which we will discuss here.

Fixing a Riemannian manifold (M, g), geodesic trajectories can be viewed as zero-acceleration curves in the Levi–Civita connection ∇, governed by the equation $\nabla_{\dot\gamma}\dot\gamma = 0$. A way to consider other flows can be done by adding a mechanically motivated force field, characterized by a bundle map $F: TM \to TM$ for which the trajectories evolve under Newton's second law

$$\nabla_{\dot\gamma}\dot\gamma = F(\gamma, \dot\gamma).$$

If, in addition, this force field is skew-Hermitian, then the quantity $|\dot\gamma|$ is preserved along trajectories, and we obtain a flow ψ_t on SM again, whose generator $G := \frac{d}{dt}\psi_t|_{t=0}$ can be shown to take the form $G = X + \lambda V$ (e. g., in two dimensions), where $\lambda : SM \to \mathbb{R}$ incorporates information about the force field. One may then define associated ray transforms of functions and tensor fields via solving transport equations of the form $Gu = -f$. Many objects then depend on the flow under consideration, namely: the kernel of the ray transform over tensor fields, the notion of convexity at the boundary, the notion of conjugate points, etc... One may also lift this flow to a bundle $E \to M$ with a connection A and Higgs field Φ and consider the associated notions of scattering data $C_{A,\Phi}$ and x-ray transforms $\mathcal{I}_{A,\Phi}$. In this context, the following results have been derived.

4.3.6.1 Magnetic ray transforms

A magnetic field on M is a closed two-form Ω, and this gives rise to the magnetic force $F : TM \to TM$ uniquely defined by

$$\Omega_x(\xi, \eta) = g(F_x(\xi), \eta), \quad x \in M, \ \xi, \eta \in T_xM,$$

see [2, 15, 61]. Here, the function $\lambda(x, v) = -g(F_x(v), v_\perp)$ in fact does not depend on v. One can then define the concept of a magnetically convex boundary, and being "simple" with respect to the magnetic flow, and consider Problems 1–4.

The first results for such transforms were given in [15], where energy methods (formulated there in the language of semi-basic tensor fields) are used in [15, Section 5] to prove injectivity of the magnetic ray transform over functions, one-forms [15, Theorem 5.3] and two-tensors [15, Theorem 5.4] under certain curvature conditions. Further results are established (generic injectivity, magnetic boundary rigidity) using methods discussed later in this article.

On simple magnetic surfaces, positive answers to Problems 3 (tensor tomography problem) and 4 (determination up to gauge of (A, Φ) from their scattering data) are obtained in [2, Theorems 1.2, 1.4, 1.5] in the presence of a unitary connection and a skew-Hermitian Higgs field, as in Section 4.3.5. The schemes of proof of [2, Theorems 1.2, 1.4] follow [81], where the new key step is to derive Pestov identities for the operator $V(X + \lambda V + A + \Phi)$.

4.3.6.2 Thermostat ray transforms on surfaces

Another example of external field is given by a Gaussian thermostat, characterized by a smooth vector field E on M; see [8, 61]. The force field in this case is given by

$$F(\gamma, \dot{\gamma}) = E(\gamma) - \frac{g_\gamma(E(\gamma), \dot{\gamma})}{|\dot{\gamma}|^2}\dot{\gamma} = \frac{g_\gamma(E(\gamma), \dot{\gamma}_\perp)}{|\dot{\gamma}|^2}\dot{\gamma}_\perp$$

and the function $\lambda(x, v) = -g(F_x(v), v_\perp)$ is now linear in v. Upon defining an associated thermostat ray transform over tensor fields, and defining a notion of terminator value β_{Ter} with respect to a thermostat-Jacobi equation, it is proved in [8, Theorem 1.5] that the thermostat ray transform is injective (up to natural obstruction) over m-tensors if $\beta_{\mathrm{Ter}} \ge \frac{m+1}{2}$. In particular, for this notion of terminator value, we again have $\beta_{\mathrm{Ter}} = \infty$ if the thermostat curvature is nonpositive, in which case the previous result holds for any tensor order m; see [8, Corollary 1.6]. The proofs are based on deriving Pestov identities for the operator $V(X + \lambda V)$. Associated results in the case of closed surfaces without boundary are given there as well, see [8, Theorem 1.2, Corollaries 1.3, 1.4].

It is worth pointing out that in the geodesic case, the condition on β_{Ter} is not necessary (namely $\beta_{\mathrm{Ter}} > 1$ implies the result for any m) because one can use holomorphic integrating factors for scalar connections to move from any harmonic level to any other. It may be of interest to seek a similar construction here.

4.4 Inversion formulas and another route to injectivity in two dimensions

In this section, we present constructive inversion approaches for various integrands and geometric contexts.

4.4.1 Pestov–Uhlmann inversion formulas on simple surfaces

Recall the scattering relation $\mathcal{S}: \partial SM \to \partial SM$ as follows: if $(x,v) \in \partial_+ SM$, $\mathcal{S}(x,v) := \varphi_{\tau(x,v)}(x,v) \in \partial_- SM$; if $(x,v) \in \partial_- SM$, $\mathcal{S}(x,v) := \varphi_{-\tau(x,-v)}(x,v) \in \partial_+ SM$. Recall the following definitions of $A_\pm: L^2_\mu(\partial_+ SM) \to L^2_{|\mu|}(\partial SM)$ and their adjoints:

$$A_\pm w(x,v) = \begin{cases} w(x,v) & (x,v) \in \partial_+ SM, \\ \pm w \circ \mathcal{S}(x,v) & (x,v) \in \partial_- SM, \end{cases} \qquad A_\pm^* u := (u \pm u \circ \mathcal{S})|_{\partial_+ SM}.$$

Recall also the definition of the *fiberwise Hilbert transform* $H: L^2(SM) \to L^2(SM)$, defined on the fiberwise harmonic decomposition by

$$Hu_k = -i\,\mathrm{sign}(k) u_k, \quad u_k \in \Omega_k \text{ with the convention } \mathrm{sign}(0) = 0.$$

Introducing this transform, Pestov and Uhlmann obtained the following formulas in [92] (written with slight updates as in [66, Proposition 2.2]), inverting the ray transform over functions and solenoidal vector fields:

$$\begin{aligned} f + W^2 f &= \frac{1}{8\pi} I_\perp^* (A_+^* H A_-) I_0 f, \quad f \in L^2(M), \\ h + W_\perp^2 h &= \frac{-1}{8\pi} I_0^* (A_+^* H A_-) I_\perp h, \quad h \in H_0^1(M), \end{aligned} \tag{14}$$

where the operators W, $W_\perp$ are $L^2(M) \to L^2(M)$-adjoints, and compact smoothing. Such equations take the form of *filtered-backprojection* formulas, where $A_+^* H A_-: L^2(\partial_+ SM) \to L^2(\partial_+ SM)$ is a continuous "filter" and the adjoints I_0^* or $I_\perp^*$ are viewed as "backprojection" operators; see Figures 4.3–4.4 for an example.

Remark 4.4.1. Formulas (14) hint at us that the classical *filtered-backprojection* formula inverting the 2D Radon transform $f = \frac{1}{4\pi} R^* H \partial_s R f$, contains two formulas for the price of one: indeed, $\partial_s Rf$ can also be viewed as $R_\perp f := R(X_\perp f)$, where $X_\perp f = \sin\theta \partial_x f - \sin\theta \partial_y f$ is the restriction to SM of the solenoidal one-form $-\partial_y f\, dx + \partial_x f\, dy$, generated by the solenoidal potential f. In that case, the solenoidal potential $X_\perp f$ can be reconstructed by first reconstructing f via $f = \frac{1}{4\pi} R^* H R_\perp f$.

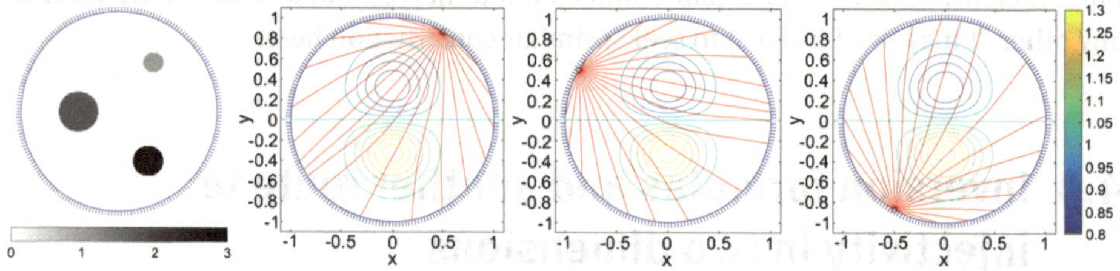

Figure 4.3: Example of a function f defined on the unit disk (left). The domain is endowed with a scalar metric $g = c^{-2} Id$ where the "sound speed" c is contour-plotted, and some geodesics are superimposed on three pictures on the right.

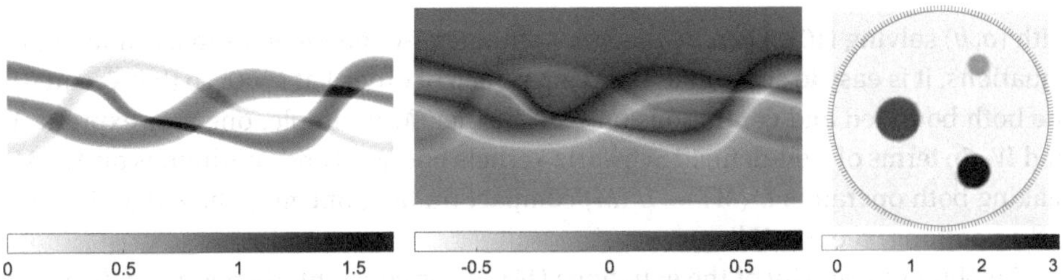

Figure 4.4: An example of approximate inversion. Left: $I_0 f$; Middle: $A_+^* H A_- I_0 f$ (the "filtering" step); Right: $-(1/8\pi) I_\perp^* A_+^* H A_- I_0 f$ (the "backprojection" step). On the left two pictures, the horizontal axis describes the boundary point from where geodesics are cast; the vertical axis describes the shooting direction. See [63] for details of the implementation.

The proof of (14) can be found in [92] and in the recent form in [66], and is based on the interaction of the fiberwise Hilbert transform with the transport equation (5), in particular relying heavily on the commutator formula below, first derived in [93]

$$[H, X]u = (X_\perp u)_0 + X_\perp u_0, \quad u \in C^\infty(SM). \tag{15}$$

This is a commutator formula involving a distinguished nonlocal zeroth order ΨDO, and is used in a similar fashion to the local commutator formulas of Lemma 4.3.1. No higher dimensional analogue for this nonlocal formula is known.

4.4.1.1 Analysis of W and $W_\perp$

Equations (14) in fact make sense on any nontrapping surface. On the other hand, one must further assume that the surface is simple to establish that W, $W_\perp$ are compact. In general, as is the case with other operators emanating from this context, the operators W, $W_\perp$ are integral operators with Schwartz kernels naturally written in exponential coordinates. Namely, such operators take the form

$$Kf(x) = \frac{1}{2\pi} \int_{S_x} \int_0^{\tau(x,v)} k(x, v, t) f(\gamma_{x,v}(t))\, dt\, dS(v),$$

where $k(x, v, t)$ is the Schwartz kernel of K in exponential coordinates. In the absence of conjugate points, the mapping $(v, t) \mapsto \gamma_{x,v}(t)$ is a global diffeomorphism onto M with Jacobian $dM_x = b(x, v, t)\, dt\, dS(v)$ and inverse denoted $M \ni x' \mapsto (v_x(x'), t = d_g(x, x'))$, so that the actual Schwartz kernel of K is, up to a constant, $\mathcal{K}(x, x') = \frac{k(x, v_x(x'), d_g(x, x'))}{b(x, v_x(x'), d_g(x, x'))}$. A sufficient condition for K to be compact is if $\mathcal{K} \in L^2(M \times M)$.

In the setting of equations (14), one may show (see, e. g., [92]) that W, $W_\perp$ have respective kernels

$$w(x, v, t) = V\left(\frac{a}{b}\right)(x, v, t), \quad w_\perp(x, v, t) = V\left(\frac{1}{b}\right)(x, v, t),$$

with (a, b) solving (10). Then by theory of parameter-dependent ordinary differential equations, it is easy to show that w and $w_\perp$ vanish of order 1 at $t = 0$, so that $\frac{w}{b}$ and $\frac{w_\perp}{b}$ are both bounded and continuous on $\mathcal{D}$. If further, M is simple, one may express W and $W_\perp$ in terms of well-defined Schwartz kernels bounded and continuous on $M \times M$, making both operators $L^2(M) \to L^2(M)$ compact (in fact, one may show that they are C^∞ smoothing, see [37, 92]).

To obtain injectivity of the equations (14), if κ is constant, then $w, w_\perp \equiv 0$, and if $\sup_M |d\kappa|$ is small enough, then W, $W_\perp$ are contractions (see [53]), and recovery of f or h from (14) can be done via Neumann series. In such cases, this in fact gives another mechanism to prove injectivity of I_0 and $I_\perp$ than energy identities, though for now, it is open as to whether $Id + W^2$ and $Id + W_\perp^2$ are injective for all simple surfaces. A quantitative bound estimating the norm of W in a neighborhood of constant curvature, simple surfaces was given in [69, Appendix A]: if τ_∞, $\operatorname{Vol} M$ denote diameter and volume of M, and if (M, g) is simple with constants (C_1, C_2) in the sense that

$$C_1 t \le |b(x, v, t)| \le C_2 t, \quad (x, v, t) \in \mathcal{D}, \tag{16}$$

then one may obtain the estimate

$$\|W\|_{L^2 \to L^2} \le \frac{C_2^3 \tau_\infty^2}{24 C_1^{5/2}} \left(\frac{\operatorname{Vol} M}{2\pi} \right)^{\frac{1}{2}} \|d\kappa\|_\infty.$$

4.4.2 Transforms over *k*-differentials on simples surfaces

Following the template outlined above, the second author generalized in [64] the inversion of I_0 and $I_\perp$ to inversion formulas for the recovery of sections of Ω_k and their horizontal derivatives, with $k \in \mathbb{Z}$ fixed for this paragraph. Namely, for $u \in \Omega_k$ (locally of the form $f(x, y) e^{ik\theta}$ in isothermal coordinates), one may define

$$I_k u := \mathcal{I}[u], \quad I_{k,\perp} u := \mathcal{I}[X_\perp u],$$

and upon introducing a *shifted Hilbert transform* $H_{(k)}$ by the formula

$$H_{(k)} u_\ell := -i \operatorname{sign}(\ell - k) u_\ell, \quad u_\ell \in \Omega_\ell,$$

one may prove a commutator relation $[H_{(k)}, X] u = (X_\perp u)_k + X_\perp u_k$ which allows Fredholm equations of the form (14) pseudo-inverting I_k and $I_{k,\perp}$, modulo compact error operators W_k and $W_{k,\perp} = W_k^*$. The kernel of W_k in exponential coordinates is given by

$$w_k(x, \theta, t) = \left(-\partial_\theta \left(\frac{a}{b} \right) + ik \frac{a-1}{b} \right) e^{ik(\alpha_{x,\theta}(t) - \theta)},$$

where $\alpha_{x,\theta}(t)$ is the angle of $\dot\gamma_{x,\theta}(t)$ with ∂_x in isothermal coordinates. Further study of these kernels gives exact reconstructions via Neumann series in [64, Corollary 5.9] under certain assumptions on the curvature.

4.4.3 Transforms with nonunitary connections on simple surfaces

A case encompassing both previous ones is to consider transforms with connections. Namely, given $M \times \mathbb{C}^n \to M$ the trivial[8] vector bundle of rank n and a connection A given by a $n \times n$ matrix of one-forms on M, and $f \in L^2(SM, \mathbb{C}^n)$, one may now solve the equivalent of a coupled system of transport equations for $u: SM \to \mathbb{C}^n$ by

$$Xu + A_x(v)u = -f \quad (SM), \quad u|_{\partial\_SM} = 0,$$

and define $\mathcal{I}_A f := u|_{\partial_+ SM}$, the ray transform of f with connection A. Note in what follows that we will abuse notation $A(x, v) \equiv A_x(v) \in \mathbb{C}^{n\times n}$. Naturally, $\mathcal{I}_A$ is continuous in the $L^2(SM, \mathbb{C}^n) \to L^2_\mu(\partial_+ SM, \mathbb{C}^n)$ setting, though a dimension count shows that one may not recover all of f from $\mathcal{I}_A f$, and must therefore restrict to certain classes of f. In particular, one may define

$$\begin{aligned} I_{A,0} f &:= \mathcal{I}_A[f \circ \pi], \quad f \in L^2(M, \mathbb{C}^n), \\ I_{A,\perp} h &:= \mathcal{I}_A[(X_\perp - A_V)(h \circ \pi)], \quad h \in H^1_0(M, \mathbb{C}^n), \end{aligned}$$

the restrictions of $\mathcal{I}_A$ to functions and certain one-forms (of the form $\star dh - A_V h$). In the same way that curvature of (M, g) has an impact on x-ray transforms, the curvature of the connection $F_A := dA + A \wedge A$ (i. e., a matrix of 2-forms with components $(F_A)_{ij} = dA_{ij} + \sum_{k=1}^n A_{ik} \wedge A_{kj}$) will have an impact on x-ray transforms via the function $\star F_A: M \to \mathbb{C}^{n\times n}$.

Such transforms generalize the case of symmetric differentials because if ϕ denotes a nonvanishing section of Ω_1 (in isothermal coordinates, take $e^{i\theta}$), then every element of Ω_k can be uniquely written as $f\phi^k$ for some function $f : M \to \mathbb{C}$, and the transport equation $Xu = -f\phi^k$ (with $u|_{\partial_- SM=0}$), upon setting $v = \phi^{-k} u$, is equivalent to the transport equation $Xv + k(\phi^{-1} X\phi)v = -f$ (with $v|_{\partial_- SM} = 0$) on the trivial bundle $SM \times \mathbb{C}$ with connection $k\phi^{-1} d\phi$. In particular, we have

$$I_{k,0}[f\phi^k] = u|_{\partial_+ SM} = \phi^k|_{\partial_+ SM} v|_{\partial_+ SM} = \phi^k|_{\partial_+ SM} I_{k\phi^{-1} d\phi, 0} f,$$

and the two transforms are strictly equivalent for injectivity and inversion purposes.

Upon introducing the fiberwise Hilbert transform, acting this time on each component of $\mathbb{C}^n$, one may derive the commutator formula (see [81])

$$[H, X + A]u = (X_\perp - A_V)u_0 + ((X_\perp - A_V)u)_0, \quad u \in C^\infty(SM, \mathbb{C}^n),$$

and derive the inversion formulas (see [69, Theorem 1]):

$$\begin{aligned} f + W_A^2 f &= \frac{1}{8\pi} I^*_{-A^*,\perp} B_{A,+} H Q_{A,-} I_{A,0} f, \quad f \in C^\infty(M, \mathbb{C}^n), \\ h + W_{A,\perp}^2 h &= \frac{-1}{8\pi} I_{-A^*,0} B_{A,+} H Q_{A,-} I_{A,\perp} h, \quad h \in C_0^\infty(M, \mathbb{C}^n), \end{aligned} \tag{17}$$

8 All vector bundles are trivial when M is simply connected.

extendible by density to $f \in L^2(M,\mathbb{C}^n)$ and $h \in H_0^1(M,\mathbb{C}^n)$, and where the error operators admit respective kernels

$$w_A(x,v,t) = \left(X_\perp - A_V + \frac{a}{b}V + V\left(\frac{a}{b}\right)\right)E_A^{-1}(x,v,t),$$

$$w_{A,\perp}(x,v,t) = E_A^{-1}(x,v,t)\left(V\left(\frac{1}{b}\right) - A_V(\varphi_t)\right) + \frac{1}{b}V(E_A^{-1}(x,v,t)),$$

where $E_A(x,v,t)$ is the attenuation matrix solving the (x,v)-dependent ODE

$$\frac{d}{dt}E_A(x,v,t) + A(\varphi_t(x,v))E_A(x,v,t) = 0, \quad (x,v,t) \in \mathcal{D}, \quad E_A(0,x,v) = I_n.$$

Note that $w_A, w_{A,\perp} : \mathcal{D} \to \mathbb{C}^{n\times n}$ are again such that $\frac{w_A}{b}$ and $\frac{w_{A,\perp}}{b}$, composed with the inverse of the exponential map $(v,t) \mapsto \gamma_{x,v}(t)$ are uniformly bounded on $M \times M$, in particular the kernels of W_A, $W_{A,\perp}$ belong to $L^2(M\times M)$ so that W_A, $W_{A,\perp}$ are compact.

By the Fredholm alternative, this implies that $\ker I_{A,0} \subset \ker(Id + W_A^2)$ and $\ker I_{A,\perp} \subset \ker(Id + W_{A,\perp}^2)$ are finite-dimensional. Varying connections with a complex parameter, one may combine this with analytic Fredholm theory to enlarge the known cases of injective ray transforms with connection. The steps go as follows.

Theorem 4.4.2 ([69, Theorem 3]). *For any analytic $C^1(M,(\Lambda^1)^{n\times n})$-valued family of connections $\lambda \mapsto A_\lambda$, the corresponding $L^2(M,\mathbb{C}^n) \to L^2(M,\mathbb{C}^n)$-valued families of operators $\lambda \mapsto W_{A_\lambda}$ and $\lambda \mapsto W_{A_\lambda,\perp}$ are analytic.*

For $A \in C^1(M,(\Lambda^1)^{n\times n})$ a fixed connection, applying this to a family of the form $\lambda \mapsto \lambda A$ with A implies that if there is $\lambda_0 \in \mathbb{C}$ such that $Id + W_{\lambda_0 A}^2$ is injective, this remains so for all $\lambda \in \mathbb{C}$ except possibly over a discrete set of λ, and implies the same conclusions for the ray transforms $I_{\lambda_0 A,0}$. An obvious choice is $\lambda = 0$, which reconducts the question to a transform with no connection. In that case, we are left inquiring whether $Id + W^2$ is injective.

Theorem 4.4.3 ([69, Theorem 4]). *Let (M,g) be a simple Riemannian surface with constants C_1, C_2 as in (16) and Gaussian curvature $\kappa(x)$. Given the C^1 connection A with curvature F_A, let us denote $\alpha_A = \sup_{(x,v)\in SM}\{\|(A+A^*)/2\|(x,v)\}$ and τ_∞ the diameter of M. There exist constants C, C' depending on $(n, C_1, C_2, \tau_\infty, \alpha_A)$ such that*

$$\|W_A\|_{L^2\to L^2}, \|W_{A,\perp}\| \le \left(\frac{\operatorname{Vol} M}{2\pi}\right)^{\frac{1}{2}} \sqrt{C\| \star F_A\|_\infty^2 + C'\|d\kappa\|_\infty^2}. \tag{18}$$

This gives us a few settings where transforms with connections defined over any structure group are injective.

Theorem 4.4.4 ([69, Theorem 5]). *Let (M,g) be a simple surface and A a C^1 connection. Then the following conclusions hold:*

(i) *If κ is constant and A is flat, the operators W_A and $W_{A,\perp}$ vanish identically and* (17) *implies that the transforms $I_{A,0}$, $I_{A,\perp}$, $I_{-A^*,0}$, and $I_{-A^*,\perp}$ are all injective, with explicit, one-shot inversion formulas.*
(ii) *Injectivity still holds if $(n, C_1, C_2, \tau_\infty, \alpha_A, \| \star F_A\|_\infty, \|d\kappa\|_\infty, \mathrm{Vol}\, M)$ are such that the right-hand side of* (18) *is less than 1, with a Neumann series type inversion.*
(iii) *If (M, g) is such that the operator $Id + W^2$ in* (14) *is injective, then for every $\lambda \in \mathbb{C}$ outside a discrete set, the transforms $I_{\lambda A,0}$, $I_{\lambda A,\perp}$, $I_{-\lambda A^*,0}$ and $I_{-\lambda A^*,\perp}$ are all injective.*

The fact that the conclusions apply jointly to connections A and $-A^*$ comes from exploring the symmetries in the inversion formulas (17) and establishing for instance that $(W_A)^* = W_{-A^*,\perp}$; see [69, Lemma 13].

4.4.4 Surfaces with no conjugate points and hyperbolic trapping

As mentioned in Section 4.2.3, some of the arguments can be adapted to situations when hyperbolic trapping is present, a special case of which is surfaces with negative curvature. In [37], Guillarmou and the second author provided reconstruction formulas for functions and solenoidal vector fields from knowledge of their ray transform; in the case of surfaces with convex boundary, no conjugate points and hyperbolic trapping, see [37, Theorem 1.1]. The derivation of the formulas is similar to obtaining (14), as it relies on the commutator formula (15) which holds locally and independently of the presence of trapping. The added technicalities are then: the control of the regularity and wavefront sets when writing transport equations as explained in Section 4.2.3; the presence of infinite-length geodesics in the kernel of the error operator, requiring further control for the sake of continuity estimates. In particular, one can write explicit estimates in the neighborhood of constant negative curvature metrics, making (14) invertible via Neumann series.

Theorem 4.4.5 ([37, Theorem 1.2]). *Let (M, g_0) be a manifold with strictly convex boundary and constant negative curvature $-\kappa_0$ and trapped set K, and let $\delta = \frac{1}{2}(\dim_{\mathrm{Haus}}(K) - 1) \in [0, 1)$.*[9] *Then for each $\lambda_1, \lambda_2 \in (0, 1)$ so that $1 \ge \lambda_1\lambda_2 > \max(\delta, \frac{1}{2})$, there is an explicit constant $A(\delta, \lambda_1, \lambda_2) > 0$ depending only on δ, λ_1, λ_2 such that for all metrics g on M with strictly convex boundary and Gauss curvature $\kappa(x)$ satisfying*

$$\lambda_1^2 g_0 \le g \le \lambda_1^{-2} g_0, \quad \kappa(x) \le -\lambda_2^2\kappa_0, \quad |d\kappa|_\infty \le A(\delta, \lambda_1, \lambda_2)\kappa_0^{3/2},$$

the remainder operator W in (14) *is a $L^2(M, g)$-contraction, and hence f and h are reconstructible from* (14) *via a convergent series. When $\delta < 1/2$, the constant $A(\delta, \lambda_1, \lambda_2)$ does not depend on δ.*

9 In constant negative curvature, the trapped set K is a fractal set with $\dim_{\mathrm{Haus}}(K) \in [1, 3)$ when $K \neq \emptyset$.

Let us mention in passing that proving Theorem 4.4.5 requires writing continuity estimates for the normal operator $I_0^* I_0$ on a surface of a constant negative curvature, and this leads to a striking expression of $I_0^* I_0$ in terms of the (negative) Laplacian on that surface Δ_M, given by

$$I_0^* I_0 = 4\frac{\Gamma(\frac{1}{4} - S)\Gamma(\frac{1}{4} + S)}{\Gamma(\frac{3}{4} - S)\Gamma(\frac{3}{4} + S)}, \qquad \text{where } S := \frac{i\sqrt{-\Delta_M - \frac{1}{4}}}{2}, \tag{19}$$

$\Gamma(\cdot)$ the Euler Gamma function, and with the convention that $\sqrt{s(1-s) - \frac{1}{4}} = i(s - 1/2)$ when $s(1-s) \in (0, \frac{1}{4})$ is a L^2-eigenvalue of $-\Delta_M$ with $s \in (1/2, 1)$; see [37, Lemma A.1] for more detail. This is to be contrasted with the Euclidean case where $I_0^* I_0 = \sqrt{-\Delta}$, a fact which is often also stated at the level of principal symbols in Riemannian settings, though *global* relations such as (19) show that the relation between $I_0^* I_0$ and the Laplace–Beltrami operator can be intricate.

4.4.5 Reconstruction of higher-order tensor fields

To reconstruct tensor fields from their x-ray transform, it was realized in [65] that the solenoidal representative of a tensor field may not lead to the most efficient reconstructions. To this end, we first describe a different gauge than the solenoidal one, which we call the Killing gauge here. The first appearance known to the authors is in [16, Theorem 1.5], stating that every m-tensor $f \in H^k(S^m(T^*M))$ ($m \geq 0$, $k \geq 1$) can be uniquely represented in the form

$$f = \sigma\nabla v + \sigma(g \otimes \lambda) + \tilde{f},$$

where $v \in H^{k+1}(S^{m-1}(T^*M))$ is trace-free and vanishes at ∂M, λ belongs to $H^k(S^{m-2}(T^*M))$ and $\tilde{f} \in H^k(S^m(T^*M))$ is trace-free and divergence-free, and the decomposition is continuous in the appropriate spaces. When restricting all tensors to SM, one equivalently has $f \in H^k(\Omega_m) \cap \ker X_-$. On the data side, we immediately have $\mathcal{I}[\sigma\nabla v] = 0$, and thus

$$\mathcal{I}f = \mathcal{I}[\sigma(g \otimes \lambda)] + \mathcal{I}[\tilde{f}] = \mathcal{I}[\lambda] + \mathcal{I}[\tilde{f}].$$

An advantage of this decomposition for inversion purposes, found in [65], is that in the Euclidean case, the two components in the last right-hand side are orthogonal for the $L^2(\partial_+ SM)$ topology. While not orthogonal in the usual target space $L^2_\mu(\partial_+ SM)$, this still gives us a direct decomposition, allowing one to break down the tensor tomography problem into smaller pieces, both for inversion and range characterization purposes. To see the gist of the method on the Euclidean unit disk, given a $2m$-tensor field f

restricted to SM with L^2 components, one may iterate the decomposition above to find a $2m$-tensor g with L^2 components such that $\mathcal{I}f = \mathcal{I}g$, and g is of the form

$$g = g_0 + \sum_{k=1}^{m} g_{2k}, \quad g_0 \in L^2(M), \quad g_{2k} \in L^2(\Omega_{2k}) \cap \ker X_-, \quad 1 \le k \le m.$$

Moreover, the ray transforms of each component are $L^2(\partial_+SM)$-orthogonal; see [65, Theorems 2.1, 2.2]. Then each g_{2k} can be reconstructed explicitly via Cauchy-type integrals (its components are complex-analytic or anti-analytic in the base coordinates), while g_0 (a full $L^2(M)$ function) can be inverted via applying I_0^{-1} to the data; see [65, Theorems 2.4]. A similar story holds for odd-order tensors. In the case of the Euclidean disk, the approach was generalized to transforms with arbitrary position-dependent attenuation in [67], leading to explicit and efficient inversions of the attenuated x-ray transform over tensor fields of arbitrary order. To reconstruct the g_{2k} components, special care must be paid in constructing special invariant distributions with specified harmonic content. This is the topic of the next section.

4.5 Tensor tomography and special invariant distributions

4.5.1 An equivalence principle

As providing a potential alternate route toward proving tensor tomography on simple manifolds, the following theorem was provided in [89]. In the statement, L_m denotes the L^2-L^2 adjoint of the operator ℓ_m defined in (6).

Theorem 4.5.1 ([89, Theorem 1.2]). *Let M a compact simple Riemannian manifold, then the following are equivalent:*

(1) *I_m is s-injective on $C^\infty(S^m(T^*M))$;*
(2) *for every $u \in L^2(S^m_{\text{sol}}(T^*M))$, there exists $f \in H^{-1}(\partial_+SM)$ such that $u = I_m^*\varphi$;*
(3) *for every $u \in L^2(S^m_{\text{sol}}(T^*M))$, there exists $f \in H^{-1}(SM)$ satisfying $Xf = 0$ and $u = L_m f$;*
(4) *for every $u \in C^\infty(S^m_{\text{sol}}(T^*M))$, there exists $\varphi \in C^\infty_\alpha(\partial_+SM)$ such that $u = I_m^*\varphi$;*
(5) *for every $u \in C^\infty(S^m_{\text{sol}}(T^*M))$, there exists $f \in C^\infty(SM)$ with $Xf = 0$ such that $L_m f = u$.*

In the theorem above, f is an invariant distribution for the geodesic flow, and the condition $L_m f = u$ is a prescription on its harmonic content. Constructing such invariant distributions not only provides another way to look at the problem, but their explicit construction provides an immediate way to reconstruct explicit features of the unknown tensor: if u is an unknown solenoidal m-tensor and one knows an invariant distribution $f = I^*\varphi$ such that $L_m f = v$ for some $v \in C^\infty(S^m_{\text{sol}}(T^*M))$, then the inner product $\langle u, v\rangle_{S^m(T^*M)}$ is known from the chain of equalities

$$\begin{aligned}(u,v)_{S^m(T^*M)} &= (u, L_m f)_{S^m(T^*M)} = (\ell_m u, f)_{SM} \\ &= (\ell_m u, I^*\varphi)_{SM} = (I_m u, \varphi)_{\partial_+ SM}.\end{aligned}$$

The construction of such invariant distributions, or equivalently the surjectivity of adjoints to integral transforms has appeared in several places as explained in the next paragraphs.

4.5.2 Surjectivity of adjoints using microlocal arguments

On simple surfaces, the surjectivity of $I_0^*\colon C_\alpha^*(\partial_+ SM) \to C^\infty(M)$ was proved in [93, Theorem 1.4] via microlocal arguments, where

$$C_\alpha^\infty(\partial_+ SM) := \{h \in C^\infty(\partial_+ SM),\ h_\psi \in C^\infty(SM)\}.$$

Specifically, the normal operator $I_0^* I_0$ is elliptic on a simple open neighborhood of M, and can be extended into an invertible operator on the closed "double" of M. After appropriate restriction, this allows to construct a right-inverse for I_0^*. This scheme of proof for simple surfaces was further used in the following contexts:

- In [101, Lemma 4.5], it is proved that the operator $I_\perp^*\colon C_\alpha^\infty(\partial_+ SM) \to C^\infty(M)$ is surjective for M a simple surface.
- In [86, Theorem 5.4], it is proved that the operator $I_{0,A}^*\colon \mathcal{S}^\infty(\partial_+ SM, \mathbb{C}^n) \to C^\infty(M, \mathbb{C}^n)$ is surjective for M a simple surface and A a unitary connection, where the space $\mathcal{S}^\infty(\partial_+ SM, \mathbb{C}^n)$ is a natural generalization of $C_\alpha^\infty(\partial_+ SM)$ to the case with connection.

4.5.3 Iterated Beurling series

A building block toward fulfilling (3) or (5) in Theorem 4.5.1 amounts to seeking an invariant distribution of the form $w := w_{k_0} + w_{k_0+2} + w_{k_0+4} + \cdots$ such that $w_{k_0} = u \in H_{k_0}$ is prescribed and $Xw = 0$. In light of (8), this implies

$$X_- w_{k_0} = 0, \quad \text{and} \quad X_- w_{k_0+2(m+1)} = -X_+ w_{k_0+2m}, \quad m = 0, 1, \ldots.$$

The first equation is a requirement on u, while the second family of equations suggests to construct w_{k_0+2} from w_{k_0}, then w_{k_0+4} from w_{k_0+2}, etc. provided that one can "invert" X_- on $X_+\Omega_p$ in some sense. This is the purpose of the *Beurling transform*, defined for any $p \geq 0$ as

$$B\colon \Omega_p \to \Omega_{p+2}, \quad f_p \mapsto f_{p+2},$$

where $f_{p+2} \in \Omega_{p+2}$ is the unique solution to the equation $X_- f_{p+2} = -X_+ f_p$ that is orthogonal to $\ker^{p+2} X_-$ (equivalently, the unique solution with minimal L^2 norm). That this is well-defined follows from [85, Lemma 11.1]. With B defined as above, the following theorem is an example where the construction of such invariant distributions is well understood, done via the formal iterated Beurling series.

Let $C_n(m)$ denote the constant

$$C_n(m) = \begin{cases} \sqrt{2}, & m = 0 \text{ and } n = 2, \\ [1 + \frac{1}{(2m+1)(m+2)^2}]^{1/2}, & m \geq 0 \text{ and } n = 3, \\ 1, & \text{otherwise.} \end{cases}$$

Theorem 4.5.2 ([85, Theorem 11.4]). *Let (M, g) be a compact manifold with boundary, and assume that the sectional curvatures are nonpositive. Then*

$$\|X_- u\|_{L^2} \leq C_n(m) \|X_+ u\|_{L^2}, \quad u \in \Omega_m, \quad u|_{\partial SM} = 0, \quad m \geq 0.$$

The Beurling transform satisfies

$$\|Bf\|_{L^2} \leq C_n(m) \|f\|_{L^2}, \quad f \in \Omega_m, \ m \geq 0.$$

If $k_0 \geq 0$ and if $f \in \Omega_{k_0}$ satisfies $X_- f = 0$, then there exists a solution of $Xw = 0$ in SM such that $w_{k_0} = f$, given by $w = \sum_{k=0}^{\infty} B^k f$. One has $w \in L_x^2 H_v^{-\frac{1}{2}-\varepsilon}(SM)$ for any $\varepsilon > 0$, and the Fourier coefficients of w satisfy

$$\|w_k\|_{L^2} \leq \|f\|_{L^2}, \quad k \geq k_0.$$

In the theorem above, we have introduced the mixed norm spaces

$$\begin{aligned} L_x^2 H_v^s(SM) &= \{u \in \mathcal{D}'(SM) \colon \|u\|_{L_x^2 H_v^s} < \infty\}, \\ \|u\|_{L_x^2 H_v^s}^2 &:= \sum_{p=0}^{\infty} \langle p \rangle^{2s} \|u_p\|_{L^2}^2, \end{aligned} \tag{20}$$

where as usual $\langle p \rangle = (1 + p^2)^{1/2}$.

We point out that the estimate $\|X_- u_k\|_{L^2} \lesssim \|X_+ u_k\|_{L^2}$ can be useful and true even in settings where the Beurling transform itself is not needed.

4.5.4 Explicit constructions on the Euclidean disk

The constructions above via formal Beurling series is explicit if one understands the Beurling transform explicitly, which is only well understood in some cases yet little documented. Another approach toward building such invariant distributions is done in the case of the Euclidean disk, provided by the second author in [67, Theorem 5.2]. There, it is enough to consider finding invariant distribution with prescribed complex-analytic, $L^2(\mathbb{D})$ average, and this serves as a crucial building block to write an explicit inversion of the attenuated tensor tomography problem over tensors of any order.

When $M = \mathbb{D}$, parameterizing $\partial_+ SM$ in fan-beam coordinates $(\beta, \alpha) \in \mathbb{S}^1 \times (-\pi/2, \pi/2)$, borrowing notation from [67], we define for $k = 0, 1, \dots$

$$Z_k(x,y) := \frac{\sqrt{k+1}}{2\pi^2}(x+iy)^k, \quad (x,y) \in \mathbb{D},$$
$$W_k(\beta,\alpha) := (-1)^k \frac{\sqrt{k+1}}{2\pi\sqrt{2}} e^{ik\beta}(e^{i(2k+1)\alpha} + (-1)^k e^{-i\alpha}), \quad (\beta,\alpha) \in \partial_+ SM.$$

Then [67, Proposition 4] shows that for every k, $((\frac{W_k}{\cos\alpha})_\psi)_0 = Z_k$. Moreover, we have the following theorem, reformulated here in terms of invariant distributions.

Theorem 4.5.3 ([67, Theorem 5.2]). *For any $f \in L^2(\ker\bar\partial)$, given by $f = \sum_{k=0}^\infty (f, Z_k)_{SM} Z_k$, the function $W_f \in \ell^2(\{W_k\}_{k=0}^\infty)$ given by*

$$W_f := \sum_{k=0}^\infty (f, Z_k)_{SM} W_k, \quad \textit{satisfies} \quad \left(\left(\frac{W_f}{\cos\alpha}\right)_\psi\right)_0 = f.$$

Moreover, the distribution $(\frac{W_f}{\cos\alpha})_\psi$ is fiberwise holomorphic and orthogonal to $\ker^k \eta_-$ for any $k > 0$.

The last claim implies that such invariant distributions have minimal norm in some sense. In addition, one may show that such distributions make sense in $L^2_x H_v^{-1/2-\varepsilon}(SM)$ (as defined in (20)) for every $\varepsilon > 0$, and that this is sharp. Notice also that each invariant function $(\frac{W_k}{\cos\alpha})_\psi$ belongs to $C^\infty(SM)$. For $k > 1$, since $\ker^k X_- = \ker^k \eta_- + \ker^{-k} \eta_+$, the theorem above can be viewed as a building block to construct invariant distribution with prescribed kth moment in $\ker^k X_-$.

4.5.5 Anosov flows on closed manifolds

While not covered in detail in this review, invariant distributions also play a role in solving integral geometric problems on a closed *Anosov*[10] Riemannian manifold (M, g). Namely, there is a notion of x-ray transform, where one integrates a function or tensor field over all possible closed geodesics, which appears when considering the linearization of spectral rigidity questions ("is a metric uniquely determined up

10 The geodesic flow is Anosov if there is a continuous invariant splitting $T(SM) = \mathbb{R}X \oplus E^u \oplus E^s$ and constants $C > 0$ and $0 < \rho < 1 < \eta$ such that for all $t > 0$, $\|d\varphi_{-t}|_{E^u}\| \le C\eta^{-t}$, and $\|d\varphi_t|_{E^s}\| \le C\rho^t$.

to gauge, from the spectrum of its Laplace–Beltrami operator? or from its so-called marked length spectrum?"); see, e. g., [39].

In this context, injectivity of the ray transform considered is again linked in [83] to the existence of certain invariant distributions. Specifically, in the context of Anosov surfaces in [83, Theorems 1.4, 1.5, 1.6], the authors establish the existence of distributional solutions of $Xw = 0$ with prescribed zeroth moment in $C^\infty(M)$ ([83, Theorem 1.4]) and first moment as a prescribed smooth solenoidal one-form ([83, Theorem 1.4]). Under additional conditions on β_{Ter}, [83, Theorem 1.6] establishes the existence of invariant distributions with prescribed fiberwise moments on order 2. This is the case of interest for the nonlinear problem as it relates to the x-ray transform over second-order tensors.

All distributions mentioned above live in the space $H^{-1}(SM) = (H^1(SM))'$, where $H^1(SM)$ is the completion of $C^\infty(SM)$ with respect to the norm

$$\|u\|_{H^1}^2 := \|u\|^2 + \|Xu\|^2 + \|X_\perp u\|^2 + \|Vu\|^2.$$

This is to be contrasted to the previous paragraphs (involving an $L_x^2 H_v^{-1/2-\varepsilon}(SM)$ norm) which may indicate that the norm can be sharpened. However the present method, based on Pestov identities (and duality arguments *à la* Hahn–Banach), is not yet amenable to fractional Sobolev norms.

The results above, namely regarding the existence of invariant distributions with prescribed zeroth or first fiberwise moments, were generalized in [8, Theorems 1.7, 1.8] to the case of Anosov thermostat flows (see also Section 4.3.6) on closed Riemannian surfaces.

4.6 Microlocal methods

This section is devoted to microlocal methods, first introduced in integral geometry by Guillemin. Here, the results are based on the description of a ray transform I (over functions or tensor fields of a fixed degree) or its corresponding normal operator I^*I in certain classes of operators, namely the first one as a Fourier Integral Operator (FIO), and the second one as a pseudo-differential operator (ΨDO) when the family of curves is simple, or in more general classes if the geometry is more complex. Here, microlocal analysis helps one to prove finiteness theorems or Fredholmess (invertibility of the problem up to a finite-dimensional kernel made of smooth ghosts) under geometric restrictions, to study how to recover the singularities[11] of the unknown from the singularities of the data, and to describe how and when this is not possible. Such analysis ultimately provides stability estimates, allows to work with weights in the transform,

11 By "singularity," here we mean element of the wavefront set.

and to study partial data problems. One can also prove injectivity results in the case where the family of curves and weights is real-analytic, using analytic microlocal analysis.

4.6.1 Results for complete families of curves

A sufficient-but-not-necessary microlocal condition for stability and sometimes injectivity, first formulated in [113], is that the family of geodesics be *geodesically complete*, in the sense that for every $x \in M$ and every $\omega \in T^*M$, there exists a geodesic γ free of conjugate points passing through x and normal[12] to ω. Such a condition ensures that integrals over geodesics in a neighborhood of γ allow to resolve possible singularities at ω without creating artifacts elsewhere. If the metric is analytic, one may use analytic microlocal analysis to prove injectivity of the ray transform over any complete complex of geodesics; see [113, Theorem 1]. Then using stability estimates which remain true under C^k-perturbation of the metric for k large enough, one may promote this to a generic injectivity result for functions but also for solenoidal tensor fields of order up to two; see [113, Theorems 2, 3].

Such results have been generalized to the case of magnetic flows in [15] and for generic general families of curves and weights including analytic ones in [28]. For instance in [28], the completeness condition can also be written for a general family of curves, upon defining simple curves and assuming that the conormal bundle of all simple curves covers T^*M, and the analysis can be made robust to transforms with smooth weight $\phi\colon SM \to \mathbb{R}$

$$I_{\phi,0}f(\gamma) = \int_\gamma f(\gamma(t))\phi(\gamma(t),\dot\gamma(t))\,dt, \quad \gamma \in \mathcal{G}.$$

If the curves are geodesics and the metric is simple, the associated normal operator $I_{\phi,0}^* I_{\phi,0}$ is a ΨDO of order −1, with principal symbol

$$a_{-1}(x,\eta) = 2\pi \int_{\ker \eta} |\phi(x,v)|^2\, dS(v), \quad (x,\eta) \in T^*M,$$

where we have defined $\ker\eta := \{v \in S_xM, \eta(v) = 0\}$. In particular, if ϕ is *admissible* in the sense that for every $x \in M$ and $\eta \in T_x^*M$, there exists $v \in S_xM$ such that $\eta(v) = 0$ and $\phi(x,v) \neq 0$, the normal operator is elliptic, therefore, the problem is Fredholm and Hölder-stable on a complement of the (at most finite-dimensional) kernel; see also [111, 112]. If I_ϕ is defined over tensor fields instead, the normal operator is elliptic over divergence-free tensors in the interior of M.

12 Specifically, there exists t such that $\gamma(t) = x$ and $\omega(\dot\gamma(t)) = 0$.

4.6.2 Mapping properties of the normal operator and an uncertainty quantification result on simple manifolds

4.6.2.1 Mapping properties of $I_0^* I_0$ in the simple case

In the case of simple manifolds, recent sharp mapping properties of the normal operator $I_0^* I_0$ were obtained in [68]. Specifically, calling d_M any positive C^∞ function that equals $\mathrm{dist}(\cdot, \partial M)$ near the boundary, the following mapping properties were derived.

Theorem 4.6.1 ([68, Theorem 4.4]). *The operator $I_0^* I_0$ is an isomorphism in the following functional settings:*

$$
\begin{aligned}
&I_0^* I_0 \colon d_M^{-1/2} C^\infty(M) \to C^\infty(M), \\
&I_0^* I_0 \colon H^{-1/2(s)}(M) \to H^{s+1}(M), \quad s > -1 \quad (\textit{bi-continuous}).
\end{aligned} \tag{21}
$$

The theorem above also applies to x-ray transforms with attenuation whenever they are injective. Looking at equation (21), we see that it is in fact natural for the integrand to have a certain blow-up at the boundary. The spaces $H^{-1/2(s)}(M)$, so-called Hörmander $\frac{-1}{2}$-transmission spaces, are Hilbert spaces whose elements are morally H^s inside of M, with some special behavior near the boundary; see [31, 32]. The proof of Theorem 4.6.1 is based on the fact that an extension P of $I_0^* I_0$ satisfies a so-called μ-transmission condition[13] with $\mu = -1/2$ with respect to ∂M in the sense that

$$
\partial_x^\beta \partial_\xi^\alpha p_j(x, \nu_x) = e^{\pi i(m-2\mu-j-|\alpha|)} \partial_x^\beta \partial_\xi^\alpha p_j(x, -\nu_x), \quad x \in \partial M,\ j \geq 0,
$$

for all multi-indices α, β, where $p \sim \sum_{j=0}^\infty p_j$ is the full symbol of P, of order $m = -1$ so that $p_j(x, t\xi) = t^{-1-j} p_j(x, \xi)$. Such a condition makes the operator P Fredholm in the functional settings above, after which one proves that the kernel and cokernel are trivial.

4.6.2.2 Statistical interlude: regularization of noisy inversions and uncertainty quantification

Theorem 4.6.1 allows to give the first rigorous statistical approach to regularization in the case of inversions with noisy data in a Riemannian context. The setting is as follows: given $\psi \in C^\infty(M)$, suppose that one attempts at estimating a "smooth aspect" $\langle f, \psi \rangle_{L^2(M)}$ from noisy measurements

$$
Y = I_0 f + \varepsilon \mathbb{W},
$$

13 A generalization of Boutet de Monvel's *transmission condition* which corresponds to the case $\mu = 0$.

where $\varepsilon > 0$ is the noise level and W is a Gaussian white noise on the space $L^2_\mu(\partial_+ SM)$. One then adopts a Bayesian approach where, upon a choice of Gaussian prior distribution and noise model, Bayes' formula gives the density of the posterior random variable $f|Y$. One may then attempt to understand where posterior densities concentrate depending on the choice of prior and, under the assumption that a true f_0 generated the noisy data, whether the posterior distributions concentrate their mass near f_0.

In this regard, [68, Theorems 2.5, 2.7] provide a limiting behavior of the posterior distribution that is independent of the choice of Gaussian prior for f, chosen among a rather loosely constrained family (including all priors modeling Sobolev smoothness), corrected at the boundary to account for the boundary behavior as in equation (21). Upon choosing such a prior, if $\bar{f}$ is the mean of the posterior distribution and f_0 the ground truth, not only do we have convergence in distribution[14]

$$\frac{1}{\varepsilon}\langle \bar{f} - f_0, \psi\rangle_{L^2(M)} \to Z \sim \mathcal{N}(0, \|I_0(I_0^* I_0)^{-1}\psi\|^2_{L^2_\mu(\partial_+ SM)}), \quad \text{as } \varepsilon \to 0,$$

but in fact the posterior distribution of $\langle f, \psi\rangle_{L^2(M)}$ is approximately a normal density centered at $\langle \bar{f}, \psi\rangle_{L^2(M)}$ with variance $\|I_0(I_0^* I_0)^{-1}\psi\|^2_{L^2_\mu(\partial_+ SM)}$. This variance is optimal and its expression justifies why sharp mapping properties of $I_0^* I_0$ were required. Note that since the posterior law is Gaussian here, the posterior mean $\bar{f}$ agrees with the Maximum A Posteriori (i. e., the argmax of the posterior distribution) which can also be obtained by Tychonov regularization.

The result above tells us how the mass of the posterior distribution concentrates about the mean, and allows one to use *Bayesian credible intervals* (intervals of the posterior distribution centered at the mean and containing $1-\alpha$ of the total mass with $\alpha \in (0,1)$ a fixed threshold) as approximate *frequentist confidence sets* (sets which contain the ground truth with some given probability). While the latter are usually hard to compute, the former can be obtained by visualizing ensembles of posterior draws. For uncertainty quantification purposes, the results above tell us that one may use the former intervals to infer where the true unknown lies with a certain probability.

4.6.3 Microlocal analysis of cases with conjugate points

In the presence of conjugate points, $I_0^* I_0$ is no longer an elliptic ΨDO, and the presence of a conjugate locus can destroy stability. One may view this microlocally by seeing that the Schwarz kernel of $I_0^* I_0$ develops singularities away from the diagonal, and this generates artifact singularities in reconstructions. In some case, they can be removed, either because they are "weaker" than the original singularity which generated them, or because the geometry allows to reconstruct the original singularity using integrals

14 $\mathcal{N}(\mu, \sigma^2)$ denotes a Gaussian law (or "normal density") of mean μ and variance σ^2.

over other curves than the ones which generate the artifacts. The latter case only occurs in dimensions greater than three, where the problem is overdetermined and some geometries with conjugate points still allow to restrict the data in a way that fulfills the completeness condition, thereby restoring stability. While this is a somewhat global consideration, the aim of the following sections is to locally describe the ray transform when defined on a neighbourhood of a curve containing conjugate points, and possibly study what these local descriptions imply globally.

The first result in this direction appeared in [114] in the case of fold caustics, and the analysis was done by studying the structure of the normal operator $I_0^* I_0$. Following this approach, the results were extended to more general types of conjugate points in [44]. In two dimensions, the results in [114] were refined in [43, 70] by studying the operator I_0 as a FIO directly. We present these results in the next few sections.

4.6.3.1 FIOs and the clean composition calculus

One way to make our way up to this description is by expressing $\mathcal{I}$ and I_0 as composites of push-forwards and pull-backs by smooth submersions, following [44, 114], or equivalently, exploiting the double fibration structure below:

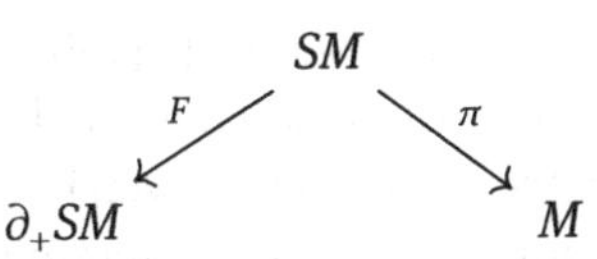

Namely, if (M, g) is nontrapping with strictly convex boundary, the canonical projection $\pi\colon SM \to M$ and basepoint map $F\colon SM \to \partial_+ SM$ defined by $F(x,v) = \varphi_{-\tau(x,-v)}(x,v)$ are both smooth submersions with π proper, and as such define pull-backs $\pi^*\colon C_c^\infty(M) \to C_c^\infty(SM)$ ($\pi^* f(x,v) := f(x)$) and $F^*\colon C_c^\infty(\partial_+ SM) \to C^\infty(SM)$ ($F^* g(x,v) = g(F(x,v))$), as well as push-forwards $\pi_*\colon C_c^\infty(SM) \to C_c^\infty(M)$ and $F_*\colon C_c^\infty(SM) \to C_c^\infty(\partial_+ SM)$ via the relations

$$\int_{SM} (\pi^* f) g \, d\Sigma^3 = \int_M f(\pi_* g) \, d\,\mathrm{Vol}_g,$$

$$\int_{SM} (F^* h) g \, d\Sigma^3 = \int_{\partial_+ SM} h\,(F_* g)\mu \, d\Sigma^2,$$

to be true for all $f \in C_c^\infty(M)$, $g \in C_c^\infty(SM)$, and $h \in C_c^\infty(\partial_+ SM)$. Note in particular that $\mathcal{I} = F_*$. As explained in [44], π_* and π^* are FIOs of order $\frac{1-n}{4}$ with canonical relations

$$\mathcal{C}_{\pi_*} = \{(\omega, d\pi|_{(x,v)}^T \omega) : (x,v) \in SM,\ \omega \in T_x^* M \setminus \{0\}\} \subset T^* M \times T^* SM,$$

$$\mathcal{C}_{\pi^*} = \{(d\pi|_{(x,v)}^T \omega, \omega) : (x,v) \in SM,\ \omega \in T_x^* M \setminus \{0\}\} \subset T^* SM \times T^* M,$$

while F_* and F^* are FIOs of order $\frac{-1}{4}$ with canonical relations $\mathcal{C}_{F_*} \subset T^*(\partial_+ SM) \times T^*SM$ and $\mathcal{C}_{F^*} \subset T^*SM \times T^*(\partial_+ SM)$, respectively, given by

$$\mathcal{C}_{F_*} = \{(\eta, dF|_{(x,v)}^T \eta) : (x,v) \in SM,\ \eta \in T^*_{F(x,v)}\partial_+ SM \setminus \{0\}\},$$
$$\mathcal{C}_{F^*} = \{(dF|_{(x,v)}^T \eta, \eta) : (x,v) \in SM,\ \eta \in T^*_{F(x,v)}\partial_+ SM \setminus \{0\}\}.$$

Note that $\mathcal{I} = F_*$ and as such one can tell which singularities are "collapsed": given $\eta \in T^*_\xi(\partial_+ SM)$, all the singularities $dF|_{\varphi_t(\xi)}^T \eta$ for $t \in (0, \tau(\xi))$ are collapsed into η. Now, to understand I_0 as a FIO, one may view I_0 as $I_0 = F_* \circ \pi^*$ and use the *clean composition calculus* of Duistermaat and Guillemin to obtain the canonical relation of I_0 by computing the composition $\mathcal{C}_{I_0} = \mathcal{C}_{F_*} \circ \mathcal{C}_{\pi^*} \subset T^*(\partial_+ SM) \times T^*M$.

Traditionally, similar to geodesic completeness, it is of interest that the so-called *Bolker condition* be satisfied in the sense that on the microlocal diagram below

$$\begin{array}{ccc} & \mathcal{C}_{I_0} & \\ \swarrow & & \searrow \\ T^*(\partial_+ SM) & & T^*M \end{array}$$

the projection $\mathcal{C}_{I_0} \to T^*(\partial_+ SM)$ is an injective immersion. When this is the case (for I_0 or any other integral operator of interest see, e. g., [26]), one may use the clean composition calculus again to compute the associated normal operator and show that this is in fact a ΨDO, the most favorable scenario for inversion purposes. For geodesic x-ray transforms, the presence of conjugate points is precisely what invalidates the Bolker condition, and further analysis is required. Similar issues occur in [23, 115, 124]. The next two sections summarize what can be said in cases with conjugate points; first, in dimension two, then three and higher.

4.6.3.2 Two dimensions

In two dimensions, the manifolds M and $\partial_+ SM$ have the same dimension, and some statements can be made more precise. In this section, we present results from [43, 70, 114]. The results in [43, 70] are presented in the context of the double fibration of the point-geodesic relation, though for consistency of the present article, we will present the ideas using the *SM* notation.

Let $(X, X_\perp, V)$ the canonical frame of *SM* and $(X^\flat, X_\perp^\flat, V^\flat)$ its dual co-frame with respect to the Sasaki metric.[15] A basis of $T_{(x,v)}(\partial_+ SM)$ is given by $(V_{(x,v)}, T_{(x,v)} := \frac{1}{\langle v, \nu_x \rangle} \nabla_T|_{(x,v)})$, where $\nabla_T|_{(x,v)} = \langle -(\nu_x)_\perp, \nabla \rangle$ (horizontal derivative along the tangent

15 It corresponds to $(\alpha, -\beta, \gamma)$ in [61].

vector). Let us denote $(V^\flat, T^\flat)$ the dual coframe to (V, T) on $T^*_{(x,v)}\partial_+ SM$. Then the composition $\mathcal{C}_{I_0} = \mathcal{C}_{F_*} \circ \mathcal{C}_{\pi^*}$ can be computed explicitly as follows (the proof is relegated to the Appendix).

Lemma 4.6.2. *The composition $\mathcal{C}_{I_0} = \mathcal{C}_{F_*} \circ \mathcal{C}_{\pi^*} \subset T^*(\partial_+ SM) \times T^*M$ is given by*

$$\begin{aligned} \mathcal{C}_{F_*} \circ \mathcal{C}_{\pi^*} &= \{(\lambda\, \eta_{x,v,t}, \lambda\, \omega_{x,v,t}),\ (x,v) \in \partial_+ SM,\ t \in (0, \tau(x,v)),\ \lambda \in \mathbb{R}\}, \\ \textit{where} \quad \eta_{x,v,t} &:= -b(x,v,t) V^\flat_{(x,v)} + a(x,v,t) T^\flat_{(x,v)} \in T^*_{(x,v)}\partial_+ SM, \\ \omega_{x,v,t} &:= (\dot\gamma_{x,v}(t))^\flat_\perp \in T^*_{\gamma_{x,v}(t)} M. \end{aligned} \tag{22}$$

Checking the Bolker condition now is just to ask whether, along a geodesic $\varphi_t(x,v)$, two (or more) singularities of the form $\lambda_1 \omega_{x,v,t_1}$ and $\lambda_2 \omega_{x,v,t_2}$ are mapped to the same $\eta \in T^*_{(x,v)}(\partial_+ SM)$? This is equivalent to asking whether the mapping $t \mapsto \eta_{x,v,t}$ in (22) is injective. Such a condition is violated precisely when $a(t_2)b(t_1) - a(t_1)b(t_2) = 0$ for some $0 < t_1 < t_2 < \tau(x,v)$, but then this occurs precisely when the nontrivial Jacobi field

$$J(t) := (a(x,v,t)b(x,v,t_1) - b(x,v,t)a(x,v,t_1))\dot\gamma_{x,v}(t)_\perp,$$

vanishes at t_1 and t_2, i. e., when $\varphi_{t_1}(x,v)$ and $\varphi_{t_2}(x,v)$ are conjugate; see also [70, Theorem 4.1].

In such a case, upon defining $\omega_1 := \lambda_1 \omega_{x,v,t_1}$, $\omega_2 := \lambda_1 \frac{a(t_1)}{a(t_2)} \omega_{x,v,t_2}$ and $\eta \in T^*_{(x,v)}(\partial_+ SM)$ uniquely defined by $\eta_V = -b(t_1)\lambda_1$ and $\eta_T = a(t_1)\lambda_1$, we have that

$$(\eta, \omega_1) \in \mathcal{C}_{I_0}, \quad \text{and} \quad (\eta, \omega_2) \in \mathcal{C}_{I_0},$$

describing how both singularities at ω_1 and ω_2 collapse into η upon applying I_0.

Cancellation of singularities and artifact-generating operators

Next, one may exploit that $\mathcal{C}_{I_0}$ is a local graph to construct artifact-generating operators which will produce a proof of global instability of the problem. In the setting above, there exists a conical neighborhood V of η in $T^*(\partial_+ SM)$ and $V_{1,2}$, conical neighborhoods of $\omega_{1,2}$ in T^*M such that $\mathcal{C}_{1,2} := \mathcal{C} \cap (V \times V_{1,2})$ are diffeomorphisms. Then $\mathcal{C}_{21} := \mathcal{C}_2^{-1} \circ \mathcal{C}_1$ is a canonical relation itself, and a diffeomorphism, and it provides the basis for constructing an artifact-generating FIO $F_{21}\colon H^s(V_1) \to H^s(V_2)$ (for all $s \in \mathbb{R}$) with canonical relation $\mathcal{C}_{21}$; see [70, Theorem 4.3]. The operator F_{21} will be such that given f_1 a compactly supported distribution with wavefront set included in V_1, one may construct the artifact $F_{21}f_1$, whose wavefront set is included in V_2, and such that

$$I(f_1 - F_{21}f_1) \in C^\infty(\partial_+ SM),$$

see also [70, Corollary 4.1]. Such an approach therefore allows to construct distributions whose image is a smooth function, thereby removing any hopes for a global stability estimate on the Sobolev scale, of the form

$$\|f\|_{H^{s_1}(M)} \le C\|If\|_{H^{s_2}(\partial_+ SM)} + C'\|f\|_{H^{s_3}(M)},$$

no matter the choice of indices s_1, s_2, s_3. In practice, if ω_1 sits above x_1 and ω_2 sits above x_2, upon defining $U_{1,2}$ simple neighborhoods of $x_{1,2}$ in M, and denoting $r_{U_{1,2}}: L^2(M) \to L^2(U_{1,2})$ the restriction operators, the transforms $I_0 r_{U_1}$ and $I_0 r_{U_2}$ are injective and explicitly invertible (if the domain is small enough, the error operators in (14) become contractions, and thus the Fredholm equations are explicitly invertible), and F_{21} may be constructed as $F_{21} := (I_0 r_{U_2})^{-1} \circ I_0 r_{U_1}$. One may also define $F_{12} := (I_0 r_{U_1})^{-1} \circ I_0 r_{U_2}$. To express the fact that such operators generates artifacts "of the same strength," it was further proved in [43, Theorem 2.1] that the operators $F_{21}: H^{-1/2}(V^1) \to H^{-1/2}(V^2)$ and $F_{12}: H^{-1/2}(V^2) \to H^{-1/2}(V^1)$ were *principally unitary*, in the sense that $F_{12}F_{21} - Id$ and $F_{21}F_{12} - Id$ are smoothing operators.

Outcomes of the Landweber iteration

In cases where the operator I_0 is proved unstable in theory, one may wonder what happens if we run an adjoint-based inversion (such as Landweber's iteration) to the data. Let us recall that given $\mathcal{L}: \mathcal{H}_1 \to \mathcal{H}_2$ a bounded operator between two Hilbert spaces, and given data $m = \mathcal{L}f$, the Landweber iteration consists in choosing a constant $\gamma > 0$ and iterating the following scheme:

$$f^{(0)} = 0, \quad f^{(k)} = f^{(k-1)} - \gamma \mathcal{L}^* (\mathcal{L} f^{(k-1)} - m), \quad k = 1, 2, \ldots.$$

One can use $\mathcal{L} = I_0$, though $\mathcal{L} = (-\Delta_g)^{1/2} \chi I_0^* I_0$ is an equivalent choice which removes the smoothing properties of I_0 and may speed up convergence ($\mathcal{L}^* \mathcal{L}$ is a ΨDO of order zero while $I_0^* I_0$ is not).

The analysis made in [43], specifically regarding the principal unitarity of the operators F_{12} and F_{21}, allows to draw conclusions on what happens to the iterations in the presence of conjugate points. To be specific, considering the situation where f is supported on $U_1 \cup U_2$ with $WF(f) \subset V^1 \cup V^2$, we may identify f with the couple $(f_1, f_2) := (r_{U_1} f, r_{U_2} f)$. Then the operator $\mathcal{L}^* \mathcal{L}$ can be regarded as a 2×2 matrix of operators involving the operators F_{12} and F_{21}, namely, modulo smoother operators, one may write

$$\mathcal{L}^* \mathcal{L} = \begin{pmatrix} Id + F_{12}F_{21} & 2F_{12} \\ 2F_{21} & Id + F_{21}F_{12}. \end{pmatrix} \quad \text{mod } \Psi^{-1}.$$

Picking $f_1 \in L^2(U_1)$ arbitrary and $f_2 = 0$, and writing the Landweber iteration at leading order, the iterations converge to

$$\text{Landweber solution} = (Id - P)f_1 + F_{21} P f_1, \quad P := (Id + F_{21}^* F_{21}),$$

see [43, Section 3.2]. This means that, starting from an input initially supported on U_1, the Landweber iteration reconstructs a portion of the input, but also generates the artifact $F_{21} P f_1$, supported on U_2. This is not so much a defect of the method than a common issue that any adjoint-based method will share. Indeed f_1 and $F_{21} f_1$, while having disjoint wavefront sets, lead to the same singularities in data. Without additional prior knowledge, any linear combination of the two would be an acceptable reconstruction given the data.

Attenuated transforms

So far, the statements were made for transforms without weight, although the analysis carries over to the case of a transform with weight of the form

$$I_\phi f(x,v) := \int_0^{\tau(x,v)} f(\gamma_{x,v}(t))\phi(\varphi_t(x,v))\,dt, \quad (x,v) \in \partial_+ SM,$$

where $\phi : SM \to \mathbb{R}$ is a smooth weight, nonvanishing for simplicity of exposition. A specific example is that of the *attenuated x-ray transform*, where the weight is given by $\phi(x,v) = \exp(-\int_0^{\tau(x,v)} a(\gamma_{x,v}(t))\,dt)$, with a some *attenuation* function. The presence of a weight can salvage stability in certain scenarios, even in the presence of conjugate points. Namely, the following results were established in [70]:

- If conjugate points occur at most in pairs, and for any conjugate pair (x_1, v_1), (x_2, v_2), we have $\det\left[\begin{smallmatrix} \phi(x_1,v_1) & \phi(x_2,v_2) \\ \phi(x_1,-v_1) & \phi(x_2,-v_2) \end{smallmatrix}\right] \neq 0$, then stability is possible.
- If the condition above does not hold at some conjugate pair, or if some conjugate points occur in triples (i. e., there exists a Jacobi field vanishing more than twice along some geodesic), then the problem is unstable.

The work in [43, 70] also covers the case of attenuations, namely in constructing corresponding artifact-generating operators in the unstable case, and studying the long-term behavior of the Landweber iteration when applied to the attenuated x-ray transform. A characteristic example of the locality of the discussion regarding the interplay between the geometry and the presence of a weight can be found in Figure 4.5 (these pictures also appear in [43]).

4.6.3.3 Three dimensions and higher

In higher dimensions, M has dimension n while $\partial_+ SM$ has dimension $2n-2 > n$, and the analysis from above is no longer formally determined. The analysis will also depend on the order of conjugate points, which can be of any order between 1 and $n-1$. Following pioneering work in [114], the most complete result known to date is given in [44], and applies to general weighted transforms of the form $\mathcal{I}_{\phi,0} := F_* \circ \phi \circ \pi^*$, where $\phi: SM \to \mathbb{R}$ is admissible in the sense of Section 4.6.1. We identify ϕ with the corresponding multiplication operator $f \mapsto \phi f$. As seen earlier, a natural approach to study $I_{\phi,0}$ is to study the normal operator

$$N_\phi := I_{\phi,0}^* I_{\phi,0} = \pi_* \circ \phi \circ F^* \circ F_* \circ \phi \circ \pi^*,$$

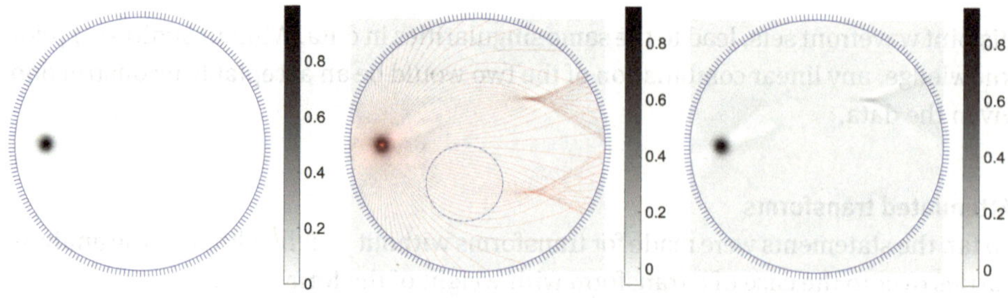

Figure 4.5: Left: a sharply peaked Gaussian f defined on the unit disk. Middle: some geodesics superimposed, showing where the conjugate locus of the (approximate) wavefront set of f is. A positive attenuation a supported inside the dashed circle is included in the transform. Right: reconstruction of f from $I_a f$ after convergence of Landweber's iteration. Comments: In the absence of attenuation, one would expect that the Landweber iteration generates artifacts at both top and bottom conjugate loci. On the other hand, adding a positive attenuation supported inside the dashed circle on the middle picture "stabilizes" the reconstruction of certain singularities by removing the bottom artifact at convergence.

using again the clean composition calculus of four FIOs. As mentioned in Section 4.6.1, in the absence of conjugate points, N_ϕ is an elliptic ΨDO of order −1. In the case of conjugate points, two clean compositions can be carried out easily, while the third creates multiple connected components in the canonical relation, corresponding to differing orders of conjugate points. Before mentioning the main theorems, let us briefly introduce notation about the decomposition of the conjugate locus on M.

Two vectors on the same geodesic, say (x, v) and $\varphi_t(x, v)$ for $t \neq 0$, are *conjugate of order* k ($1 \le k \le n-1$) if

$$\dim(\mathcal{V}(x,v) \cap d\varphi_{-t}|_{\varphi_t(x,v)}\mathcal{V}(\varphi_t(x,v))) = k,$$

and we write $C \subset SM \times SM$ the set of all conjugate pairs.[16] The set C splits into $C = \bigcup_{k=1}^{n-1} C_{R,k} \cup C_S$ where $C_{R,k}$ consists of the *regular conjugate pairs of order* k (those that have a neighborhood in $SM \times SM$ such that any pair is either nonconjugate or conjugate of order k), and C_S (singular conjugate points) contains the rest. Each $C_{R,k}$ is an embedded submanifold of $SM \times SM$ of dimension $2n-1$ (see [44, Theorem 3]), and one may construct a vector bundle $J_{R,k} \subset T(SM) \times T(SM)$ of dimension k on $C_{R,k}$, consisting of the pairs $((\gamma(t_1), \dot\gamma(t_1), \frac{dJ}{dt}(t_1)), (\gamma(t_2), \dot\gamma(t_2), \frac{dJ}{dt}(t_2)))$ whenever $\gamma(t_1)$, $\gamma(t_2)$ are conjugate along γ and J is any Jacobi field along γ vanishing at t_1 and t_2. Each such bundle maps into $C_k(J_{R,k}) \subset T^*M \times T^*M$, obtained by mapping each argument of $J_{R,k}$ using the connection map and the musical isomorphism, given at any point by $K_{(x,v)}: \mathcal{V}(x,v) \to \{v\}^\perp$ (where $\{v\}^\perp \subset T_xM$) and $\flat_g|_x: T_xM \to T_x^*M$, respectively.

16 This corresponds to the more traditional definition that on some geodesic γ, two points $x = \gamma(t_1)$ and $x' = \gamma(t_2)$ are conjugate of order k along γ if the space of Jacobi fields along γ vanishing at t_1 and t_2 has dimension k, see [44, Lemma 3].

The main result is as follows.

Theorem 4.6.3 ([44, Theorem 4]). *Suppose M has no self-intersecting geodesics and that $C_S = \emptyset$. Then the sets*

$$C_{A_k} = \mathcal{C}_k(\mathcal{J}_{R,k}) \subset T^*M \times T^*M$$

are either empty of are local canonical relations made out of M_k connected components. On the level of operators, we have the decomposition

$$N_\phi = \Upsilon + \sum_{k=1}^{n-1} A_k, \quad A_k = \sum_{m=1}^{M_k} A_{k,m},$$

where Υ is a ΨDO of order -1, elliptic where ϕ is admissible, and for each k, either

$$A_{k,m} \in \mathcal{I}^{-(1+(n-1-k)/2)}(M \times M, C'_{A_{k,m}}; \Omega^{1/2}_{M\times M}),$$

where $C_{A_{k,m}} \subset C_{A_k}$ for each m, or $M_k = 1$ and $A_{k,1} = 0$ if $C_{A_k} = \emptyset$.

From this result, we see that the components A_k for $1 \le k < n-1$ are FIOs of order strictly less than -1. This hints at us that when $C_S = C_{R,n-1} = \emptyset$, and if one can prove Sobolev mapping properties of the form $A_k : H^s \to H^{s+1+(n-1-k)/2}$ for $1 \le k \le n-2$, the problem may still be stable in the sense that, when solving for f the equation $N_\phi f = g$, writing $A = \sum_{k=1}^{n-2} A_k$ and applying a parametrix Q for Υ such that $Q\Upsilon = I + K$ with K a smoothing operator, we arrive at the equation $f + Kf + QAf = Qg$, where QA is also smoothing therefore compact. The problem is then Fredholm again, as such, solvable modulo a finite-dimensional kernel of smooth ghosts.

The main challenge at this point is to prove said mapping properties. In [44], the argument is brought to completion under the following additional assumptions: suppose that there are only conjugate pairs of order 1 and that no two points are conjugate along more than one geodesic, implying $M_1 = 1$; suppose in addition that C_{A_1} is a local canonical graph. Under these additional assumptions, we give the second main theorem of [44].

Theorem 4.6.4 ([44, Theorem 5]). *Suppose $n \ge 3$, let $(\widetilde{M}, \tilde{g})$ a smooth extension of M with no self-intersecting geodesic, conjugate pairs of order at most 1, and such that C_{A_1} is a local canonical graph. Let $\phi : SM \to [0, \infty)$ smooth and admissible at every $x \in M$. Then the kernel of $I_{\phi,0}$ acting on $L^2(\Omega^{1/2}_M)$ is at most finite dimensional and contained in $C_c^\infty(\Omega^{1/2}_M)$, and for any $f \in L^2(\Omega^{1/2}_M)/\ker I_{\phi,0}$,*

$$\|\mathcal{I}_\phi f\|_{L^2(\Omega_{\partial_+SM})} \sim \|f\|_{H^{-1/2}(\Omega^{1/2}_M)}.$$

The hypothesis that C_{A_1} be a local canonical graph was first formulated in [114, equation (4.4)]. Some conditions on the geometry guaranteeing the graph condition were formulated in [44, 114]. To the authors' knowledge, there are no examples yet of metrics satisfying that condition, but other families that do are given in [114]. Other counterexamples were provided there, hinting that conjugate loci which either contain pairs of higher order, or those that violate the graph condition, require further analysis. In this context, some open questions are formulated in Section 4.9.

4.7 Methods exploiting convexity

4.7.1 Heuristics

Let us first give local considerations. Let a smooth hypersurface $S = \{\rho = 0\}$ with p a point on it and U a neighborhood of p. Then $U \setminus S$ splits into two components $U_\pm = \{\pm\rho > 0\}$. Considering a family of smooth curves $\mathcal{G}$ across U which are near tangential to S, call $\mathcal{G}_+$ those that pass through U_+ and $\mathcal{G}_-$ the other ones. Then p is a convex point on S (as viewed from U_+, or concave as viewed from U_-) if for all curves in $\mathcal{G}$, we have $\frac{d^2}{dt^2}\rho(\gamma(t)) \le -c_0 < 0$ for some constant $c_0 > 0$. This imposes curves in $\mathcal{G}_+$ to be short, poking through S twice, coming from and returning to U_-; see Figure 4.6. In this type of geometry, the problem of reconstructing f supported in U from $If|_\mathcal{G}$ admits the following triangular structure:

$$\begin{bmatrix} If|_{\mathcal{G}_-} \\ If|_{\mathcal{G}_+} \end{bmatrix} = \begin{bmatrix} A_1 & 0 \\ * & A_2 \end{bmatrix} \begin{bmatrix} f|_{U_-} \\ f|_{U_+} \end{bmatrix}.$$

This implies that one may consider reconstructing $f|_{U_-}$ first, from $If|_{\mathcal{G}_-}$, then consider the reconstruction of $f|_{U_+}$ later. This is the key idea behind *support theorems* (does $If|_{\mathcal{G}_-} = 0$ imply that f is supported in U_+?), and several old and new inversion approaches. Such a condition makes the problem local in nature, and becomes global if one may exhaust the entire manifold in that way,[17] as one may reconstruct the function f layer after layer.

Convexity was initially assumed in [42, 123] in the case of radial, scalar metrics of the form $g = c^{-2}(r)Id$ where the sound speed c satisfies $\frac{d}{dr}(r/c(r)) > 0$, arising as a natural condition for invertibility. Such a condition is natural in that if $\frac{d}{dr}(r/c(r))$ vanishes

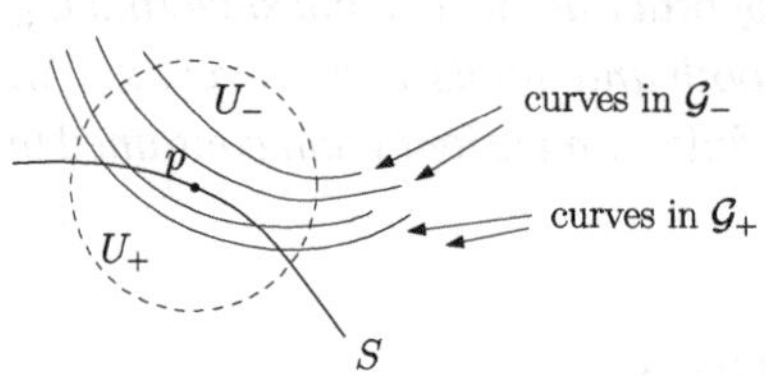

Figure 4.6: A locally convex setting.

17 The exhaustion can exclude a small set (zero measure, empty interior, or other) corresponding to the function space at hand.

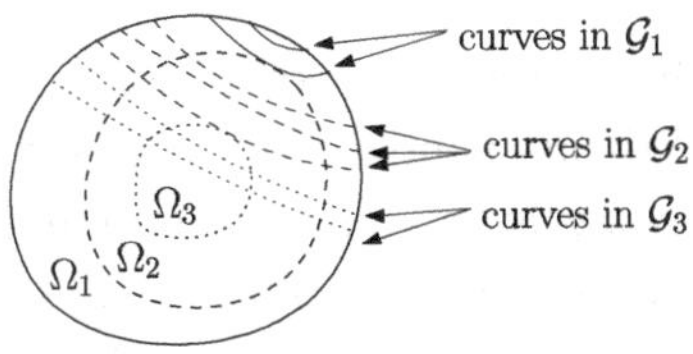

Figure 4.7: Layer stripping in action: reconstruct $f_1 = f|_{\Omega_1}$ from $If|_{\mathcal{G}_1}$ first, then $f_2 = f|_{\Omega_2}$ from $(If - If_1)|_{\mathcal{G}_2}$, then $f_3 = f|_{\Omega_3}$ from $(If - If_1 - If_2)|_{\mathcal{G}_3}$.

at r_0, then the sphere $\{|x| = r_0\}$ is totally geodesic, i. e., neither convex nor concave. Convexity conditions were exploited further in [97, Section 1.8] when the metric only depends on a single (not necessarily radial) variable, and in [104] for radial metrics to produce a positive answer to the tensor tomography problem. It was first used in a general context without symmetries in [120].

In order to understand *layer stripping* when a global foliation holds, split Ω into $\Omega_1 \cup \Omega_2 \cup \Omega_3$ as in Figure 4.7, and for $1 \le j \le 3$, denote $\mathcal{G}_j \subset \mathcal{G}$ the geodesics intersecting Ω_j but not Ω_{j+1}, so that $\mathcal{G} = \mathcal{G}_1 \cup \mathcal{G}_2 \cup \mathcal{G}_3$. Then the convex foliation allows to establish that the forward operator has the triangular structure

$$\begin{bmatrix} If|_{\mathcal{G}_1} \\ If|_{\mathcal{G}_2} \\ If|_{\mathcal{G}_3} \end{bmatrix} = \begin{bmatrix} A_1 & 0 & 0 \\ * & A_2 & 0 \\ * & * & A_3 \end{bmatrix} \begin{bmatrix} f|_{\Omega_1} \\ f|_{\Omega_2} \\ f|_{\Omega_3} \end{bmatrix},$$

where for each $1 \le j \le 3$, the geodesics in $\mathcal{G}_j$ are geodesically complete over Ω_j (so the operators A_j are Fredholm, i. e., injective up to a finite-dimensional space of smooth ghosts, and stable). In addition, adding more intermediate slabs preserves the triangular structure, and if the slabs are thin enough, then the error operators within the results of [120] becomes contractions, so that the operators A_j are actually injective.

In the following sections, we describe in more detail the literature based on exploiting convexity, first reviewing support theorems in Section 4.7.2 then injectivity and reconstruction approaches in geometries with and without symmetries in the remaining sections.

4.7.2 Support theorems

A version of Helgason's support theorem states that if a compactly supported smooth function $f: \mathbb{R}^n \to \mathbb{R}$ integrates to zero over all lines that avoid a compact, convex obstacle $K \subset \mathbb{R}^n$, then $f = 0$ in $\mathbb{R}^n \setminus K$.

There are a number of possible extensions of this result to Riemannian manifolds. The mentioned Euclidean result is global: it concerns all rays that avoid a compact obstacle. Typical support theorems on manifolds are local: a boundary point $x \in \partial M$ has a neighborhood U so that if a sufficiently regular function $f: U \to \mathbb{R}$ integrates to zero over all maximal geodesics that stay in U, then f vanishes in U. If f is a tensor field, then one can only expect it to vanish up to gauge, meaning that $f = \sigma\nabla h$ for a tensor field f of one order lower.

On real analytic simple Riemannian manifold a local support theorem was obtained by Krishnan for both scalar [55] fields and by Krishnan and Stefanov for second order tensor [54] fields using analytic microlocal analysis, an approach initiated by Boman and Quinto [9]. The technique uses the complex stationary phase by Sjöstrand as applies for the first time in [28]. The result was later extended to tensor fields of all orders [1]. A result without real analyticity was obtained by Uhlmann and Vasy in dimensions three and higher [120].

On radially symmetric manifolds it is natural to consider a spherical layer instead of a small neighborhood. The support theorem holds in any dimension when the boundary is strictly convex, as this guarantees that the Herglotz condition is satisfied locally [17, 104].

All of these results on Riemannian manifolds reproduce Helgason's theorem when applied to Euclidean geometry. Support theorems have also been obtained in Lorentzian geometry [95, 109].

As explained in the previous section, local results of this kind lead to global injectivity results if one assumes a foliation condition compatible with the local support theorem. A typical combination is a strictly convex foliation and a local support theorem near strictly convex boundary points.

4.7.3 Metrics dependent on a single variable

4.7.3.1 Parallel layers

V. G. Romanov studied integral geometric problems in [97, Section I.4] on the slab $S = \mathbb{R}^{n-1} \times \{0 \le x_n \le H\}$, integrating a function along a family of curves which is translation-invariant in the $\mathbb{R}^{n-1}$ factor, and with a diving behavior in the x_n variable. Specifically, a crucial assumption is that through every point $x = (x', x_n) \in S$ and direction v with $v_n = 0$, there is a curve γ passing through (x, v), with both endpoints at $\{x_n = 0\}$, and such that x is the farthest point to $\{x_n = 0\}$ on the curve γ. Together with additional curvature and smoothness conditions, this nothing but formulates that the planes $\{x_n = \text{const.}\}$ form a convex foliation of S.

A way to achieve this with a geodesic flow is to consider a metric of the form $g = e^{2\lambda(x_n)} Id$, with $\lambda' < 0$, a form of Herglotz condition adapted to parallel layers.

In this case, a continuous function with compact support can be reconstructed from its integrals (see [97, Theorem 1.5, Section I.4]), allowing for certain weights as well in [97, Theorem 1.6, Section I.4]. To prove this, one notices that upon Fourier-transforming in x', the problem diagonalizes frequency-wise, and for each frequency, the integral geometry problem looks like a Volterra equation of the second kind (integral equation with causal kernel). The case where curves are parabolas is also treated in [10, Chapter 4.3] using Volterra operator equations.

4.7.3.2 Spherical layers (radial metrics)

When the manifold is a ball equipped with a radial, scalar metric of the form $c^{-2}(r)Id$, the sphere of radius r and center 0 is strictly convex if and only if the Herglotz condition $\frac{d}{dr}\frac{r}{c(r)} > 0$ is satisfied. A rotation invariant Riemannian metric on an annulus may be written in this form [18, Proposition C.1]. In this context, Sharafutdinov has obtained a positive answer to the tensor tomography problem [104].

Theorem 4.7.1 ([104, Theorem 1.1]). *Let g a Riemannian metric on the spherical layer*

$$D = \{x \in \mathbb{R}^n \mid \rho_0 \leq |x| \leq \rho_1\} \quad (0 < \rho_0 < \rho_1,\ n \geq 2).$$

Assume g invariant under all orthogonal transformations of $\mathbb{R}^n$ *and such that the sphere* $S_\rho = \{x \mid |x| = \rho\}$ *is strictly convex for every* $\rho \in [\rho_0, \rho_1]$. *Let* $G = S_{\rho_1}$. *If a symmetric tensor field* $f \in C^\ell(S^m(T^*D))$ $(\ell \geq 1, m \geq 0)$ *lies in the kernel of the ray transform* I_G, *then* $f = dv$ *for some* $v \in C^\ell(S^{m-1}(T^*D))$ *satisfying* $v|_G = 0$.

The same result for scalar fields was proven in lower regularity in [17], assuming only $f \in L^2$ and $g \in C^{1,1}$. The metric may even have jump discontinuities.

Theorem 4.7.2 ([17]). *Consider the spherical layer D as in Theorem 4.7.1. Suppose* $c\colon [\rho_0, \rho_1] \to (0, \infty)$ *is piecewise* $C^{1,1}$ *and satisfies the Herglotz condition. At jump discontinuities, the Herglotz condition is interpreted as positivity of the distributional derivative* $\frac{d}{dr}\frac{r}{c(r)} > 0$, *meaning* $c(r-) > c(r+)$ *when c is not continuous at r. If* $f \in L^2(D)$ *integrates to zero over all geodesics between points on the outer boundary* S_{ρ_1}, *then* $f = 0$.

The two proofs are similar. The problem is first reduced to the two-dimensional case. In two dimensions, the ray transform is block-diagonalized by the angular Fourier transform. More precisely, the function can be written in polar coordinates as $f(r, \theta)$ and expanded as a Fourier series in $\theta \in S^1$. Identifying a point in the annulus with the geodesic whose minimal radius is at that point, one can regard $If(r, \theta)$ as a function on the annulus as well. The kth Fourier component of $If(r, \theta)$ only depends on the kth Fourier component of $f(r, \theta)$. This one-dimensional dependence is encoded in Abel-type integral transforms depending on the index k. Once one proves that the Abel transform is injective for all k, injectivity of the x-ray transform follows. In the case of tensor fields, the radial problem for each k becomes a system of integral equations due to the gauge freedom.

The Herglotz condition does not prohibit conjugate points. At least in two dimensions, this allows cases where the problem is definitely unstable yet the ray transform is solenoidally injective in all orders.

4.7.4 Geometric method for piecewise constant functions

Using a layer stripping argument associated with a foliation is somewhat technical when there is no explicit symmetry. One can relax the geometrical assumptions on the underlying manifold if one restricts to a smaller class of functions. The argument can be used to show that a piecewise constant function in the kernel of the x-ray transform has to vanish on any manifold of dimension two or higher, provided that there is a strictly convex foliation [48]. One has to be careful with the wording, since the set of piecewise constant functions is not a vector space unless the partition is fixed. In two dimensions, this is not covered by any existing result, and in higher dimensions this provides a simpler proof than the scattering calculus approach described next.

4.7.5 The Uhlmann–Vasy method by scattering calculus

All the methods above rely to some extent on a "local" problem near a convex boundary point, where convexity allows to write integral equations with causal kernels as in, e. g., [97, 104], or to set up an invertible linear problem in the case of [48]. Another method, first introduced by Uhlmann and Vasy in [120] to study the local problem in dimensions three and higher, is to express a post-processed version of the x-ray transform into a "normal-like" operator with good ellipticity properties within Melrose's scattering calculus. This yields local injectivity of x-ray transforms over functions, which we first discuss including a sketch of the approach. We then outline the many generalizations (to higher-order tensor fields, other flows and weighted transforms), for which the approach's handling of each case proves very robust.

4.7.5.1 Geodesic x-ray transform over functions

The main result hinges on a local integral geometric problem near a concave hypersurface $\{x = 0\}$, and suppose that on the slab $S = \{0 \le x \le c\}$, every level set $\{x = x_0\}$ is geodesically concave when viewed from the super-level set $\{x \ge x_0\}$. Let us denote a general point $z = (x, y)$, and coordinatize a unit tangent vector $v \in T_z S$ as $v = \lambda \partial_x + \omega$ with $g_z(\partial_x, \omega) = 0$ and $g_z(\omega, \omega) = 1 - \lambda^2$; see Figure 4.8.

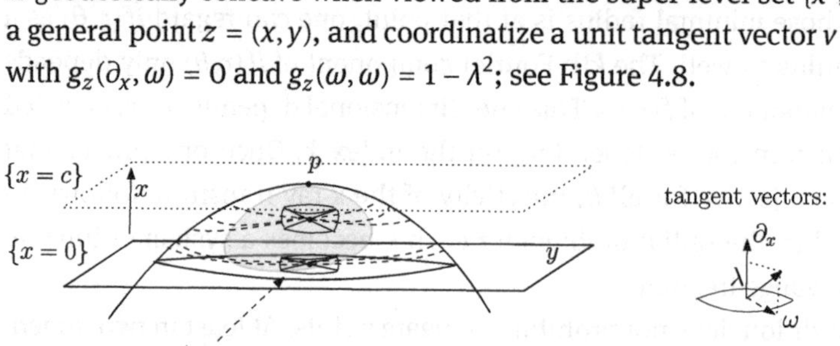

Figure 4.8: A local setting amenable to scattering calculus.

Back to our case of a manifold with boundary, if p is a convex point of ∂M, then there exists a neighborhood $O = M \cap \{0 < x \leq c\}$ of p open in M which can be described by a convex foliation as above. We denote by $\mathcal{M}_O$ the set of O-local geodesics. In this context, the main local result in [120] is the following.

Theorem 4.7.3. *With the notation as above, if c is small enough, the local transform I is injective on $H^s(O)$ for any $s \geq 0$. Upon defining for $F > 0$,*

$$H^s_F(O) = e^{F/x} H^s = \{f \in H^s_{\mathrm{loc}}(O) : e^{F/x} f \in H^s(O)\}.$$

For any $s \geq 0$, there exists $C > 0$ such that for all $f \in H^s_F(O)$,

$$\|f\|_{H^{s-1}_F(O)} \leq C \|If|_{\mathcal{M}_O}\|_{H^s(\mathcal{M}_O)}.$$

The main idea behind the theorem is to define a "truncated" normal operator

$$Af(z) := x^{-1} \int_{S_z} If(\gamma_v) \chi(\lambda/x) \, d\mu(v),$$

and to show that for $F > 0$,

$$A_F = x^{-1} e^{-F/x} A e^{F/x} \in \Psi^{-1,0}_{\mathrm{sc}}(\{x \geq 0\}),$$

and is elliptic within Melrose's scattering algebra $\Psi^{\cdot,\cdot}_{\mathrm{sc}}(\{x \geq 0\})$, a $\mathbb{Z} \times \mathbb{Z}$-graded algebra of pseudo-differential operators defined on a manifold with boundary, whose grading has two indices: a classical one measuring the growth rate in the momentum variable at "fiber infinity," and another measuring the rate of vanishing near the boundary.[18]

To prove ellipticity, one must compute the principal symbol of A_F, and this accounts for much of the work. The Schwartz kernel $K_{A_F}(z, z')$ is easy to compute via exponential coordinates; the main singularity of K_{A_F} is on the diagonal $z = z'$, and upon rewriting K_{A_F} as a function of $(z, \xi := z' - z)$, the full symbol may be computed via Fourier transform $\sigma_{A_F}(z, \zeta) \sim \mathcal{F}_{\xi \to \zeta} K_{A_F}(z, \zeta)$. Once such an amplitude is computed, one must find the leading behavior in terms of x near $x = 0$ and in terms of ζ near $\zeta \to \infty$, and show that both leading-order terms vanish neither on $\{x = 0\} \times \overline{\mathbb{R}^n_\zeta}$, nor on $\overline{O} \times \{|\zeta| = \infty\}$ ($\{|\zeta| = \infty\}$ is understood as the boundary of the radial compactification of $\mathbb{R}^n_\zeta$).

Once ellipticity is proved, this shows Fredholm properties of such an operator, namely the existence of a B such that $BA_F = Id + K$, where K is compact. In addition, if c is small enough, then K becomes a contraction, and the inversion of $Id + K$ can be done by Neumann series so that A_F is in fact injective and Hölder-stable. If the manifold can

---

18 The boundary of the base is also called "base infinity," as the geometric model associated is usually such that geodesics do not attain the boundary in finite time.

be globally foliated by convex hypersurfaces, with convexity constants bounded away from 0, then compactness arguments allow to upgrade the local Theorem 4.7.3 into a global one; see, e. g., [120, Corollary].

Before discussing the latest results, let us mention some generalizations of this approach, the first two of which are covered in the topical review [121]. These generalizations attest to the robustness of the method for integral geometric problems in dimensions three and higher.

4.7.5.2 General families of curves

In the Appendix to [120], Zhou generalized the results above to ray transforms of functions, where integration is performed along general families of curves on a Riemannian manifold (M, g). Such curves are generated by fixing $(x, v) \in TM$ and solving the differential equation

$$\nabla_{\dot\gamma}\dot\gamma = G(\gamma, \dot\gamma), \quad \gamma(0) = x, \quad \dot\gamma(0) = v \quad (\gamma := \gamma_{x,v}),$$

where ∇ is the Levi–Civita connection and G is a fixed smooth bundle map from TM to itself. The local result is first considered near a convex boundary point p, where by convexity here we mean that if ρ is a boundary defining function near p (such that $\rho = 0$ on ∂M and $\rho > 0$ inside M), we have $\frac{d^2}{dt^2}\rho(\gamma_{p,v}(t))|_{t=0} < 0$ for any $v \in T_p(\partial M)$.

Related results are described in Section 4.6.1.

4.7.5.3 Injectivity over tensor fields and judicious choices of gauges

The approach was then generalized by Stefanov–Uhlmann–Vasy in [116] to prove solenoidal injectivity of geodesic x-ray transforms over tensor fields of order 1 and 2 (as a stepping stone toward proving boundary rigidity, see Section 4.8.1 below). In this case, the transform has a natural kernel, and given that one is working with two distinct metrics (the initial one and the scattering one), one must make a judicious choice of gauge. Upon choosing the solenoidal gauge with respect to the scattering metric, the work is then to show full ellipticity (up to gauge) of a certain operator defined out of appropriately restricted x-ray transforms. Ellipticity is here defined in the sense of a scattering algebra of operators defined on sections of bundles, namely, $\Psi_{\mathrm{sc}}^{\cdot,\cdot}(X; {}^{\mathrm{sc}}T^*M, {}^{\mathrm{sc}}T^*M)$ and $\Psi_{\mathrm{sc}}^{\cdot,\cdot}(X; \mathrm{Sym}^2({}^{\mathrm{sc}}T^*M), \mathrm{Sym}^2({}^{\mathrm{sc}}T^*M))$. Though not written, the results are expected to generalize to tensor fields of arbitrary order.

The solenoidal gauge used above gives rise to an elliptic problem, which is an advantage to study Fredholmness in inverse problems, but a curse when considering local problems and extension issues, since elliptic gauges require solving a global problem in order to be computed. To gain more flexibility in the local problem and in fact tackle the nonlinear ones (boundary and lens rigidity, see Section 4.8.1.1), the same authors improved the results above in [118], by working with tensor fields which are in the *normal gauge* at the boundary (one-forms in the normal gauge only have tangential components, 2-tensors only have tangential-tangential components). One must then construct "normal-like" operators which land in this gauge instead of the solenoidal one. Such operators are only partially elliptic, and their Fredholm properties are obtained after a refined study of the characteristic directions. In the case of one-forms, the operator has real principal type with radial points, and those can be directly dealt with. The case of two-tensors is more delicate, as one must deal with double characteristics (i. e., second-order vanishing of the principal symbol at the characteristic set). The authors circumvent this by reducing the inversion to that in the elliptic gauge and working out the effect of the change of gauge. Such results are compatible with having microlocal weights in the ray transform, a necessary step toward using pseudo-linearization identities for the nonlinear problem.

4.7.5.4 Magnetic flows

Injectivity over tensor fields was also established in [127] in the case of magnetic flows. The added technical step is that the kernel of the magnetic ray transform couples pairs of tensor fields of consecutive order and, therefore, invertibility is proved by jointly considering pairs (function, one-form) and pairs (one-form, symmetric two-tensor). A local injectivity result requires a "magnetic convexity" condition near a boundary point; a global injectivity result (over a gauge representative similar to a solenoidal gauge defined on pairs) can then be derived when the manifold can be foliated by magnetically convex hypersurfaces. The ellipticity of the system is established by working in the scattering algebras $\Psi_{sc}^{\cdot,\cdot}(M, {}^{sc}T^*M \times M, {}^{sc}T^*M \times M)$ and $\Psi_{sc}^{\cdot,\cdot}(M, \mathrm{Sym}^2({}^{sc}T^*M) \times {}^{sc}T^*M, \mathrm{Sym}^2({}^{sc}T^*M) \times {}^{sc}T^*M)$. Though not written, the results are expected to generalize to tensor fields of arbitrary order.

4.7.5.5 Transform with matrix weights, connections, and Higgs fields

Another generalization of the approach is to work with transforms with connections and Higgs fields, as done in [87], or in the most general form, on transforms with a matrix weight. Namely, given a (known) weight function $W \in C^\infty(SM, GL(N,\mathbb{C}))$, one may define the x-ray transform $\mathcal{I}_W \colon C^\infty(SM,\mathbb{C}^N) \to C^\infty(SM,\mathbb{C}^N)$

$$\mathcal{I}_W h(x,v) = \int_0^{\tau(x,v)} W(\varphi_t(x,v)) h(\varphi_t(x,v))\,dt, \quad (x,v) \in \partial_+ SM.$$

Special examples of weights are those that arise from a pair (A,Φ) (connection, Higgs field) on the bundle $M\times C^N$, where the weight is assumed to solve the transport equation

$$XW = W(A+\Phi) \quad (\text{on } SM\times\mathbb{C}^N), \quad W|_{\partial_- SM} = Id,$$

in which case we exactly have $\mathcal{I}_W = \mathcal{I}_{A,\Phi}$. In [87], the authors study the invertibility of $\mathcal{I}_W$ in the following cases: (i) W is arbitrary and h is a $\mathbb{C}^N$-valued function on M; (ii) W arises from a pair (A,Φ), and h is the sum of $\mathbb{C}^N$-valued functions and one-forms.

The main resting assumption is that the manifold (M,g) where the transform $\mathcal{I}_W$ is defined admits a strictly convex function. This implies the existence of a strictly convex foliation, allowing a successful study of the local problem by scattering calculus (see [87, Theorem 1.5]) followed by a global argument to prove injectivity of $\mathcal{I}_W$ in both settings described above (see [87, Theorems 1.1, 1.6]). In addition, an extended discussion is provided in [87, Section 2] regarding which manifolds admit a strictly convex function. The injectivity result for $\mathcal{I}_{A,\Phi}$ over functions and one-forms also implies a positive answer to the nonlinear Problem 3, as described in Section 4.8.2 below.

4.8 From linear to nonlinear results

While linearization of nonlinear inverse problems is a common approach to obtaining nonlinear results in a neighborhood of "favorable" case, integral geometric problems have enjoyed striking identities, out of which one may derive a *global* nonlinear uniqueness result which ultimately relies on the injectivity of an x-ray transform.

We give two examples of such identities below, one with applications to boundary/lens rigidity, the other with applications to inverse problems for connections and Higgs fields.

4.8.1 From Problem 2 to Problem 1

In the case of inverse problems for metrics, a good example of a "pseudo-linearization" identity is given below, first appearing in [110]. Let N a manifold and V, $\widetilde{V}$ two vector fields on N. We denote $X(s,X^{(0)})$ the solution of $\dot{X} = V(X)$, similarly for $\widetilde{X}$ in terms of $\widetilde{V}$.

Lemma 4.8.1 ([110, 117]). *For any $t > 0$ and any initial condition $X^{(0)}$, if $\widetilde{X}(\cdot,X^{(0)})$ and $X(\cdot,X^{(0)})$ exist on the interval $[0,t]$, then*

$$\begin{aligned}\widetilde{X}(t,X^{(0)}) &- X(t,X^{(0)})\\ &= \int_0^t \frac{\partial\widetilde{X}}{\partial X^{(0)}}(t-s,X(s,X^{(0)}))(\widetilde{V}-V)(X(s,X^{(0)}))\,ds. \qquad (23)\end{aligned}$$

See [117] for a proof. To see how this relates to the lens rigidity problem (see Problem 1), fix two metrics g and $\tilde{g}$ on M and suppose they have same lens data.[19] Both metrics give rise to geodesic vector fields V and $\widetilde{V}$ on T^*M, each of which uniquely characterizes g and $\tilde{g}$. If g and $\tilde{g}$ have the same scattering relation at the boundary, then the left-hand side of (23) vanishes for $X^{(0)}$ at the boundary and for $t > 0$ is the length of the geodesic emanating from $X^{(0)}$ for either metric. Then this implies the vanishing of a weighted x-ray transform of $V - \widetilde{V}$ along the geodesics of V. If the weight has good properties and if the ray transform is injective, this solves the nonlinear problem.

Such an approach has been documented in the review [121, Section 5.4.1] and the interested reader is invited to refer to it for earlier results. We now cover a couple of further recent uses of this approach.

4.8.1.1 Boundary and lens ridigity

The works [117] and [118] give the latest progress on boundary rigidity results to date. Specifically, [118, Theorem 1.1] states that on a Riemannian manifold (M, g) of dimension three and higher, if ∂M is strictly convex for two metrics whose boundary distance functions agree, then these metrics are gauge equivalent in a neighborhood of the boundary. Under a global foliation condition, a lens rigidity result is also established in [118, Theorem 1.3], namely: if (M, g) has a strictly convex foliation, and if $\hat{g}$ is another metric whose lens data agrees with that of g, then g and $\hat{g}$ are gauge-equivalent. The results in [117] were first established by the same authors in the case of the recovery of conformal a factor, a problem which does not require addressing ray transforms over tensor fields.

To find the range of applicability of the results above, one must then study which manifolds admit strictly convex foliations, and this is discussed at length in [87, Section 2]. Note that it is still open whether simple manifolds of dimension three and higher admit strictly convex foliations, therefore, the earlier result [112], establishing boundary rigidity for generic simple manifolds using microlocal methods, may cover cases which are not treated by [117, 118]. A simple surface does admit a strictly convex foliation.

4.8.1.2 Lens rigidity for Yang–Mills fields

In [88, Theorem 1.2], the authors prove the unique identifiability modulo gauge of a Yang–Mills potential from its scattering relation on manifolds of dimension three and higher satisfying a certain foliation condition. In this example, the scattering relation

19 I. e., same boundary distance function and same scattering relation.

is defined in terms of the flow of a coupled dynamical system for a particle in SM and a Lie algebra-valued "color charge," and the nature of the coupling is driven by the Yang–Mills potential. The approach combines the tools of Section 4.7.5, with a pseudo-linearization identity similar to Lemma 4.8.1. A similar problem for magnetic fields was considered earlier by Zhou in [126].

4.8.1.3 An inverse problem from condensed matter physics

In [56], the authors consider the recovery of a potential from the dynamical behavior of vortex dipoles in an inhomogeneous Gross–Pitaevskii equation in the plane, a problem with applications to condensed matter physics. The inverse problem can be viewed as a lens rigidity problem where the measurements resemble a scattering relation for a flow perturbed by the unknown potential, and the recovery of the potential in [56, Theorem 2]) uses ideas such as Lemma 4.8.1.

4.8.2 From Problem 4 to Problem 3

The solenoidal injectivity of ray transforms $I_{A,\Phi}$ over sums of functions and one-forms implies reconstructibility of a connection and a Higgs field from their scattering data up to gauge, i. e., a positive answer to Problem 3.

Theorem 4.8.2 ([81, Theorem 1.5]). *Assume M is a compact simple surface, let A and B be two Hermitian connections, and let Φ and Ψ be two skew-Hermitian Higgs fields. Then $C_{A,\Phi} = C_{B,\Psi}$ implies that there exists a smooth $U: M \to U(n)$ such that $U|_{\partial M} = Id$ and $B = U^{-1}dU + U^{-1}AU$, $\Psi = U^{-1}\Phi U$.*

While similar theorems exist in a perturbative context (see references in [81]), Theorem 4.8.2 is a striking example of how injectivity of the linearized problem implies *global* injectivity (modulo gauge) of the nonlinear operator $(A, \Phi) \mapsto C_{A,\Phi}$. The argument is as short as it is powerful and we repeat it here.

Proof of Theorem 4.8.2. The equality $C_{A,\Phi} = C_{B,\Psi}$ implies that the fundamental matrix solutions $U_{A,\Phi}, U_{B,\Psi}: SM \to U(n)$, satisfying

$$(X + A + \Phi)U_{A,\Phi} = 0, \quad (X + B + \Psi)U_{B,\Psi} = 0,$$
$$U_{A,\Phi}|_{\partial_+ SM} = U_{B,\Psi}|_{\partial_+ SM} = Id,$$

agree on ∂SM. Then the proof consists in showing that $U := U_{A,\Phi}(U_{B,\Psi})^{-1}$, which is smooth by construction, only depends on the basepoint, and thus fulfills the conclusion of the theorem. Looking at $W := U - Id$, W is a matrix solution, vanishing at ∂SM, of the transport equation on SM

$$XW + AW - WB + \Phi W - W\Psi = B - A + \Psi - \Phi.$$

This equation can be viewed as a transport equation on the bundle $M \times \mathbb{C}^{n\times n}$ with (Hermitian) connection $\hat{A}(R) := AR - RB$ and (skew-Hermitian) Higgs field $\hat{\Phi}(R) := \Phi R - R\Psi$, and the vanishing of W at ∂SM expresses that $I_{\hat{A},\hat{\Phi}}(B - A + \Psi - \Phi) = 0$. Theorem 4.3.13 then implies that W is only a function on the basepoint, and thus $U = Id + W$ fulfills all the desired properties. □

Following these ideas, we briefly describe similar contexts where such results have been obtained.

4.8.2.1 Skew-Hermitian pairs on simple magnetic surfaces

A similar scheme of proof was used in [2] to show that a skew-Hermitian pair (A, Φ) is determined by the scattering data $C_{A,\Phi}$ defined through a simple magnetic flow; see [2, Theorem 1.4]. The proof relies on the injectivity of all magnetic ray transforms with skew-Hermitian pairs, proved in this context via energy identities as described in Section 4.3.

4.8.2.2 Skew-Hermitian pairs on manifolds with negative sectional curvature

The conclusion of Theorem 4.8.2 also holds on arbitrary bundles with Hermitian connection and skew-Hermitian Higgs field over a manifold (M, g) with negative sectional curvature and strictly convex boundary, as stated in [38, Theorem 1.2]. The scheme of proof mimics that of Theorem 4.8.2 by relying on the injectivity of the linear problem as stated in Theorem 4.3.14. The added technicality is the presence of trapping, which in the case of negative sectional curvature, is hyperbolic and allows to control appropriately the regularity of transport solutions, as explained in Section 4.2.3.

4.8.2.3 General pairs on manifolds admitting a strictly convex function

In general, one may notice that the scheme of proof above also works for pairs (A, ϕ) which are not necessarily skew-Hermitian. Namely, on a fixed Riemannian manifold of dimension at least three (say contractible with strictly convex boundary), if $\mathcal{I}_{A,\Phi}$ is injective over functions and one-forms for any smooth connection A and Higgs field Φ, then for any smooth pair (A, Φ), the scattering data $C_{A,\Phi}$ determines the pair (A, Φ) up to gauge.

In this context, on Riemannian manifolds of dimension three and higher admitting a strictly convex function, and for any smooth pair (A, Φ), $\mathcal{I}_{A,\Phi}$ is proved injective in [87], then the authors also prove there that general pairs (A, Φ) are determined (up to gauge) by their scattering data; see, e. g., [87, Theorem 1.1].

4.9 Open questions

We conclude this review with some open questions.

1. On simply connected, nontrapping Riemannian surfaces with strictly convex boundary, it is commonly conjectured that the tensor tomography problem is solvable.
2. Is the x-ray transform solenoidally injective on tensor fields of all orders on a simple Riemannian manifold of any dimension? There is a complete answer in dimension two, but only partial results in higher dimensions.
3. On a simple Riemannian manifold with boundary in dimension three and higher, the boundary rigidity problem is still open.
4. Regarding injectivity of the x-ray transform with conjugate points in Section 4.6.3:
   (a) What happens in the presence of regular conjugate points of order $n-1$ when $n \geq 3$? What about singular conjugate points? In dimensions three and higher, singular conjugate pairs (e. g., of type D_4, as defined in [6]) occur generically and correspond to singular conjugate pairs of order 2.
   (b) What happens in the presence of regular conjugate points of any order violating the graph condition? (And when is the graph condition satisfied in the first place?) Some works studying FIOs where the graph condition fails may be found in [22, 24], and examples of metrics of product type in [114] shows that failure of this condition can destroy stability in some cases. The question is however open in general.
5. On simple Riemannian surfaces, is the ray transform injective when considering it over a bundle with any connection and Higgs field with structure group $GL(n, \mathbb{C})$?
6. Is there a local support theorem for tensor fields of all orders without real analytic metrics?
7. Can a Pestov identity be used to prove an estimate which is localized in space? Localization in frequency is a recent observation [79].
8. What happens to the various linear and nonlinear integral geometry problems when the metric is not C^∞? Namely, the results that hold for smooth metrics would be expected to hold with roughly C^2 regularity.
9. In dimensions three and higher, do simple manifolds admit strictly convex foliations?
10. On which manifolds is the geodesic x-ray transform and variants thereof *not* (solenoidally) injective?

Appendix. Proof of Lemma 4.6.2

Proof of Lemma 4.6.2. Given $\omega \in T^*_{x'}M$, we compute immediately

$$d\pi|^T_{(x',v')}\omega = \omega(v')X^b_{(x',v')} - \omega(v'_\perp)X^b_{\perp,(x',v')}.$$

Given $\eta \in T^*_\xi(\partial_+ SM)$, we compute, for any $t \in (0, \tau(\xi))$,

$$dF|^T_{\varphi_t(\xi)}\eta = \eta(dF|_{\varphi_t(\xi)}(X_{\perp,\varphi_t(\xi)}))X^\flat_{\perp,\varphi_t(\xi)} + \eta(dF|_{\varphi_t(\xi)}(V_{\varphi_t(\xi)}))V^\flat_{\varphi_t(\xi)}.$$

Given $(x', (x, v)) \in M \times \partial_+ SM$ and $(\omega, \eta) \in T^*_{x'}M \times T^*_{(x,v)}(\partial_+ SM)$, to find whether $(\eta, \omega) \in C_{I_0}$, the only point in SM where canonical relations can compose is (x', v'), where $x' = \gamma_{x,v}(t)$ for some $t \in (0, \tau(x, v))$, and $v' = \dot\gamma_{x,v}(t)$. There, writing the condition $d\pi|^T_{(x',v')}\omega = dF|^T_{\varphi_t(x,v)}\eta$ gives

$$\begin{aligned} 0 = \omega(v'), \quad \eta(dF|_{\varphi_t(x,v)}(X_{\perp,\varphi_t(x,v)})) &= -\omega(v'_\perp), \\ \eta(dF|_{\varphi_t(x,v)}(V_{\varphi_t(x,v)})) &= 0. \end{aligned} \tag{24}$$

This imposes $\omega = \lambda(v'_\perp)^\flat$ for some $\lambda \in \mathbb{R}$, $\lambda = \omega(v'_\perp)$. We now use the following fact, proved after the conclusion:

$$\begin{aligned} dF|_{\varphi_t(x,v)}(X_{\perp,\varphi_t(x,v)}) &= \dot a(x, v, t)V_{(x,v)} + \frac{\dot b(x, v, t)}{\langle \nu_x, v\rangle}\nabla_T|_{(x,v)}, \\ dF|_{\varphi_t(x,v)}(V_{\varphi_t(x,v)}) &= a(x, v, t)V_{(x,v)} + \frac{b(x, v, t)}{\langle \nu_x, v\rangle}\nabla_T|_{(x,v)}, \end{aligned} \tag{25}$$

and upon writing $\eta_V = \eta(V_{(x,v)})$ and $\eta_T = \eta(T_{(x,v)})$, the linear system (24) has a unique solution

$$\eta_V = -b(x, v, t)\lambda, \quad \eta_T = a(x, v, t)\lambda,$$

hence the proof.

To prove (25), we want to compute

$$dF|_{\varphi_t(x,v)}(X_{\perp,\varphi_t(x,v)}) \quad \text{and} \quad dF|_{\varphi_t(x,v)}(V_{\varphi_t(x,v)}).$$

A basis of $T_{(x,v)}(\partial_+ SM)$ is given by $V_{(x,v)}$ and $\nabla_T|_{(x,v)} = \langle T_x, \nabla\rangle$ (horizontal derivative along the tangent vector), where $T_x := -(\nu_x)_\perp$, in particular expressed in the frame $(X, X_\perp, V)$ as

$$\begin{aligned} \nabla_T|_{(x,v)} &= \langle T_x, v\rangle\langle v, \nabla\rangle + \langle T_x, v_\perp\rangle\langle v_\perp, \nabla\rangle = \langle T_x, v\rangle X_{(x,v)} - \langle T_x, v_\perp\rangle X_{\perp(x,v)} \\ &= \langle \nu_x, v_\perp\rangle X_{(x,v)} + \langle \nu_x, v\rangle X_{\perp(x,v)}. \end{aligned}$$

Now for every $(x, v) \in \partial_+ SM$ and $t \in (0, \tau(x, v))$, we have $F(\varphi_t(x, v)) = (x, v)$, so that the following identity holds:

$$dF|_{\varphi_t(x,v)}(d\varphi_t|_{(x,v)}Y) = Y, \quad Y \in T_{(x,v)}(\partial_+ SM). \tag{26}$$

In addition, we can compute directly that

$$d\varphi_t|_{(x,v)}(V_{(x,v)}) = -b(x, v, t)X_{\perp,\varphi_t(x,v)} + \dot b(x, v, t)V_{\varphi_t(x,v)},$$

$$d\varphi_t|_{(x,v)}(\nabla_T|_{(x,v)}) = \langle v_x, v_\perp\rangle X_{\varphi_t(x,v)} + \cdots$$
$$+ \langle v_x, v\rangle \big(a(x,v,t)X_{\perp,\varphi_t(x,v)} - \dot a(x,v,t)V_{\varphi_t(x,v)}\big).$$

Applying $dF|_{\varphi_t(x,v)}$, using (26) and the fact that $dF|_{\varphi_t(x,v)}(X_{\varphi_t(x,v)}) = 0$, we obtain the relations

$$dF|_{\varphi_t(x,v)}\big(-b(x,v,t)X_{\perp,\varphi_t(x,v)} + \dot b(x,v,t)V_{\varphi_t(x,v)}\big) = V_{(x,v)},$$
$$dF|_{\varphi_t(x,v)}\big(a(x,v,t)X_{\perp,\varphi_t(x,v)} - \dot a(x,v,t)V_{\varphi_t(x,v)}\big) = \frac{1}{\langle v_x, v\rangle}\nabla_T|_{(x,v)},$$

which, via linear combinations and using $a\dot b - b\dot a = 1$, yields (25). □

Bibliography

[1] A. Abhishek and R. K. Mishra. Support theorems and an injectivity result for integral moments of a symmetric m-tensor field. arXiv preprint, 2017. arXiv:1704.02010.

[2] G. Ainsworth. The attenuated magnetic ray transform on surfaces. *Inverse Probl. Imaging*, 7:27–46, 2013.

[3] G. Ainsworth and Y. M. Assylbekov. On the range of the attenuated magnetic ray transform for connections and Higgs fields. *Inverse Probl. Imaging*, 9:317–335, 2015. arXiv:1311.4582.

[4] Y. E. Anikonov. *Some Methods for the Study of Multidimensional Inverse Problems for Differential Equations*. Nauka Sibirsk. Otdel., Novosibirsk, 1978.

[5] Y. E. Anikonov and V. Romanov. On uniqueness of determination of a form of first degree by its integrals along geodesics. *J. Inverse Ill-Posed Probl.*, 5:487–490, 1997.

[6] V. I. Arnol'd. Normal forms for functions near degenerate critical points, the Weyl groups of A_k, D_k, E_k and Lagrangian singularities. *Funct. Anal. Appl.*, 6:254–272, 1972.

[7] Y. M. Assylbekov, F. Monard, and G. Uhlmann. Inversion formulas and range characterizations for the attenuated geodesic ray transform. *J. Math. Pures Appl.*, 111:161–190, 2018.

[8] Y. M. Assylbekov and H. Zhou. Invariant distributions and tensor tomography for Gaussian thermostats. *Commun. Anal. Geom.*, 25:895–926, 2017. arXiv:1410.7006.

[9] J. Boman, E. T. Quinto, et al. Support theorems for real analytic Radon transforms. Stockholms Universitet, Matematiska Institutionen, 1987.

[10] A. L. Bukhgeym. *Introduction to the Theory of Inverse Problems*, volume 19. Walter de Gruyter, 2000.

[11] W.-S. Chiang, J. M. LaManna, A. Katsevich, D. S. Hussey, D. L. Jacobson, M. Frenkel, D. T. Georgi, Q. Sun, Y. Liu, and J.-H. Chen. Joint X-ray and neutron tomography of shale cores for characterizing structural morphology, 2017. http://www.jgmaas.com/SCA/2017/SCA2017-019.pdf.

[12] C. B. Croke and V. A. Sharafutdinov. Spectral rigidity of a compact negatively curved manifold. *Topology*, 37:1265–1273, 1998.

[13] N. Dairbekov and V. Sharafutdinov. Some problems of integral geometry on anosov manifolds. *Ergod. Theory Dyn. Syst.*, 23:59–74, 2003.

[14] N. S. Dairbekov. Integral geometry problem for nontrapping manifolds. *Inverse Probl.*, 22:431–445, 2006.

[15] N. S. Dairbekov, G. P. Paternain, P. Stefanov, and G. Uhlmann. The boundary rigidity problem in the presence of a magnetic field. *Adv. Math.*, 216:535–609, 2007.

[16] N. S. Dairbekov and V. Sharafutdinov. On conformal killing symmetric tensor fields on Riemannian manifolds. *Sib. Adv. Math.*, 21:1–41, 2011.
[17] M. V. de Hoop and J. Ilmavirta. Abel transforms with low regularity with applications to X-ray tomography on spherically symmetric manifolds. *Inverse Probl.*, 33:124003, 2017. arXiv:1702.07625.
[18] M. V. de Hoop, J. Ilmavirta, and V. Katsnelson. Spectral rigidity for spherically symmetric manifolds with boundary, 2017. arXiv:1705.10434.
[19] N. M. Desai and W. R. Lionheart. An explicit reconstruction algorithm for the transverse ray transform of a second rank tensor field from three axis data. *Inverse Probl.*, 32:115009, 2016.
[20] G. Eskin. On the non-Abelian Radon transform. *Russ. J. Math. Phys.*, 11:391–408, 2004. arXiv:math/0403447.
[21] G. Eskin. Inverse boundary value problems in domains with several obstacles.
[22] R. Felea. Composition of fourier integral operators with fold and blowdown singularities. *Commun. Partial Differ. Equ.*, 30:1717–1740, 2005.
[23] R. Felea. Displacement of artefacts in inverse scattering. *Inverse Probl.*, 23:1519, 2007.
[24] R. Felea, R. Gaburro, and C. J. Nolan. Microlocal analysis of SAR imaging of a dynamic reflectivity function. *SIAM J. Math. Anal.*, 45:2767–2789, 2013.
[25] D. Finch. The attenuated X-ray transform: recent developments. In *Inside Out: Inverse Problems and Applications*. MSRI Publications. Cambridge University Press, 2004.
[26] D. Finch, I.-R. Lan, and G. Uhlmann. Microlocal analysis of the x-ray transform with sources on a curve. In *Inside Out: Inverse Problems and Applications*, volume 47, pages 193–218, 2003.
[27] D. Finch and G. Uhlmann. The X-ray transform for a non-Abelian connection in two dimensions. *Inverse Probl.*, 17:695–701, 2001.
[28] B. Frigyik, P. Stefanov, and G. Uhlmann. The X-ray transform for a generic family of curves and weights. *J. Geom. Anal.*, 18:81–97, 2008.
[29] P. Funk. Über eine geometrishe Anwendung der Abelschen Integralgleichnung. *Math. Ann.*, 77:129–135, 1916.
[30] C. R. Graham, C. Guillarmou, P. Stefanov, and G. Uhlmann. X-ray transform and boundary rigidity for asymptotically hyperbolic 2-manifolds. arXiv preprint, 2017. arXiv:1709.05053.
[31] G. Grubb. Local and nonlocal boundary conditions for μ-transmission and fractional elliptic pseudodifferential operators. *Anal. PDE*, 7:1649–1682, 2014.
[32] G. Grubb. Fractional Laplacians on domains, a development of Hörmander's theory of μ-transmission pseudodifferential operators. *Adv. Math.*, 268:478–528, 2015.
[33] C. Guillarmou. Invariant distributions and x-ray transform for Anosov flows. *J. Differ. Geom.*, 105:177–208, 2017.
[34] C. Guillarmou. Lens rigidity for manifolds with hyperbolic trapped sets. *J. Am. Math. Soc.*, 30:561–599, 2017.
[35] C. Guillarmou and M. Mazzucchelli. Marked boundary rigidity for surfaces. *Ergod. Theory Dyn. Syst.*, 38:1459–1478, 2018. arXiv:1602.02946.
[36] C. Guillarmou, M. Mazzucchelli, and L. Tzou. Boundary and lens rigidity for non-convex manifolds. arXiv preprint, 2017. arXiv:1711.10059.
[37] C. Guillarmou and F. Monard. Reconstruction formulas for x-ray transforms in negative curvature. *Ann. Inst. Fourier*, 67:1353–1392, 2017.
[38] C. Guillarmou, G. P. Paternain, M. Salo, and G. Uhlmann. The x-ray transform for connections in negative curvature. *Commun. Math. Phys.*, 343:83–127, 2016.
[39] V. Guillemin and D. Kazhdan. Some inverse spectral results for negatively curved 2-manifolds. *Topology*, 19:301–312, 1980.
[40] V. Guillemin and D. Kazhdan. Some inverse spectral results for negatively curved n-manifolds. In *Proc. Symp. Pure Math.*, volume 36, pages 153–180. Amer. Math. Soc., 1980.

[41] S. Helgason. *The Radon Transform*. Birkäuser, 2nd edition, 1999.

[42] G. Herglotz. Über die Elastizität der Erde bei Berücksichtigung ihrer variablen Dichte. *Z. Angew. Math. Phys.*, 52:275–299, 1905.

[43] S. Holman, F. Monard, and P. Stefanov. The attenuated geodesic X-ray transform. *Inverse Probl.*, 34:064003, 2018.

[44] S. Holman and G. Uhlmann. On the microlocal analysis of the geodesic x-ray transform with conjugate points. Preprint, 2015.

[45] A. Homan and H. Zhou. Injectivity and stability for a generic class of generalized radon transforms. *J. Geom. Anal.*, 27:1515–1529, 2017.

[46] J. Ilmavirta. Boundary reconstruction for the broken ray transform. *Ann. Acad. Sci. Fenn., Math.*, 39:485–502, 2014.

[47] J. Ilmavirta. X-ray transforms in pseudo-Riemannian geometry. *J. Geom. Anal.*, 28:606–626, 2018. arXiv:1608.03063.

[48] J. Ilmavirta, J. Lehtonen, and M. Salo. Geodesic x-ray tomography for piecewise constant functions on nontrapping manifolds. *Math. Proc. Camb. Philos. Soc.*, 2017 (to appear).

[49] J. Ilmavirta and G. Paternain. Broken ray tensor tomography with one reflecting obstacle, 2018. arXiv:1805.04947.

[50] J. Ilmavirta and M. Salo. Broken ray transform on a Riemann surface with a convex obstacle. *Commun. Anal. Geom.*, 24:379–408, 2016. arXiv:1403.5131.

[51] A. G. Kosovichev. Tomographic imaging of the Sun's interior. *Astrophys. J. Lett.*, 461:L55, 1996.

[52] M. Kramar, H. Lin, and S. Tomczyk. Direct observation of solar coronal magnetic fields by vector tomography of the coronal emission line polarizations. *Astrophys. J. Lett.*, 819:L36, 2016.

[53] V. Krishnan. On the inversion formulas of Pestov and Uhlmann for the geodesic ray transform. *J. Inverse Ill-Posed Probl.*, 18:401–408, 2010.

[54] V. Krishnan and P. Stefanov. A support theorem for the geodesic ray transform of symmetric tensor fields. *Inverse Probl. Imaging*, 3:453–464, 2009. arXiv:0803.3062.

[55] V. P. Krishnan. A support theorem for the geodesic ray transform on functions. *J. Fourier Anal. Appl.*, 15:515–552, 2009.

[56] R.-Y. Lai, R. Shankar, D. Spirn, and G. Uhlmann. An inverse problem from condensed matter physics. *Inverse Probl.*, 33:11, 2017.

[57] J. Lehtonen. The geodesic ray transform on two-dimensional Cartan–Hadamard manifolds, 2016. arXiv:1612.04800.

[58] J. Lehtonen, J. Railo, and M. Salo. Tensor tomography on Cartan–Hadamard manifolds. *Inverse Probl.*, 34:4, 2017.

[59] W. R. Lionheart and P. J. Withers. Diffraction tomography of strain. *Inverse Probl.*, 31:045005, 2015.

[60] R. B. Melrose. *Geometric Scattering Theory*, volume 1. Cambridge University Press, 1995.

[61] W. J. Merry and G. P. Paternain. Lecture notes: inverse problems in geometry and dynamics, March 2011.

[62] R. Michel. Sur la rigidité imposée par la longueur des géodésiques. *Invent. Math.*, 65:71–83, 1981.

[63] F. Monard. Numerical implementation of two-dimensional geodesic X-ray transforms and their inversion. *SIAM J. Imaging Sci.*, 7:1335–1357, 2014.

[64] F. Monard. On reconstruction formulas for the X-ray transform acting on symmetric differentials on surfaces. *Inverse Probl.*, 30:065001, 2014.

[65] F. Monard. Efficient tensor tomography in fan-beam coordinates. *Inverse Probl. Imaging*, 10:433–459, 2016.

[66] F. Monard. Inversion of the attenuated geodesic X-ray transform over functions and vector fields on simple surfaces. *SIAM J. Math. Anal.*, 48:1155–1177, 2016.

[67] F. Monard. Efficient tensor tomography in fan-beam coordinates. II: attenuated transforms. *Inverse Probl. Imaging*, 12:433–460, 2018. arXiv:1704.08294.

[68] F. Monard, R. Nickl, and G. P. Paternain. Efficient nonparametric Bayesian inference for x-ray transforms. *Ann. Stat.*, 47:1113–1147, 2019.

[69] F. Monard and G. P. Paternain. The geodesic X-ray transform with a $GL(n, \mathbb{C})$-connection. *J. Geom. Anal.*, 2017 (online). https://doi.org/10.1007/s12220-017-9931-z.

[70] F. Monard, P. Stefanov, and G. Uhlmann. The geodesic X-ray transform on Riemannian surfaces with conjugate points. *Commun. Math. Phys.*, 337:1491–1513, 2015.

[71] R. Mukhometov. Inverse kinematic problem of seismic on the plane. Math. problems of geophysics. *Akad. Nauk. SSSR, Sibirsk. Otdel., Vychisl. Tsentr, Novosibirsk*, 6:243–252, 1975.

[72] R. Mukhometov. The reconstruction problem of a two-dimensional Riemannian metric, and integral geometry. *Dokl. Akad. Nauk SSSR*, 232:32–35, 1977 (Russian).

[73] W. Munk and C. Wunsch. Ocean acoustic tomography: a scheme for large scale monitoring. *Deep-Sea Res., A, Oceanogr. Res. Pap.*, 26:123–161, 1979.

[74] F. Natterer. Inversion of the attenuated radon transform. *Inverse Probl.*, 17:113–119, 2001.

[75] F. Natterer. *The Mathematics of Computerized Tomography*. SIAM, 2001.

[76] R. G. Novikov. An inversion formula for the attenuated x-ray transformation. *Ark. Mat.*, 40:147–167 (2002). (Rapport de recherche 00/05–3 Université de Nantes, Laboratoire de Mathématiques).

[77] R. G. Novikov. On determination of a gauge field on R^d from its non-Abelian Radon transform along oriented straight lines. *J. Inst. Math. Jussieu*, 1:559–629, 2002.

[78] R. G. Novikov. On the range characterization for the two-dimensional attenuated x-ray transformation. *Inverse Probl.*, 18:677–700, 2002.

[79] G. Paternain and M. Salo. Carleman estimates for geodesic x-ray transforms, 2018. arXiv:1805.02163.

[80] G. P. Paternain. Inverse problems for connections. In *Inside Out II*, 2012.

[81] G. P. Paternain, M. Salo, and G. Uhlmann. The attenuated ray transform for connections and Higgs fields. *Geom. Funct. Anal.*, 22:1460–1480, 2012.

[82] G. P. Paternain, M. Salo, and G. Uhlmann. Tensor tomography on surfaces. *Invent. Math.*, 193:229–247, 2013.

[83] G. P. Paternain, M. Salo, and G. Uhlmann. Spectral rigidity and invariant distributions on Anosov surfaces. *J. Differ. Geom.*, 98:147–181, 2014.

[84] G. P. Paternain, M. Salo, and G. Uhlmann. Tensor tomography in two dimensions: progress and challenges. *Chin. Ann. Math., Ser. B*, 35:399–428, 2014. arXiv:1303.6114.

[85] G. P. Paternain, M. Salo, and G. Uhlmann. Invariant distributions, Beurling transform and tensor tomography in higher dimensions. *Math. Ann.*, 363:305–362, 2015.

[86] G. P. Paternain, M. Salo, and G. Uhlmann. On the range of the attenuated ray transform for unitary connections. *Int. Math. Res. Not.*, 2013 (online).

[87] G. P. Paternain, M. Salo, G. Uhlmann, and H. Zhou. The geodesic x-ray transform with matrix weights. Preprint, 2016.

[88] G. P. Paternain, G. Uhlmann, and H. Zhou. Lens rigidity for a particle in a Yang–Mills field. *Commun. Math. Phys.*, 366:681–707, 2019.

[89] G. P. Paternain and H. Zhou. Invariant distributions and the geodesic ray transform. *Anal. PDE*, 9:1903–1930, 2016.

[90] L. Pestov. *Questions of Well-Posedness of the Ray Tomography Problems*. Sib. Nauch. Izd., Novosibirsk, 2003 (Russian).

[91] L. Pestov and V. Sharafutdinov. Integral geometry of tensor fields on a manifold of negative curvature. *Sib. Math. J.*, 29:427–441, 1988.

[92] L. Pestov and G. Uhlmann. On the characterization of the range and inversion formulas for the geodesic X-ray transform. *Int. Math. Res. Not.*, 80:4331–4347, 2004.

[93] L. Pestov and G. Uhlmann. Two-dimensional compact simple Riemannian manifolds are boundary distance rigid. *Ann. Math.*, 161:1093–1110, 2005.

[94] C. W. Plaza, J. del Toro Iniesta, B. R. Cobo, V. M. Pillet, B. Lites, and A. Skumanich. Optical tomography of a sunspot. II. Vector magnetic field and temperature stratification. *Astrophys. J.*, 547:1130, 2001.

[95] S. RabienaHaratbar. Support theorem for the light-ray transform of vector fields on Minkowski spaces. *Inverse Probl. Imaging*, 12:293–314, 2018.

[96] J. Radon. Über die Bestimmung von Funktionen durch ihre Integralwerte längs gewisser Mannigfaltigkeiten. *Ber. Verh. K. Sächs. Akad. Wiss. Leipz., Math.-Phys. Kl.*, 69:262–277, 1917.

[97] V. Romanov. *Integral Geometry and Inverse Problems for Hyperbolic Equations*. Springer-Verlag, 1974.

[98] K. Sadiq, O. Scherzer, and A. Tamasan. On the X-ray transform of planar symmetric 2-tensors. *J. Math. Anal. Appl.*, 442:31–49, 2016.

[99] M. Sales, M. Strobl, T. Shinohara, A. Tremsin, L. T. Kuhn, W. R. Lionheart, N. M. Desai, A. B. Dahl, and S. Schmidt. Three dimensional polarimetric neutron tomography of magnetic fields. arXiv preprint, 2017. arXiv:1704.04887.

[100] M. Salo. The Calderón problem and normal forms. Preprint, 2017. arXiv:1702.02136.

[101] M. Salo and G. Uhlmann. The attenuated ray transform on simple surfaces. *J. Differ. Geom.*, 88:161–187, 2011.

[102] D. Schwarz, P. Vontobel, E. H. Lehmann, C. A. Meyer, and G. Bongartz. Neutron tomography of internal structures of vertebrate remains: a comparison with x-ray computed tomography. *Palaeontol. Electronica*, 8:30A, 2005.

[103] V. A. Sharafutdinov. *Integral Geometry of Tensor Fields*. VSP, Utrecht, The Netherlands, 1994.

[104] V. A. Sharafutdinov. Integral geometry of tensor fields on a surface of revolution. *Sib. Math. J.*, 38, 1997.

[105] V. A. Sharafutdinov. *Ray Transform on Riemannian Manifolds*. Eight lectures on integral geometry, 1999.

[106] V. A. Sharafutdinov. On the inverse problem of determining a connection on a vector bundle. *J. Inverse Ill-Posed Probl.*, 8:51–88, 2000.

[107] V. A. Sharafutdinov. Variations of Dirichlet-to-Neumann map and deformation boundary rigidity on simple 2-manifolds. *J. Geom. Anal.*, 17:147–187, 2007.

[108] V. A. Sharafutdinov and J.-N. Wang. Tomography of small residual stresses. *Inverse Probl.*, 28:065017, 2012.

[109] P. Stefanov. Support theorems for the light ray transform on analytic Lorentzian manifolds. Proc. Am. Math. Soc., 145:1259–1274, 2017.

[110] P. Stefanov and G. Uhlmann. Rigidity for metrics with the same lengths of geodesics. *Math. Res. Lett.*, 5:83–96, 1998.

[111] P. Stefanov and G. Uhlmann. Stability estimates for the X-ray transform of tensor fields and boundary rigidity. *Duke Math. J.*, 123:445–467, 2004.

[112] P. Stefanov and G. Uhlmann. Boundary rigidity and stability for generic simple metrics. *J. Am. Math. Soc.*, 18:975–1003, 2005.

[113] P. Stefanov and G. Uhlmann. Integral geometry of tensor fields on a class of non-simple Riemannian manifolds. *Am. J. Math.*, 130:239–268, 2008.

[114] P. Stefanov and G. Uhlmann. The geodesic X-ray transform with fold caustics. *Anal. PDE*, 5:219–260, 2012.

[115] P. Stefanov and G. Uhlmann. Is a curved flight path in SAR better than a straight one? *SIAM J. Appl. Math.*, 73:1596–1612, 2013.

[116] P. Stefanov, G. Uhlmann, and A. Vasy. Inverting the local geodesic X-ray transform on tensors, 2014. arXiv:1410.5145.

[117] P. Stefanov, G. Uhlmann, and A. Vasy. Boundary rigidity with partial data. *J. Am. Math. Soc.*, 29:299–332, 2016.

[118] P. Stefanov, G. Uhlmann, and A. Vasy. Local and global boundary rigidity and the geodesic x-ray transform in the normal gauge. Preprint, 2017.

[119] R. Stotzka, J. Wuerfel, T. O. Mueller, and H. Gemmeke. Medical imaging by ultrasound computer tomography. In *Medical Imaging 2002*, pages 110–119. International Society for Optics and Photonics, 2002.

[120] G. Uhlmann and A. Vasy. The inverse problem for the local geodesic ray transform. *Invent. Math.*, 205:83–120, 2016.

[121] G. Uhlmann and H. Zhou. Journey to the center of the earth. In *Proceedings of the International Congress of Mathematical Physics*, 2017 (to appear).

[122] L. Vertgeim. Integral geometry with a matrix weight, and a nonlinear problem of recovering matrices. *Sov. Math. Dokl.*, 44:132–135, 1992.

[123] E. Wiechert and K. Zoeppritz. Über Erdbebenwellen. *Nachr. Koenigl. Ges. Wiss. Gött.*, 4:415–549, 1907.

[124] Y. Zhang. Artifacts in the inversion of the broken ray transform in the plane. arXiv preprint, 2017. arXiv:1711.08575.

[125] H. Zhou. Appendix to "The inverse problem for the local geodesic ray transform", by G. Uhlmann and A. Vasy. *Invent. Math.*, 2015.

[126] H. Zhou. Lens rigidity with partial data in the presence of a magnetic field. Preprint, 2016.

[127] H. Zhou. The local magnetic ray transform of tensor fields. Preprint, 2016. arXiv:1609.03929.

[128] H. Zhou. Generic injectivity and stability of inverse problems for connections. *Commun. Partial Differ. Equ.*, 42:780–801, 2017. arXiv:1610.02185.

R. G. Novikov

5 Non-Abelian Radon transform and its applications

Abstract: Considerations of the non-Abelian Radon transform were started in [Manakov, Zakharov, 1981] in the framework of the theory of solitons in dimension 2+1. On the other hand, the problem of inversion of transforms of such a type arises in different tomographies, including emission tomographies, polarization tomographies, and vector field tomography. In this article, we give a short review of old and recent results on this subject. This article is an extended version of the talk given at the conference "100 Years of the Radon Transform," Linz, 27–31 March 2017.

Keywords: Non-Abelian Radon transform, tomography, integral geometry, inverse connection problem, Yang–Mills fields, inverse scattering, theory of solitons

MSC 2010: 44A12, 53C65, 78A46, 81U40, 37K15

5.1 Introduction

We consider the transport equation

$$\theta\partial_x\psi + A(x,\theta)\psi = 0, \quad x \in \mathbb{R}^d,\ \theta \in \mathbb{S}^{d-1}, \tag{1}$$

where $\theta\partial_x = \sum_{j=1}^{d}\theta_j\partial/\partial x_j$ and A is a sufficiently regular function on $\mathbb{R}^d \times \mathbb{S}^{d-1}$ with sufficient decay as $|x| \to \infty$.

We assume that A and ψ take values in $M(n,\mathbb{C})$ that is in $n \times n$ complex matrices.

For equation (1), we consider the "scattering" matrix S:

$$S(x,\theta) = \lim_{s\to+\infty}\psi^+(x+s\theta,\theta), \quad (x,\theta) \in T\mathbb{S}^{d-1}, \tag{2}$$

where

$$T\mathbb{S}^{d-1} = \{(x,\theta) \in \mathbb{R}^d \times \mathbb{S}^{d-1} : x\theta = 0\} \tag{3}$$

and $\psi^+(x,\theta)$ is the solution of (1) such that

$$\lim_{s\to-\infty}\psi^+(x+s\theta,\theta) = I, \quad x \in \mathbb{R}^d,\ \theta \in \mathbb{S}^{d-1}, \tag{4}$$

where I is the identity matrix.

R. G. Novikov, CNRS (UMR 7641), Centre de Mathématiques Appliquées, École Polytechnique, Université Paris-Saclay, 91128 Palaiseau, France; and IEPT RAS, 117997 Moscow, Russia, e-mail: novikov@cmap.polytechnique.fr

https://doi.org/10.1515/9783110560855-005

We interpret $T\mathbb{S}^{d-1}$ as the set of all rays in $\mathbb{R}^d$. As a ray γ, we understand a straight line with fixed orientation. If $\gamma = (x,\theta) \in T\mathbb{S}^{d-1}$, then $\gamma = \{y \in \mathbb{R}^d : y = x + t\theta, t \in \mathbb{R}\}$ (up to orientation) and θ gives the orientation of γ.

We say that S is the non-Abelian Radon transform along oriented straight lines (or the non-Abelian x-ray transform) of A.

We consider the following inverse problem for $d \geq 2$:

Problem 1. Given S, find A.

Note that S does not determine A uniquely, in general. One of the reasons is that S is a function on $T\mathbb{S}^{d-1}$, whereas A is a function on $\mathbb{R}^d \times \mathbb{S}^{d-1}$ and

$$\dim \mathbb{R}^d \times \mathbb{S}^{d-1} = 2d - 1 > \dim T\mathbb{S}^{d-1} = 2d - 2.$$

In particular, for Problem 1 there is a gauge-type nonuniqueness, that is, S is invariant with respect to the gauge transforms

$$\begin{aligned} &A \to A', \\ &A'(x,\theta) = g^{-1}(x,\theta)A(x,\theta)g(x,\theta) + g^{-1}(x,\theta)\theta\partial_x g(x,\theta), \end{aligned} \tag{5}$$

where g is a sufficiently regular $GL(n,\mathbb{C})$-valued function on $\mathbb{R}^d \times \mathbb{S}^{d-1}$ and $g \to I$ sufficiently fast as $|x| \to \infty$.

In addition, in particular, for Problem 1 there are Boman-type nonuniqueness (see [6, 19]) and nonuniqueness related with solitons (see [35]).

Equation (1), the "scattering" matrix S and Problem 1 arise, for example, in different tomographies (see Sections 5.2–5.6, 5.8), in differential geometry (see Section 5.7) and in the theory of the Yang–Mills fields (see Section 5.9). In Sections 5.2–5.9 we give a short review of old and recent results on this subject.

Finally, note that in the present article we consider the non-Abelian Radon transform (and its various reductions) along real lines only. In connection with related transforms along complex lines and their applications in the theory of the Yang–Mills fields and in complex geometry, we refer to [23] and references therein.

5.2 Classical x-ray transmission tomography

Problem 1 arises as a problem of the classical x-ray transmission tomography in the framework of the following reduction:

$$n = 1, \quad A(x,\theta) = a(x), \quad x \in \mathbb{R}^d,\ \theta \in \mathbb{S}^{d-1}, \tag{6}$$

$$S(\gamma) = \exp[-Pa(\gamma)], \quad Pa(\gamma) = \int_{\mathbb{R}} a(x + s\theta)\, ds, \quad \gamma = (x,\theta) \in T\mathbb{S}^{d-1}, \tag{7}$$

where a is the x-ray attenuation coefficient of the medium, P is the classical Radon transform along straight lines (classical x-ray transform), $S(\gamma)$ describes the x-ray photograph along γ.

In this case, for $d \geq 2$,

$$S|_{TS^1(Y)} \text{ uniquely determines } a|_Y, \tag{8}$$

where Y is an arbitrary two-dimensional plane in $\mathbb{R}^d$, $TS^1(Y)$ is the set of all oriented straight lines in Y. In addition, this determination can be implemented via the Radon inversion formula for P in dimension $d = 2$; see [47]. In connection with this formula, see also Remark 2 in Section 5.3.

For more information on the classical x-ray transmission tomography and on the classical x-ray transform, see, e. g., [17, 34], and references therein.

5.3 Single-photon emission computed tomography (SPECT)

In SPECT, one considers a body containing radioactive isotopes emitting photons. The emission data p in SPECT consist in the radiation measured outside the body by a family of detectors during some fixed time (where expected p is described by $P_a f$ defined below). The basic problem of SPECT consists in finding the distribution f of these isotopes in the body from the emission data p and some a priori information concerning the body. Usually, this a priori information consists in the photon attenuation coefficient a in the points of body, where this coefficient is found in advance by the methods of the classical x-ray transmission tomography (mentioned in Section 5.2).

Problem 1 arises as a problem of SPECT in the framework of the following reduction [35]:

$$n = 2, \quad A_{11} = a(x), \quad A_{12} = f(x), \quad A_{21} = 0, \quad A_{22} = 0, \quad x \in \mathbb{R}^d, \tag{9}$$

$$S_{11} = \exp[-P_0 a], \quad S_{12} = -P_a f, \quad S_{21} = 0, \quad S_{22} = 1, \tag{10}$$

where a is the photon attenuation coefficient of the medium, f is the density of radioactive isotopes, $P_0 = P$ is defined in (7), P_a is the attenuated Radon transform along oriented straight lines (attenuated ray transform), $P_a f$ describes the expected emission data,

$$P_a f(\gamma) = \int_{\mathbb{R}} \exp[-Da(x + s\theta, \theta)] f(x + s\theta)\, ds, \quad \gamma = (x, \theta) \in T\mathbb{S}^{d-1}, \tag{11}$$

$$Da(x, \theta) = \int_0^{+\infty} a(x + s\theta)\, ds, \quad x \in \mathbb{R}^d, \ \theta \in \mathbb{S}^{d-1}, \tag{12}$$

where D is the divergent beam transform.

In this case (as well as for the case of the classical x-ray transmission tomography), for $d \geq 2$,

$$S|_{TS^1(Y)} \text{ uniquely determines } a|_Y \text{ and } f|_Y, \tag{13}$$

where Y is an arbitrary two-dimensional plane in $\mathbb{R}^d$, $TS^1(Y)$ is the set of all oriented straight lines in Y. In addition, this determination can be implemented via the following inversion formula [36]:

$$f = P_a^{-1} g, \quad \text{where } g = P_a f, \tag{14}$$

$$P_a^{-1} g(x) = \frac{1}{4\pi} \int_{\mathbb{S}} \theta^\perp \partial_x (\exp[-Da(x, -\theta)] \tilde{g}_\theta(\theta^\perp x))\, d\theta,$$

$$\begin{aligned} \tilde{g}_\theta(s) = {} & \exp(A_\theta(s)) \cos(B_\theta(s)) H(\exp(A_\theta) \cos(B_\theta) g_\theta)(s) \\ & + \exp(A_\theta(s)) \sin(B_\theta(s)) H(\exp(A_\theta) \sin(B_\theta) g_\theta)(s), \end{aligned} \tag{15}$$

$$A_\theta(s) = (1/2) P_0 a(s\theta^\perp, \theta), \quad B_\theta(s) = H A_\theta(s), \quad g_\theta(s) = g(s\theta^\perp, \theta),$$

$$Hu(s) = \frac{1}{\pi} p.v. \int_{\mathbb{R}} \frac{u(t)}{s-t}\, dt, \tag{16}$$

$$x \in \mathbb{R}^2, \quad \theta^\perp = (-\theta_2, \theta_1) \quad \text{for } \theta = (\theta_1, \theta_2) \in \mathbb{S}^1,\ s \in \mathbb{R}.$$

Remark 1. The assumptions on a and f in (13)–(15) can be specified as follows:

$$\begin{aligned} & a, f \text{ are real-valued}, \quad a, f \in L^\infty(\mathbb{R}^2), \\ & a, f = O(|x|^{-\sigma}) \quad \text{as } |x| \to \infty \text{ for some } \sigma > 1, \end{aligned} \tag{17}$$

where Y is identified with $\mathbb{R}^2$ in (13).

Remark 2. For $a \equiv 0$, formulas (14), (15) are reduced to the classical Radon inversion formula for P defined in (7) for $d = 2$.

For more information on SPECT and for more results on the attenuated ray transform P_a, we refer to [1, 2, 4, 10, 14, 21, 22, 26, 27, 33, 34, 36, 37, 40, 46], and references therein.

5.4 Tomographies related with weighted Radon transforms

We consider the weighted Radon transforms P_W (along oriented straight lines) defined by the formula

$$P_W f(x, \theta) = \int_{\mathbb{R}} W(x + s\theta, \theta) f(x + s\theta)\, ds, \quad (x, \theta) \in T\mathbb{S}^{d-1}, \tag{18}$$

where $W = W(x,\theta)$ is the weight, $f = f(x)$ is a test function. The assumptions on W can be specified as follows:

$$\begin{aligned} W &\in L^\infty(\mathbb{R}^d \times \mathbb{S}^{d-1}), \\ W &= \bar{W}, \quad 0 < c_0 \le W \le c_1, \\ \lim_{s\to\pm\infty} W(x+s\theta,\theta) &= w_\pm(x,\theta), \quad (x,\theta) \in T\mathbb{S}^{d-1}. \end{aligned} \tag{19}$$

If $W = 1$, then P_W is reduced to the classical x-ray transform P defined in (7).

If

$$W(x,\theta) = \exp(-Da(x,\theta)), \tag{20}$$

where Da is defined by (12), then P_W is reduced to the classical attenuated ray transform P_a defined by (11), (12).

Transforms P_W with some other weights also arise in applications. For example, such transforms arise in positron emission tomography, optical tomography, fluorescence tomography; see [3, 32, 34].

The transforms $P_W f$ arise in the framework of the following reduction of the non-Abelian Radon transform S:

$$n = 2, \quad A_{11} = \theta\partial_x \ln W(x,\theta), \quad A_{12} = f(x), \quad A_{21} = 0, \quad A_{22} = 0, \tag{21}$$

$$S_{11} = \frac{w_-}{w_+}, \quad S_{12} = -\frac{1}{w_+} P_W f, \quad S_{21} = 0, \quad S_{22} = 1. \tag{22}$$

In connection with P_W and with the reduction (21), (22), we consider the following version of Problem 1, where we assume that W is known.

Problem 2. Given $P_W f$ and W, find f.

General uniqueness and reconstruction results on Problem 2 were given, in particular, in [5, 8, 13, 14, 22, 24, 26, 28, 31, 40].

For some W exact and simultaneously explicit formulas for solving Problem 2 are also known, see [9, 18, 36, 39, 47], and references therein.

Note that Problem 2 is nonoverdetermined for $d = 2$ and is overdetermined for $d \ge 3$. Indeed, $P_W f$ is a function on $T\mathbb{S}^{d-1}$, whereas f is a function on $\mathbb{R}^d$ and

$$\begin{aligned} &\dim T\mathbb{S}^{d-1} = 2d-2, \quad \dim \mathbb{R}^d = d, \\ &2d-2 = d \quad \text{for } d = 2, \quad 2d-2 > d \quad \text{for } d \ge 3. \end{aligned}$$

Nevertheless, Problem 2 is not uniquely solvable, in general, even for $d \ge 3$.

An example of nonuniqueness for Problem 2 for $d = 2$ was constructed in [6]. In this example, $W \in C^\infty(\mathbb{R}^2 \times \mathbb{S}^1)$, $f \in C_0^\infty(\mathbb{R}^2)$.

An example of nonuniqueness for Problem 2 for $d \ge 3$ was constructed in [19]. In this example, $W \in C^\alpha(\mathbb{R}^d \times \mathbb{S}^{d-1})$ for some $\alpha > 0$, $f \in C_0^\infty(\mathbb{R}^d)$.

In these examples assumptions (19) are also fulfilled. The notation C_0^∞ stands for infinitely smooth compactly supported functions.

For more information on the theory and applications of the transforms P_W, we refer to [5–8, 13, 19, 22, 24, 26, 28, 31, 34, 40], and references therein.

5.5 Neutron polarization tomography (NPT)

In NPT, one considers a medium with spatially varying magnetic field. The polarization data consist in changes of the polarization (spin) between incoming and outcoming neutrons. The basic problem of NPT consists in finding the magnetic field from the polarization data. See, e. g., [11, 29], and references therein.

Problem 1 arises as a problem of NPT in the framework of the following reduction:

$$\begin{aligned} &n = 3, \quad A_{11} = A_{22} = A_{33} = 0, \\ &A_{12} = -A_{21} = -gB_3(x), \quad A_{13} = -A_{31} = gB_2(x), \\ &A_{23} = -A_{32} = -gB_1(x), \end{aligned} \tag{23}$$

where $B = (B_1, B_2, B_3)$ is the magnetic field, g is the gyromagnetic ratio of the neutron; in addition, S for equation (1) with A given by (23) describes the polarization data (but, in general, S cannot be given explicitly in this case).

In this case, S on $T\mathbb{S}^2$ uniquely determines B on $\mathbb{R}^3$ as a corollary of items (1), (2) of Theorem 6.1 of [35]. In addition, the related 3D-reconstruction is based on local 2D-reconstructions based on solving Riemann conjugation problems (going back to [30]) and on the layer by layer reconstruction approach. The final 3D uniqueness and reconstruction results are global.

For the related 2D global uniqueness, see [12].

5.6 Electromagnetic polarization tomography (EPT)

In EPT, one considers a medium with zero conductivity, unit magnetic permeability, and small anisotropic perturbation of some known (e. g., uniform) dielectric permeability. The polarization data consist in changes of the polarization between incoming and outcoming monochromatic electromagnetic waves. The basic problem of EPT consists in finding the anisotropic perturbation of the dielectric permeability from the polarization data. See [38, 42, 49, 52], and references therein.

Problem 1 arises as a problem of EPT (with uniform background dielectric permeability) in the framework of the following reduction (see [42, 49]):

$$n = 3, \quad A(x,\theta) = -\pi_\theta f(x)\pi_\theta, \quad x \in \mathbb{R}^d,\ \theta \in \mathbb{S}^{d-1}, \tag{24}$$

where $\pi_\theta \in M(3,\mathbb{R})$, $\pi_{\theta,ij} = \delta_{ij} - \theta_i\theta_j$, f takes values in $M(3,\mathbb{C})$ and describes the anisotropic perturbation of the dielectric permeability tensor; by some physical arguments, f must be skew-Hermition, $f_{ij} = -\bar{f}_{ji}$; in addition, S for equation (1) with A given by (24) describes the polarization data (but, in general, S cannot be given explicitly in this case).

In this case, S on $T\mathbb{S}^2$ does not determine f on $\mathbb{R}^3$ uniquely, in general, (in spite of the fact that $\dim T\mathbb{S}^2 = 4 > \dim \mathbb{R}^3 = 3$), in particular, if

$$\begin{gathered} f_{11} = f_{22} = f_{33} \equiv 0, \\ f_{12}(x) = \partial u(x)/\partial x_3, \quad f_{13}(x) = -\partial u(x)/\partial x_2, \quad f_{23}(x) = \partial u(x)/\partial x_1, \\ f_{21} = -f_{12}, \quad f_{31} = -f_{13}, \quad f_{32} = -f_{23}, \end{gathered} \tag{25}$$

where u is a real smooth compactly supported function, then $S \equiv I$ on $T\mathbb{S}^2$; see [42].

On the other hand, a very natural additional physical assumption is that f is an imaginary-valued symmetric matrix: $f = -\bar{f}$, $f_{ij} = f_{ji}$. According to [38], in this case

$$S \text{ on } \Lambda \text{ uniquely determines } f, \text{ at least, if } f \text{ is sufficiently small,} \tag{26}$$

where Λ is an appropriate $3d$ subset of $T\mathbb{S}^2$, for example,

$$\begin{gathered} \Lambda = \bigcup_{i=1}^{6} \Gamma_{\omega^i}, \quad \Gamma_{\omega^i} = \{y = (x,\theta) \in T\mathbb{S}^2 : \theta\omega^i = 0\}, \\ \omega^1 = e_1, \quad \omega^2 = e_2, \quad \omega^3 = e_3, \\ \omega^4 = (e_1 + e_2)/\sqrt{2}, \quad \omega^5 = (e_1 + e_3)/\sqrt{2}, \quad \omega^6 = (e_2 + e_3)/\sqrt{2}, \end{gathered} \tag{27}$$

where e_1, e_2, e_3 is the basis in $\mathbb{R}^3$. In addition, this determination is based on a convergent iterative reconstruction algorithm.

For more information on EPT and for more results on related non-Abelian ray transforms, we refer to [38, 42, 49, 52], and references therein.

5.7 Inverse connection problem

Let

$$A(x,\theta) = a_0(x) + \sum_{j=1}^{d} \theta_j a_j(x), \quad x \in \mathbb{R}^d,\ \theta = (\theta_1, \ldots, \theta_d) \in \mathbb{S}^{d-1}, \tag{28}$$

where a_j are sufficiently regular $M(n,\mathbb{C})$-valued functions on $\mathbb{R}^d$ with sufficient decay as $|x| \to \infty$, $j = 0, 1, \ldots, d$. Then Problem 1 arises in differential geometry.

In particular, for $a_0 \equiv 0$ equation (1) with A given by (28) describes the parallel transport of the fiber in the trivial vector bundle with the base $\mathbb{R}^d$ and the fiber $\mathbb{C}^n$ and with the connection $a = (a_1, \dots, a_d)$ along the Euclidean geodesics in $\mathbb{R}^d$; in addition, $S(\gamma)$ for fixed $\gamma \in T\mathbb{S}^{d-1}$ is the operator of this parallel transport along γ (from $-\infty$ to $+\infty$ on γ); see [35, 50].

Besides, for $a_0 \not\equiv 0$ equation (1) with A given by (28) describes the parallel transport of the fiber in the trivial vector bundle with the base $\mathbb{R}^{d+1}_{1,d}$ and the fiber $\mathbb{C}^n$ and with the connection $a = (a_0, a_1, \dots, a_d)$ (independent of time) along the light rays in the Minkowski space $\mathbb{R}^{d+1}_{1,d}$; in addition, $S(\gamma)$ for fixed $\gamma = (x, \theta) \in T\mathbb{S}^{d-1}$ is the operator of this parallel transport along the light rays

$$l(\gamma, \tau) = \{(t, y) \in \mathbb{R}^{d+1} : t = 2^{-1/2}s + \tau, y = 2^{-1/2}s\theta + x, s \in \mathbb{R}\}, \quad \tau \in \mathbb{R},$$

with the orientation given by the vector $2^{-1/2}(1, \theta)$ (from $-\infty$ to $+\infty$ on $l(\gamma, \tau)$ for an arbitrary $\tau \in \mathbb{R}$); see [35].

In these cases, Problem 1 is an inverse connection problem. The determination in this problem is considered modulo gauge transforms

$$\begin{aligned} &a = (a_0, a_1, \dots, a_d) \to a' = (a'_0, a'_1, \dots, a'_d), \\ &a'_0 = g^{-1}a_0 g, \quad a'_i = g^{-1}a_i g + g^{-1}\partial_i g, \quad \partial_i g(x) = \frac{\partial g(x)}{\partial x_i}, \quad i = 1, \dots, d, \end{aligned} \tag{29}$$

where g is a sufficiently regular $GL(n, \mathbb{C})$-valued function on $\mathbb{R}^d$ and $g \to I$ sufficiently fast as $|x| \to \infty$.

Global uniqueness and reconstruction results on this inverse connection problem in dimension $d \geq 3$ were given for the first time in [35]. The related reconstruction is based on local 2D-reconstructions based on solving Riemann conjugation problems (going back to [30]) and on the layer by layer reconstruction approach.

In addition, counterexamples to the global uniqueness for the aforementioned inverse connection problem for $a_0 \equiv 0$ in dimension $d = 2$ were also given for the first time in [35]. These counterexamples use the soliton solutions constructed in [53, 54] for equation (38) mentioned below.

In addition, for the global uniqueness in dimension $d = 2$ for the case of compactly supported $a = (a_0, a_1, \dots, a_d)$, see [12].

Note that [35] was stimulated by [50], where [50] was preceded by [55].

For more information on the inverse connection problem, we refer to [12, 15, 20, 30, 35, 41, 43–45, 50], and references therein.

5.8 Vector field tomography

The inverse connection problem of Section 5.7 arises as a problem of the vector ultrasonic tomography in the framework of the following reduction:

$$n = 2, \quad a_0 = \begin{pmatrix} a(x) & 0 \\ 0 & 0 \end{pmatrix}, \quad a_j = \begin{pmatrix} 0 & u_j(x) \\ 0 & 0 \end{pmatrix}, \quad x \in \mathbb{R}^d, \ j = 1, \dots, d, \tag{30}$$

$$S_{11} = \exp[-P_0 a], \quad S_{12} = \exp[-P_a u], \quad S_{21} = 0, \quad S_{22} = 1, \tag{31}$$

where $P_0 a$ is defined as in (7), (10),

$$P_a u(\gamma) = \int_{\mathbb{R}} \exp[-Da(x + s\theta, \theta)]\theta u(x + s\theta)\, ds, \quad \gamma = (x, \theta) \in T\mathbb{S}^{d-1}, \tag{32}$$

$\theta u = \sum_{j=1}^{d} \theta_j u_j$, Da is defined as in (12), a is the attenuation coefficient, $u = (u_1, \ldots, u_d)$ is the flow velocity, $P_a u$ is the attenuated vectorial Radon transform of u along oriented straight lines.

The transform $P_a u$ for $a = 0$ is the standard vectorial Radon transform of u and is related to time-of-flight measurements or to Doppler measurements; $P_a u$ for $a \not\equiv 0$ is related to the attenuated Doppler measurements; see [48] and references therein.

In connection with mathematics of vector field tomography, we refer to [17, 25, 35, 48, 49, 51], and references therein.

5.9 Theory of the Yang–Mills fields

A. The inverse connection problem of Section 5.7 for $a_0 \equiv 0$ arises, in particular, in the framework of studies on inverse problems for the Schrödinger equation

$$\sum_{j=1}^{d} -\left(\frac{\partial}{\partial x_j} + a_j(x)\right)^2 \psi + v(x)\psi = E\psi \tag{33}$$

in the Yang–Mills field $a = (a_1, \ldots, a_d)$ at high energies E (i. e., for $E \to +\infty$); see [35] and references therein. The reason of this consists in the fact that for ψ of the form

$$\psi = e^{is\theta x}(\mu_0(x, \theta) + O(s^{-1})), \quad x \in \mathbb{R}^d,\ \theta \in \mathbb{S}^{d-1},\ s = \sqrt{E} \to +\infty, \tag{34}$$

equation (33) in its leading part is reduced to equation (1) with μ_0 in place of ψ, where A is given by (28) with $a_0 \equiv 0$.

B. The inverse connection problem of Section 5.7 for $d = 2$ arises, in particular, in the framework of integrating the self-dual Yang–Mills equations; see [16, 30, 35, 53, 54], and references therein.

Actually, Problem 1 for

$$A(x, \theta) = a_0(x) + \theta_1 a_1(x) + \theta_2 a_2(x), \quad x = (x_1, x_2) \in \mathbb{R}^2,\ \theta = (\theta_1, \theta_2) \in \mathbb{S}^1, \tag{35}$$

with $M(n, \mathbb{C})$-valued a_0, a_1, a_2 (and some linear relation between a_1 and a_2) was considered for the first time in [30] in the framework of integration by the inverse scattering method of the evolution equation

$$(\chi^{-1}\chi_t)_t = (\chi^{-1}\chi_z)_{\bar{z}}, \tag{36}$$

where $t, z, \bar{z}$ in (36) denote partial derivatives with respect to t, $z = x_1 + ix_2$, $\bar{z} = x_1 - ix_2$ and where χ is $SU(n)$-valued function. Equation (36) is a $(2+1)$-dimensional reduction of the self-duel Yang–Mills equations in $2+2$ dimensions.

To our knowledge, the terminology "non-Abelian Radon transform" was introduced namely in [30] where it was used for S in (2) corresponding to the aforementioned A of (35).

The inverse scattering transform in [30] is based on Riemann conjugation problems. Related analysis was significantly developed, in particular, in [35].

In addition, Problem 1 for

$$A(x,\theta) = \theta_2 a_2(x), \quad x = (x_1, x_2) \in \mathbb{R}^2, \ \theta = (\theta_1, \theta_2) \in \mathbb{S}^1, \tag{37}$$

with $M(n,\mathbb{C})$-valued a_2 arises in the framework of the inverse scattering method for the equation

$$(J^{-1}J_{x_1})_{x_1} - (J^{-1}J_{x_2})_t = 0, \tag{38}$$

where t, x_1, x_2 in (38) denote partial derivatives with respect to t, x_1, x_2, and where J is $SU(n)$-valued function; see [35, 53, 54], at least, for $n = 2$. Equation (38) is also a $(2+1)$-dimensional reduction of the self-duel Yang–Mills equations in $2+2$ dimensions. This reduction is different from (36).

The aforementioned counter examples to the global uniqueness for the inverse connection problem of Section 5.7 for $a_0 \equiv 0$ in dimension $d = 2$ were constructed in [35] using results of [54] and subsequent results of [53] concerning soliton solutions for equation (38).

Bibliography

[1] E. V. Arzubov, A. L. Bukhgeim, and S. G. Kazantsev. Two-dimensional tomography problems and the theory of A-analytic functions. *Sib. Adv. Math.*, 8:1–20, 1998.

[2] Y. M. Assylbekov, F. Monard, and G. Uhlmann. Inversion formulas and range characterizations for the attenuated geodesic ray transform. *J. Math. Pures Appl.*, 111:161–190, 2018.

[3] G. Bal. Inverse transport theory and applications. *Inverse Probl.*, 25:053001, 2009.

[4] G. Bal and A. Jollivet. Combined source and attenuation reconstructions in SPECT. *Contemp. Math.*, 559:13–27, 2011.

[5] G. Beylkin. The inversion problem and applications of the generalized Radon transform. *Commun. Pure Appl. Math.*, 37:579–599, 1984.

[6] J. Boman. An example of non-uniqueness for a generalized Radon transform. *J. Anal. Math.*, 61:395–401, 1993.

[7] J. Boman. Local non-injectivity for weighted Radon transforms. *Contemp. Math.*, 559:39–47 2011.

[8] J. Boman and E. T. Quinto. Support theorems for real-analytic Radon transforms. *Duke Math. J.*, 55:943–948, 1987.

[9] J. Boman and J. O. Strömberg. Novikov's inversion formula for the attenuated Radon transform: a new approach. *J. Geom. Anal.*, 14:185–198, 2004.

[10] A. V. Bronnikov. Reconstruction of attenuation map using discrete consistency conditions. *IEEE Trans. Med. Imaging*, 19:451–462, 2000.

[11] M. Dawson, I. Manke, N. Kardjilov, A. Hilger, M. Strobl, and J. Banhart. Imaging with polarized neutrons. *New J. Phys.*, 11:043013, 2009.

[12] G. Eskin. On non-Abelian Radon transform. *Russ. J. Math. Phys.*, 11:391–408, 2004.

[13] D. Finch. Uniqueness for the attenuated X-ray transform in the physical range. *Inverse Probl.*, 2:197–203, 1986.

[14] D. Finch. The attenuated x-ray transform: recent developments. In *Inside Out: Inverse Problems and Applications*. Math. Sci. Res. Inst. Publ., volume 47, pages 47–66. Cambridge Univ. Press, Cambridge, 2003.

[15] D. Finch and G. Uhlmann. The X-ray transform for a non-Abelian connection in two dimensions. *Inverse Probl.*, 17:695–701, 2001.

[16] A. S. Fokas and T. A. Ioannidou. The inverse spectral theory for the Ward equation and the $2+1$ chiral model. *Commun. Appl. Anal.*, 5:235–246, 2001.

[17] I. M. Gel'fand, S. G. Gindikin, and M. I. Graev. Integral geometry in affine and projective spaces. *J. Sov. Math.*, 18:39–167 1982.

[18] S. G. Gindikin. A remark on the weighted Radon transform on the plane. *Inverse Probl. Imaging*, 4:649–653, 2010.

[19] F. O. Goncharov and R. G. Novikov. A breakdown of injectivity for weighted ray transforms in multidimensions. arXiv:1711.06163.

[20] C. Guillarmou, G. P. Paternain, M. Salo, and G. Uhlmann. The X-ray transform for connections in negative curvature. *Commun. Math. Phys.*, 343:83–127, 2016.

[21] J.-P. Guillement and R. G. Novikov. Optimized analytic reconstruction for SPECT. *J. Inverse Ill-Posed Probl.*, 20:489–500, 2012.

[22] J.-P. Guillement and R. G. Novikov. Inversion of weighted Radon transforms via finite Fourier series weight approximations. *Inverse Probl. Sci. Eng.*, 22:787–802, 2014.

[23] G. M. Henkin and R. G. Novikov. The Yang–Mills fields, the Radon–Penrose transform and the Cauchy–Riemann equations. In *Encyclopaedia of Mathematical Science*, volume 54, pages 109–193. Springer, 1993.

[24] J. Ilmavirta. Coherent quantum tomography. *SIAM J. Math. Anal.*, 48:3039–3064, 2016.

[25] S. G. Kazantsev and A. A. Bukhgeim. Inversion of the scalar and vector attenuated X-ray transforms in a unit disc. *J. Inverse Ill-Posed Probl.*, 15:735–765 2007.

[26] L. Kunyansky. Generalized and attenuated Radon transforms: restorative approach to the numerical inversion. *Inverse Probl.*, 8:809–819, 1992.

[27] L. Kunyansky. A new SPECT reconstruction algorithm based on the Novikov's explicit inversion formula. *Inverse Probl.*, 17:293–306, 2001.

[28] M. M. Lavrent'ev and A. L. Bukhgeim. A class of operator equations of the first kind. *Funct. Anal. Appl.*, 7:290–298, 1973.

[29] W. Lionheart, N. Desai, and S. Schmidt. Nonabelian tomography for polarized light and neutrons. In *Quasilinear Equations, Inverse Problems and Their Applications*, MIPT, Russia, 30 November 2015–2 December 2015.

[30] S. V. Manakov and V. E. Zakharov. Three-dimensional model of relativistic-invariant field theory, integrable by the inverse scattering transform. *Lett. Math. Phys.*, 5:247–253, 1981.

[31] A. Markoe and E. T. Quinto. An elementary proof of local invertibility for generalized and attenuated Radon transforms. *SIAM J. Math. Anal.*, 16:1114–1119, 1985.

[32] E. X. Miqueles and A. R. De Pierro. Fluorescence tomography: reconstruction by iterative methods. In *5th IEEE International Symposium on Biomedical Imaging: From Nano to Macro*, pages 760–763, 2008.

[33] F. Monard. Inversion of the attenuated geodesic X-ray transform over functions and vector fields on simple surfaces. *SIAM J. Math. Anal.*, 48:1155–1177, 2016.

[34] F. Natterer. *The Mathematics of Computerized Tomography*. Teubner, Stuttgart, 1986.

[35] R. G. Novikov. On determination of a gauge field on $\mathbb{R}^d$ from its non-Abelian Radon transform along oriented straight lines. *J. Inst. Math. Jussieu*, 1:559–629, 2002.

[36] R. G. Novikov. An inversion formula for the attenuated X-ray transformation. *Ark. Mat.*, 40:145–167, 2002.

[37] R. G. Novikov. On the range characterization for the two-dimensional attenuated X-ray transformation. *Inverse Probl.*, 18:677–700, 2002.

[38] R. G. Novikov. On iterative reconstruction in the nonlinearized polarization tomography. *Inverse Probl.*, 25:115010, 2009.

[39] R. G. Novikov. Weighted Radon transforms for which Chang's approximate inversion formula is exact. *Russ. Math. Surv.*, 66:442–443, 2011.

[40] R. G. Novikov. Weighted Radon transforms and first order differential systems on the plane. *Mosc. Math. J.*, 14:807–823, 2014.

[41] R. G. Novikov. Détermination d'un champ de jauge sur $\mathbb{R}^3$ par sa transformée de Radon non-Abélienne. In *Séminaire "Equations aux Dérivées Partielles 2003–2004", exposé No. XVII*. Centre de Mathématiques, Ecole Polytechnique.

[42] R. G. Novikov and V. A. Sharafutdinov. On the problem of polarization tomography. I. *Inverse Probl.*, 23:1229–1257, 2007.

[43] G. P. Paternain. Inverse problems for connections. In *Inverse Problems and Applications: Inside Out. II*. Mathematical Sciences Research Institute Publications, volume 60, pages 369–409. Cambridge University Press, Cambridge, 2013.

[44] G. P. Paternain, M. Salo, and G. Uhlmann. The attenuated ray transform for connections and Higgs fields. *Geom. Funct. Anal.*, 22:1460–1489, 2012.

[45] G. P. Paternain, M. Salo, G. Uhlmann, and H. Zhou. The geodesic X-ray transform with matrix weights, 2016. arXiv:1605.07894.

[46] A. Puro and A. Garin. Cormack-type inversion of attenuated Radon transform. *Inverse Probl.*, 29:065004, 2013.

[47] J. Radon. Uber die Bestimmung von Funktionen durch ihre Integralwerte längs gewisser Mannigfaltigkeiten. *Ber. Saechs Akad. Wiss. Leipzig, Math.-Phys.*, 69:262–267, 1917.

[48] T. Schuster. 20 years of imaging in vector field tomography: a review. In Y. Censor, M. Jiang, and A. K. Louis, editors, *Mathematical Methods in Biomedical Imaging and Intensity-Modulated Radiation Therapy (IMRT)*. Series: Publication of the Scuola Normale Superiore, CRM Series, volume 7. Birkhäuser, 2008.

[49] V. A.Sharafutdinov. *Integral Geometry of Tensor Fields*. VSP, Utrecht, 1994.

[50] V. A. Sharafutdinov. On an inverse problem of determining a connection on a vector bundle. *J. Inverse Ill-Posed Probl.*, 8:51–88, 2000.

[51] V. A. Sharafutdinov. Slice-by-slice reconstruction algorithm for vector tomography with incomplete data. *Inverse Probl.*, 23:2603–2627, 2007.

[52] V. A. Sharafutdinov. The problem of polarization tomography. II. *Inverse Probl.*, 24:035010, 2008.

[53] J. Villarroel. The inverse problem for Ward's system. *Stud. Appl. Math.*, 83:211–222, 1990.

[54] R. S. Ward. Soliton solutions in an integrable chiral model in 2 + 1 dimensions. *J. Math. Phys.*, 29:386–389, 1988.

[55] L. B. Wertgeim. Integral geometry with matrix weight and a nonlinear problem of recovering matrices. *Dokl. Akad. Nauk SSSR*, 319:531–534, 1991 (in Russian).

Victor Palamodov

6 Remarks on the second century of the Funk–Radon theory

Abstract: Integral transforms of the Radon type are discussed in the survey. Papers published within five years are in focus. The following subjects are highlighted: nongeodesic Funk transform on a sphere and on an arbitrary smooth manifolds, reconstructions for photoacoustic applications, cone Radon transforms, ray transform of tensor fields, general transforms of Radon type, parametrix for Radon transforms, reconstructions in SAR geometry.

Keywords: Generalized Radon transform, nongeodesic Funk transform, cone Radon transform, photoacoustic geometry, ray transform

MSC 2010: 44A12, 92C55

6.1 Introduction

The following remarks are about papers related to the Funk and Radon transforms mostly published after 2013. A survey of the first 100 years of the history of the theory can be extracted from the bibliographic notes given in Natterer [36], Helgason [18] and in [45]. Pioneering papers of Alan Cormack (1963–1964) "Representation of a function by its line integrals, with some radiological applications" were devoted to medical applications [3]. Few years later, he wrote that "his first mathematical problem had been already solved by Radon's theory." These papers triggered numerous applications of the Radon transform to various inverse problems. The prehistory of applications to physics was described by Cormack [4].

6.2 Geodesic and nongeodesic Funk transforms

The Funk transform is applied to functions f defined on the sphere $X = \mathrm{S}^n$

$$g(\sigma) = \int_X \delta(\langle \sigma, x\rangle) f(x)\Omega = \int_{\langle\sigma,x\rangle=0} f(x)\frac{\Omega}{\langle \sigma, \mathrm{d}x\rangle}, \quad \sigma \in \Sigma = \mathrm{S}^n, \tag{1}$$

where Ω is the Euclidean volume form on the sphere. Paul Funk [10] found the explicit reconstruction formula for Hermann Minkowski's problem to determine an even

Victor Palamodov, Tel Aviv University, Tel Aviv, Israel, e-mails: palamodo@post.tau.ac.il, palamodo@tauex.tau.ac.il

https://doi.org/10.1515/9783110560855-006

function f on S^2 from data of its big circle integrals g. His method was used by Johann Radon who deduced his famous formula for the Euclidean plane [51]. Funk's formula was generalized by Helgason in 1959, 1960, for any even n; see [18]. For odd n, the inversion formula was found by Semyanistiy [56, 57][1]:

$$f(x) = \frac{1}{2(2\pi i)^{n-1}} \int_{S^n} \delta^{(n-1)}(\langle \sigma, x \rangle) g(\sigma)\Omega.$$

It looks different from the Funk–Radon–Helgason formulae for even dimensional case but similar to Lorentz's formula for $\mathbb{R}^3$; see [4]. It is easy to check that for even n no local formula exists for the classical geometries. This principle apparently is valid for all general integral transforms as in Section 6.7.

The hyperboloid $H = \{y_0^2 = y_1^2 + \cdots + y_n^2 + 1,\ y_0 > 0\}$ in the Euclidean space $\mathbb{R}^{n+1}$ is a model for the space of constant sectional curvature. See Helgason [18] for the analog of the Funk reconstructions for H. The reconstructions can also be obtained from the formulas for the unit half-sphere by the central (gnomonic) projection to H [45].

A similar integral operator called nongeodesic Funk transform was considered in [45]:

$$g(\sigma) = \int_{\xi \in S^n, \langle \xi - \alpha, \sigma \rangle = \rho} f(\xi)\Omega. \tag{2}$$

Here, $0 \le \rho < 1$ and $\alpha \in \mathbb{R}^3$, $|\alpha| \le 1$ are arbitrary, and Ω is the Euclidean volume form on the sphere. The reconstruction for $n = 2$ is given by

$$f(\xi) = -\frac{|\xi - \alpha|}{2\pi^2} \int_{S^2} \frac{g(\sigma)}{(\langle \xi - \alpha, \sigma \rangle - \rho)^2} \Omega, \quad \xi \in S^2 \tag{3}$$

provided there exists a vector $\sigma_0 \in S^2$ such that $\langle \xi - \alpha, \sigma_0 \rangle > \rho$ on $\operatorname{supp} f$. Singular integral in (3) can be explicitly written as follows. Let φ be the zenith angle of a point $\sigma \in S^2$ with the pole $\eta \doteqdot (\xi - \alpha)|\xi - \alpha|^{-1}$ and θ be the azimuth angle. Tangent field

$$\tau_\xi = \frac{1}{|x - \alpha| \sin \varphi} \frac{\partial}{\partial \varphi}$$

fulfils

$$\tau_\xi(\langle \xi - \alpha, \sigma \rangle - \rho) = 1.$$

1 V. I. Semyanistyi (1925–1984), PhD since 1964 (scientific adviser M. I. Graev), worked in the Kolomna Pedagogical College.

We have $\Omega = \sin\varphi d\varphi d\theta$ and by partial integration,

$$\int_{S^2} \frac{g(\sigma)\Omega}{(\langle \xi-\alpha,\sigma\rangle-\rho)^2} = -\int \tau_\xi\left(\frac{1}{\langle \xi-\alpha,\sigma\rangle-\rho}\right)g(\sigma)\Omega$$

$$= \int \frac{\tau_\xi g(\sigma)\Omega}{\langle \xi-\alpha,\sigma\rangle-\rho} - \frac{2\pi}{|\xi-\alpha|}\left(\frac{g(\eta)}{|\xi-\alpha|-\rho}+\frac{g(-\eta)}{|\xi-\alpha|+\rho}\right),$$

where $\tau_\xi(g)\Omega = \partial g/\partial\varphi d\varphi d\theta$ and the principal value in the right-hand side is taken. The singularity in the right-hand side can be smoothed if one more integration by parts is implemented. Suppose now $\alpha = 0, \rho = 0$, and extend f as an even function on S^2. We have $\langle \xi,\sigma\rangle = \cos\varphi$ and

$$f(\xi) = -\frac{1}{\pi}\int_0^{\pi/2} \frac{dG}{\cos\varphi} + \frac{1}{\pi}G(\eta), \quad G = \frac{1}{2\pi}\int 2g d\theta.$$

This coincides with the original Funk's reconstruction since where $2g$ is equal to the integrals of the even function f over big circles in S^2.

For the special case for $\rho = 0$ and arbitrary a, $|a| \le 1$ the reconstructions of different form were given in [53, 54], and also [49]. The further generalization is discussed in Section 6.7.

6.3 Helgason–Ludwig conditions

Ruhlandt et al. [52] described an iterative algorithm of reconstruction of parameters of 3D object illuminated by a monochromatic high frequency plane wave whose phaseless near-field is detected. The back propagation provides a rude approximation at each step of the algorithm. Helgason–Ludwig consistency conditions are used to improve reconstruction of low frequencies of images taken from different angles. This helps to retrieve the phase and improve evaluation of the attenuation and refraction coefficients of the object.

6.4 Photoacoustic reconstructions

Reconstruction formulae for photoacoustic transform

$$g(y,r) = \int_{|x-y|=r} f(x)dX, \quad y \in Z$$

are known for spheres, ellipsoids, or more complicated compact algebraic sets Z. Fawcett [8] and Anderssen [1] considered the problem for the central set $Z = \{x_1 = 0\}$

and functions f supported in the half-space. This acquisition geometry is significant for the SAR technics. Fawcett's method is based on the backprojection operation which is given in this case by the improper integral

$$\mathrm{R}^*g(x) = \int_Z g(y, |x - y|)\mathrm{d}y.$$

The integral diverges for any x such that $f(x) > 0$ if f is a continuous non-negative function since $g(y, |x - y|)$ has the positive limit as $y \to \infty$. Therefore, the backprojection does not exist in the classical sense for all such points x. Andersson's method is based on the Fourier transform of the function g. This method is also not complete since the Fourier transform $F(g)$ is not a point function [1] (p. 216) and further steps have to be approved; see Klein [26].

Haltmeier and Perversyev [16, 17] proposed reconstruction f from data on noncompact hypersurfaces Z that can be approximated by ellipsoids. The authors apply the reconstruction formula for ellipsoids Z which gives a FBP (filtration-backprojection) reconstruction. The case of elliptic cylinders Z was addressed again in [15]. The authors applied a reconstruction method of BPF type (backprojection-filtration) but the composition $\mathrm{R}^*\mathrm{R}$ is not well-defined in definition 1.2.3, equations (2) and (A.4) of [15] in the simplest case where the ellipsoid is one point. We show in Section 6.9 that the FBP method does work for these cases.

Elbau and Scherzer [7] have considered a model for photoacoustic plane sectional imaging with integrating half-cylinder shaped acoustic detector serving as a focusing detector. Kunyansky [28] proposed a method of approximate reconstruction of functions on the fundamental domain of the reflection group $\mathbb{Z}_N$ on $\mathbb{R}^2$ or the group $(\mathbb{Z}_2)^{\times 3}$ acting on $\mathbb{R}^3$ from integral data on the boundary.

6.5 Ray transform of tensor fields

The residual elastic strain in a body is described by the strain tensor which has six scalar components. Tomographic approach is applied for the reconstruction of small residual strain tensor from data of diffraction patterns under penetrated x-ray or neutron radiation. The mathematical model is the longitudinal (axial) line transform $X\varepsilon$ of the strain tensor ε. This data is gauge invariant since all the integrals vanish if $\varepsilon = Du$ for small deformations u (potential tensor). A reconstruction of a solenoidal (divergenceless) field was done in [29] from data of all line integrals. Denisjuk [5] considered reconstruction of the solenoidal part of a tensor field of arbitrary degree from ray integrals in a flat space.

According to [41], data of first derivatives of the line integrals of a first- order form are sufficient for stable reconstruction for the first-order form (vector field) if the lines meet the source curve Γ satisfying Tuy's condition. Similar results were obtained later in [25] and [23, 24]. Here, the geometric condition is: any plane P in $\mathbb{R}^3$ that meets the support of the unknown function must have at least 3 common points with Γ. This number is minimal anyway since the first-order differential form has 3 components. The polarization tomography is other method of reconstruction of a strain field in a transparent solid. The mathematical model is the line integral transform $T\varepsilon$ of the traceless normal part of the stress field ε. An analytic algorithm of complete reconstruction of an arbitrary strain tensor ε from nonredundant data.

Paternain, Salo, and Uhlmann [48] have shown that on a simple Riemannian surface the geodesic axial transform acting on solenoidal symmetric tensor fields of arbitrary order is injective. Sharafutdinov [58] develops the theory of magneto-photoelasticity for determination of dielectric tensor ε of a medium by application of the exterior magnetic field. In the linear approximation, the traceless transversal part $T\varepsilon$ of ε is evaluated in terms of the integrals of the gyration field along optical rays of the medium. It is shown in [43] that in the Euclidean 3-space the strain 2-tensor can be reconstructed from data of both integral transforms $X\varepsilon$ and $T\varepsilon$.

Denisjuk [6] describes several methods of analytic reconstruction of a function in $\mathbb{R}^n$ from data of integrals over n-dimensional family of straight lines.

6.6 Cone Radon transform

The cone of rotation in an Euclidean space $\mathbb{R}^n$ can be written in the form

$$C(\lambda) = \{x \in \mathbb{R}^n : \lambda x_1 = r\}, \quad \lambda > 0,\ r^2 = x_2^2 + \cdots + x_n^2.$$

The line $r = 0$ is the axis and $\theta \div \arctan \lambda$ is the half-opening of the cone. The integral operator

$$\mathrm{D}_\lambda^k f(y) = \int_{x \in C(\lambda)} f(y + x)|x|^{-k} \mathrm{d}S, \quad y \in \mathbb{R}^n$$

is discussed in recent publications under the name of *cone Radon* transform. Here, $\mathrm{d}S$ is the Euclidean hypersurface element. This operator is a convolution with a distribution in the case $k < n - 1$ (regular case). In the case $k = n - 1$ (singular case), the kernel is not integrable and the condition $f(y) = 0$ is necessary. The realistic model of point spread function for single-scattering optical 3D tomography is based on the photometric law of scattered radiation and is modeled by the singular cone Radon transform. This transform models multiple views of the object in form integrals of the photon source distribution. Analytic reconstructions from the regular cone transform with a restricted apex were obtained by Nguyen and Truong [37], Smith [59], Grangeat et al. [38], Maxim et al. [32], Maxim [31], Haltmeier [14], Jung and Moon [21], Terzioglu [61], Kuchment-Terzioglu [27], Louis [30], Moon [33, 34]. Papers [32], Maxim [31] contain representative numerical reconstructions. These reconstructions are based on the reduction of redundant data of all cone integrals to the classical Radon transform in several ways, for example,

$$\int_0^{\pi} \mathrm{D}_\lambda^0 f \mathrm{d}\theta = \mathrm{const}\, \mathrm{R}^* \mathrm{R} f$$

where R^* is the back projection operator for the Radon transform R. Jung and Moon [22] applied inversion formulae for the regular cone transform on $\mathbb{R}^n$ for numerical reconstruction using nonredundant data from a line of detectors and rotating axis. In [35] and [55], other reconstruction schemes were proposed based on the integral data from all cones with apices on a big sphere.

Gouia-Zarrad and Ambartsoumian [12, 13] found new reconstruction formulae for the regular cone transform on a half-space with free apex and one opening. In [46], a reconstruction from a regular cone transform D_λ^k with constant axis and opening on $\mathbb{R}^3$ is given in terms of the operator D_λ^{1-k}. This approach does not work when the weight density is not integrable and the cone integral data cannot be collected for all positions of the apex. A more complicated method is proposed in [46] for reconstruction from data of singular cone integrals with apices running a 1D set. The inversion formula for the nongeodesic Funk transform (3) was applied for this reconstruction.

6.7 General Funk–Radon transform

Integral transforms of Funk–Radon type are described in more general terms. Let X be a smooth manifold of dimension $n > 1$ and $\mathrm{d}X$ be a volume form on X, and Σ be a manifold of the same dimension n (parameter space). Let Φ be a smooth real function Φ on $X \times \Sigma$ (called generating function) satisfying the condition $\mathrm{d}\Phi \neq 0$ on the set $Z \doteqdot \{\Phi = 0\}$ which implies $\mathrm{d}_x\Phi \neq 0$ and $\mathrm{d}_\sigma\Phi \neq 0$. It follows that the set Z is a smooth manifold of dimension $2n-1$ and $Z(\sigma) \doteqdot \{\Phi(\cdot, \sigma) = 0\}$, $Z(x) \doteqdot \{\Phi(x, \cdot) = 0\}$ are smooth hypersurfaces in X, respectively in Σ. The projections $p_X : Z \to X$ and $p_\Sigma : Z \to \Sigma$ have rank n. The general Funk–Radon transform is defined in [45] by

$$\mathrm{M}f(\sigma) = \lim_{\varepsilon \to 0} \frac{1}{2\varepsilon} \int_{|\Phi(x,\sigma)| \le \varepsilon} f(x) \mathrm{d}X$$

for an arbitrary continuous function f on X with compact support and any $\sigma \in \Sigma$.

The following assumption is more strong:

Condition (I) the map $\mathrm{D} : Z \times (\mathbb{R}\backslash 0) \to T_0^*(X) \doteqdot T^*(X)\backslash(X \times 0)$ is a diffeomorphism, where $\mathrm{D}(x, \sigma, t) = (x, t\mathrm{d}_x\Phi(x, \sigma))$. This condition is equivalent to the following: the map

$$\mathrm{D}_S : Z \to \mathrm{S}^*(X),\ \mathrm{D}_S(x, \sigma) = (x, \|\mathrm{d}_x\Phi(x, \sigma)\|_{\mathrm{g}}^{-1} \mathrm{d}_x\Phi(x, \sigma))$$

is a diffeomorphism where g is an arbitrary Riemannian metric on X and $S^*(X)$ means the bundle of unit spheres in the cotangent bundle. This implies that p_X is proper and for any $x \in X$, $Z(x)$ is diffeomorphic to a $n-1$ sphere.

Proposition 1. *If Φ fulfils* (**I**), *then* M_Φ *can be extended to a bounded operator* $\mathrm{M} : H_K^\alpha(X) \to H_\Lambda^{\alpha+(n-1)/2}(\Sigma)$ *for any $\alpha \in \mathbb{R}$, an arbitrary compact set $K \subset X$ and* $\Lambda = p_\Sigma(p_X^{-1}(K))$.

This fact is well known in special cases; for a proof see, for example, [42]. Points $x, y \in X$ are called *conjugate* for a generating function Φ, if $x \neq y$, $\Phi(x,\sigma) = \Phi(y,\sigma) = 0$ and $\mathrm{d}_\sigma\Phi(x,\sigma) \wedge \mathrm{d}_\sigma\Phi(y,\sigma) = 0$ for some $\sigma \in \Sigma$.

Condition (**II**): there are no conjugate points. This condition under the name "Bolker condition" was introduced by Quinto [50] in a more general situation.

Condition (**III**): for any $x, y \in X$, $x \neq y$, equation holds $\operatorname{Re} i^n Q_n(x,y) = 0$ where

$$Q_n(x,y) = \int\limits_{Z(y)} \frac{1}{(\Phi(x,\sigma) - i0)^n} \frac{\mathrm{d}\Sigma}{\mathrm{d}_\sigma\Phi(y,\sigma)}.$$

Theorem 2 ([45]). *If a generating function Φ satisfies conditions* (**I, II, II**), *then an arbitrary function $f \in L_2(X)$ with compact support can be reconstructed from data of* $\mathrm{M}f$ *by*

$$f(x) = \frac{1}{2\mathrm{j}^{n-1}D_n(x)} \int\limits_\Sigma \delta^{(n-1)}(\Phi(x,\sigma))\mathrm{M}f(\sigma)\mathrm{d}\Sigma \tag{4}$$

for odd n, and by

$$f(x) = \frac{(n-1)!}{\mathrm{j}^n D_n(x)} \int\limits_\Sigma \frac{\mathrm{M}f(\sigma)}{\Phi(x,\sigma)^n}\mathrm{d}\Sigma \tag{5}$$

for even n where

$$D_n = \frac{1}{|S^{n-1}|} \int\limits_{Z(x)} \frac{1}{\|\mathrm{d}_x\Phi(x,\sigma)\|_g^n} \frac{\mathrm{d}\Sigma}{\mathrm{d}_\sigma\Phi}. \tag{6}$$

The integral (5) converges in mean on any compact set in X.

The Funk and Radon transforms fulfils all conditions (**I, II, III**). For the Funk transform, we can take $X = S^n/\mathbb{Z}_2$, $\Sigma = S^n/\mathbb{Z}_2$ and $\Phi = \langle x, \sigma\rangle$ where $S^n/\mathbb{Z}_2$ means the sphere with identified opposite points. For the Radon transform, $X = \mathbb{R}^n$, $\Sigma = \mathbb{R}\times S^{n-1}$, $\Phi(x,\sigma) = p - \langle x, \omega\rangle$, $\sigma = (p, \omega)$.

Example. The following result is a generalization of the Funk–Radon inversion as well as the inversion of the nongeodesic transform which considered in Section 6.2. Let E^{n+1} be an affine space with an invariant volume form $\mathrm{d}V$, Σ be an ellipsoid in E^{n+1}. Let q is the second-order polynomial in E such that $q = 1$ on Σ. Conditions (**I, II, III**) hold for the phase function

$$\Phi(x,\sigma) = \langle x - \sigma, \nabla q \rangle$$

defined on $X \times \Sigma$ provided the following condition holds for a function f to be reconstructed: any line in E^{n+1} that meets X in two points does not meet Σ. By Theorem 2 inversion formulas (4) and (5) can be applied to the integral transform $\mathrm{M}\Phi$ generated by this phase function.

Remarks. Condition (**I**) is "almost" necessary for existence of a parametrix. This means that it cannot be violated on an open set in Z. However, it is not fulfilled on a subset of dimension n for Cormack's family of confocal paraboloids; see [45], Section 5.7. The more detailed analysis is used for this family to obtain the reconstruction of the standard form. Condition (**II**) is apparently necessary for existence of a parametrix since its violation makes any pair of conjugated points indistinguishable on the microlocal level. Condition (**III**) is necessary for the exact reconstruction.

Beylkin [2] and also Homan and Zhou [19] defined the "generalized Radon transform" in a slightly different way by means of the "defining" function $\varphi = \varphi(x,\theta)$ that is positive homogeneous of degree 1 in $\theta \in \mathbb{R}^n$. This is equivalent to the above definition for the generating function $\Phi(x,\sigma) \doteqdot s - \varphi(x,\theta)$ which is positive homogeneous function on $\sigma = (s,\theta)$. The authors consider the integral transform which essentially coincides with the weighted transform $\mathrm{M}_{\Phi,w}$ and construct a parametrix. The "global Bolker condition" formulated in [19] plays the role of condition (**II**). The authors show injectivity and stability for an open, dense subset of "generalized Radon" transforms with analytic defining functions satisfying the Bolker condition. Homan and Zhou [19] define "Generalized Radon transforms" in a slightly different way. An exact inversion of the general Funk–Radon transform was obtained in the paper [47] where X is an arbitrary hypersurface in an affine space $\mathbb{R}^{n+1}$ and Σ is an ellipsoid. A density f defined on X is integrated over intersections of X with hyperplanes tangent to Σ. The only condition on X and Σ is that no three points $x, y \in \operatorname{supp} f$, $\sigma \in \Sigma$ are collinear which guarantees (**II**) and (**III**). The reconstruction formula looks like (3).

6.8 Approximate inversion and parametrix

Stefanov and Uhlmann [60] constructed an approximate time reversal operator for medium with variable sound speed in a nontrapping compact domain. The time reversal is a Fredholm operator with convergent Neumann series. It is not shown but plausible that the remainder is a smoothing operator.

Frigyik, Stefanov, and Uhlmann [9] studied the weighted curve integral transform for a family of curves on a manifold X with a boundary. They prove that the generic integral transform is injective provided that the union of conormals to the curves cover the cotangent bundle to X. The arguments include checking injectivity for analytic families of curves. Novikov [39, 40] studied inversion of the weighted line transform on a plane. He reduces the inversion problem to study of first-order differential systems on the complex plane. This method gives new injectivity and inversion results for the original problem. Goncharov [11] addressed the weighted Radon transform in dimensions $n \geq 3$. Ilmavirta [20] studied the reconstruction problem in $\mathbb{R}^3$ for the weighted line transform of unknown field of Hamiltonian 3 matrices. He considered this transform as the model for tomography of the earth from data of oscillations of neutrino emanated by the artificial sources. Ilmavirta stated uniqueness of reconstruction under some simplifying assumptions.

A construction of parametrix was given in [44] for the general Funk–Radon transform with a generating function of the form $\Phi(x;\omega,p) = \theta(x,\omega) - p$ where $p \in R$, $\omega \in S^{n-1}$.

6.9 Reconstructions for SAR geometry

Theorem 3. *Let Z be the zero set of an elliptic second order polynomial p in $\mathbb{R}^n$. Any function $f \in C^{n-1}(\mathbb{R}^n)$ with compact support in the cavity $\mathrm{H} = \{p < 0\}$ can be reconstructed by*

$$f(x) = \mathrm{j}^{1-n} p(x) \int_{\mathrm{Z}} \frac{\mathrm{d}\xi}{\mathrm{d}p} \left(\frac{1}{r} \frac{\partial}{\partial r} \right)^{n-1} \frac{\mathrm{R}f(r,\xi)}{r} \bigg|_{r=|x-\xi|} \tag{7}$$

for odd n, and by

$$f(x) = 2\mathrm{j}^{-n} p(x) \int_{\mathrm{Z}} \frac{\mathrm{d}\xi}{\mathrm{d}p} \int_0^\infty \frac{\mathrm{d}r^2}{|x-\xi|^2 - r^2} \left(\frac{1}{r} \frac{\partial}{\partial r} \right)^{n-1} \frac{\mathrm{R}f(r,\xi)}{r} \tag{8}$$

for even n from data of spherical integrals

$$\mathrm{R}f(r,\xi) = \int_{|x-\xi|=r} f(x)\mathrm{d}S, \quad \xi \in \mathrm{Z}. \tag{9}$$

For a proof see [45], p. 102.

Theorem 4. *Integrals (7) and (8) converge for any second-order polynomial p on $\mathbb{R}^n$ with nonnegative principal part and any function $f \in C^{n-1}(\mathbb{R}^n)$ with compact support in the cavity $\mathrm{H} = \{p < 0\}$. They provide reconstruction of f for odd and even n, respectively.*

Proof. For simplicity, we assume that $p = x_1$ and consider the family of polynomials

$$p_\varepsilon = x_1 - \frac{\varepsilon}{2}|x|^2, \quad \varepsilon > 0.$$

The cavity $H_\varepsilon = \{p_\varepsilon < 0\}$ is the ball of radius ε^{-1} in the upper half-space $\mathbb{R}^n_+$ with the center $x_1 = \varepsilon^{-1}$. Set $Z_\varepsilon = \{p_\varepsilon = 0\}$ is the sphere with the center $(0, \varepsilon^{-1})$ and radius ε^{-1} and is contained in the upper half-space. We have $p_\varepsilon \to p = x_1$ and $Z_\varepsilon \to Z$ as $\varepsilon \to 0$ uniformly on any ball in $\mathbb{R}^n$. We apply Theorem 3 to p_ε and show that the integral has the limit as $\varepsilon \to 0$ where

$$F(x,\xi) \doteqdot \left(\frac{1}{r}\frac{\partial}{\partial r}\right)^{n-1} \frac{Rf(r,\xi)}{r} = c_1 \frac{\partial_r^{n-1} Rf(r,\xi)}{r^n} + \cdots + c_n \frac{Rf(r,\xi)}{r^{2n-1}} \tag{10}$$

for some integers $c_1 = 1, c_2, \ldots, c_n$. By (9),

$$|\partial_r^k Rf(r,\xi)| \le C_1, \quad k \le n-1,$$

where C_1 does not depend on $r > 0$ and $\xi \in Z$. Therefore, for $r = |x - \xi|$, $x \in \operatorname{supp} f$,

$$|F(x,\xi)| \le \frac{C_2}{(|\xi| + 1)^n} \tag{11}$$

for some constant C_2 since f has compact support. On the other hand,

$$\frac{d\xi}{dp_\varepsilon} = \frac{dS}{|\nabla p_\varepsilon|} = dS$$

on Z_ε where dS is the Euclidean measure on Z_ε and $|\nabla p_\varepsilon| = 1$. Therefore,

$$\left| F(x,\xi)\frac{d\xi}{dp_\varepsilon} \right| \le C_3 \frac{dS}{(|\xi| + 1)^n}$$

on Z_ε. Set $W_\varepsilon = \{\xi \in Z_\varepsilon, \ \xi_1 > \delta\}$ where $\delta = \varepsilon^{1/n}$. We have $dS = |S^{n-2}| \sin^{n-2}\varphi d\varphi$ where φ is the zenith angle on the sphere Z_ε with the pole at the origin. By (11),

$$\left| \int_{W_\varepsilon} F(x,\xi)\frac{d\xi}{dp_\varepsilon} \right| \le C_3 \int_{W_\varepsilon} \frac{dS}{(|\xi| + 1)^n} \le C_4 \int_{\psi}^{\pi} \frac{\varepsilon \sin^{n-2}\varphi d\varphi}{(2\sin\varphi/2 + \varepsilon)^n} \tag{12}$$

where $(1 - \cos\psi) = \delta$. This yields $\sin\psi/2 = (\delta/2)^{1/2}$, hence for $\pi \ge \varphi \ge \psi$,

$$2\sin\frac{\varphi}{2} \ge 2\sin\frac{\psi}{2} = \left(\frac{\delta}{2}\right)^{1/2}$$

and by (12)

$$\left| \int_{W_\varepsilon} F(x,\xi)\frac{d\xi}{dp_\varepsilon} \right| \le C_5 \frac{\varepsilon}{\delta^{n/2}} = C_5 \varepsilon^{1/2} \to 0$$

as $\varepsilon \to 0$. Set $Z_\varepsilon \backslash W_\varepsilon$ is the intersection of the sphere Z_ε with the thin strip $\{0 \le \xi_1 \le \delta\}$. It is projected by the map $\mathbb{R}^n \to Z$ on the ball of radius $\rho \approx (2\delta\varepsilon^{-1})^{1/2} = c\varepsilon^{(1-n)/2n} \to \infty$. This implies

$$\int_{Z_\varepsilon \backslash W_\varepsilon} F(x,\xi)\frac{d\xi}{dp_\varepsilon} \to \int_Z F(x,\xi)\frac{d\xi}{dp}$$

as $\varepsilon \to 0$ since $p_\varepsilon \to p$ and reconstruction (7) follows for p. The similar arguments can be applied to (8). □

Bibliography

[1] L.-E. Andersson. On the determination of a function from spherical averages. *SIAM J. Math. Anal.*, 19(N1):214–232, 1988.

[2] G. Beylkin. The inversion problem and applications of the generalized Radon transform. *Commun. Pure Appl. Math.*, 37:579–599, 1984.

[3] A. M. Cormack. Representation of a function by its line integrals, with some radiological applications I. *J. Appl. Phys.* 34:2722–2727; II *J. Appl. Phys.* 35:2908–2912, 1963–1964.

[4] A. M. Cormack. Computed tomography: some history and recent developments. In: *Computed tomography*. Proc. Sympos. Appl. Math., volume 27, pages 35–42. AMS, Providence, 1982.

[5] A. Denisjuk. Inversion of the x-ray transform for 3D symmetric tensor fields with sources on a curve. *Inverse Probl.*, 22:399–411, 2006.

[6] A. Denisjuk. Inversion of the x-ray transform for complexes of lines in R^n. *Inverse Probl.*, 32:025007, 2016.

[7] P. Elbau and O. Scherzer. Modelling the effect of focusing detectors in photoacoustic sectional imaging. *SIAM J. Imaging Sci.*, 8:1–18, 2015.

[8] J. A. Fawcett. Inversion of N-dimensional spherical averages. *SIAM J. Appl. Math.*, 45(N2):336–341, 1985.

[9] B. Frigyik, P. Stefanov, and G. Uhlmann. The X-ray transform for a generic family of curves and weights. *J. Geom. Anal.*, 18:89–108, 2008.

[10] P. Funk. Über Flächen mit lauter geschlossenen geodätischen Linien. *Math. Ann.*, 74(2):278–300, 1913.

[11] F. O. Goncharov. An iterative inversion of weighted Radon tranforms along hyperplanes. *Inverse Probl.*, 33:124005, 2017.

[12] R. Gouia-Zarrad. Analytical reconstruction formula for n-dimensional conical Radon transform. *Comput. Math. Appl.*, 68:1016–1023, 2014.

[13] R. Gouia-Zarrad and G. Ambartsoumian. Exact inversion of the conical Radon transform with a fixed opening angle. *Inverse Probl.*, 30:045007, 2014.

[14] M. Haltmeier. Exact reconstruction formulas for a Radon transform over cones. *Inverse Probl.*, 30:035001, 2014.

[15] M. Haltmeier and S. Moon. The spherical Radon transform with centers on cylindrical surfaces. *J. Math. Anal. Appl.*, 448:567–579, 2017.

[16] M. Haltmeier and S. Pereverzyev Jr.. Recovering a function from circular means or wave data on the boundary of parabolic domains. *SIAM J. Imaging Sci.*, 8:592–610, 2015.

[17] M. Haltmeier and S. Pereverzyev Jr.. The universal back-projection formula for spherical means and the wave equation on certain quadric hypersurfaces. *J. Math. Anal. Appl.*, 429:366–382, 2015.

[18] S. Helgason. *Integral geometry and Radon transforms*. Springer-Verlag, Berlin, 2011.

[19] A. Homan and H. Zhou. Injectivity and stability for a generic class of generalized Radon transform. *J. Geom. Anal.*, 27:1515–1529, 2017.

[20] J. Ilmavirta. Coherent quantum tomography. *SIAM J. Math. Anal.*, 48:3039–3064, 2016.

[21] Ch.-Y. Jung and S. Moon. Inversion formulas for cone transforms arising in application of Compton cameras. *Inverse Probl.*, 31:015006, 2015.

[22] Ch.-Y. Jung and S. Moon. Exact Inversion of the cone transform arising in an application of a Compton camera consisting of line detectors. *SIAM J. Imaging Sci.*, 9:520–536, 2016.
[23] A. Katsevich, D. Rothermel, and Th. Schuster. An improved exact inversion formula for solenoidal fields in cone beam vector tomography. *Inverse Probl.*, 33:064001, 2017.
[24] A. Katsevich and Th. Schuster. An exact inversion formula for cone beam vector tomography. *Inverse Probl.*, 29:065013, 2013.
[25] S. G. Kazantsev and Th. Schuster. Asymptotic inversion formulas in 3D vector field tomography for different geometries. *J. Inverse Ill-Posed Probl.*, 19:769–799, 2011.
[26] J. Klein. Inverting the spherical Radon transform for physically meaningful functions, 2003.
[27] P. Kuchement and F. Terzioglu. Three-dimensional image reconstruction from Compton camera data. *SIAM J. Imaging Sci.*, 9:1708–1725, 2016.
[28] L. Kunyansky. Inversion of the spherical means transform in corner-like domains by reduction to the classical Radon transform. *Inverse Probl.*, 31:095001, 2015.
[29] S. J. Lade, D. Paganin, and M. J. Morgan. 3-D Vector tomography of Doppler-transformed fields by filtered-backprojection. *Opt. Commun.*, 253:382–391, 2005.
[30] A. Louis. Exact cone beam reconstruction formulae for functions and their gradients for spherical i flat detectors. *Inverse Probl.*, 32:115005, 2016.
[31] V. Maxim. Filtered back-projection reconstruction and redundancy in Compton camera imaging. *IEEE Trans. Image Process.*, 23:332–341, 2014.
[32] V. Maxim, M. Frandes, and R. Prost. Analytic inversion of the Compton transform using the full set of available projections. *Inverse Probl.*, 25:095001, 2009.
[33] S. Moon. On the determination of a function from its conical Radon transform with a fixed central axis. *SIAM J. Math. Anal.*, 48:1833–1847, 2016.
[34] S. Moon. Inversion of the conical Radon transform with vertices on a surface of revolution arising in an application of a Compton camera. *Inverse Probl.*, 33:065002, 2017.
[35] S. Moon and M. Haltmeier. Analytic inversion of a conical Radon transform arising in application of Compton camera on the cylinder. *SIAM J. Imaging Sci.*, 535–557, 2017.
[36] F.Natterer. *The mathematics of computerized tomography*. Teubner, Stuttgart, 1986.
[37] M. K. Nguyen and T. T.Truong. On an integral transform and its inverse in nuclear imaging. *Inverse Probl.*, 18:265–277, 2002.
[38] M. K. Nguyen, T. T. Truong, and P. Grangeat. Radon transforms on a class of cones with fixed axis direction. *J. Phys. A, Math. Gen.*, 38:8003–8015, 2005.
[39] R. G. Novikov. Weighted Radon transform for which Chang's approximate inversion formula is exact. *Russ. Math. Surv.*, 66:442–443, 2011.
[40] R. G. Novikov Weighted Radon transform and first order differential systems on the plane. *Mosc. Math. J.*, 14:807–823, 2014.
[41] V. Palamodov. Reconstruction of a differential form from Doppler transform. *SIAM J. Math. Anal.*, 41:1713–1720, 2009.
[42] V. Palamodov. Remarks on the general Funk transform and thermoacoustic tomography. *Inverse Probl. Imaging*, 4:693–702, 2010.
[43] V. Palamodov. On reconstruction of strain fields from tomographic data. *Inverse Probl.*, 31:085002, 2015.
[44] V. Palamodov. A parametrix method in integral geometry. *J. Anal. Math.*, 125:353–370, 2015.
[45] V. Palamodov. *Reconstruction from data of integrals*. CRC, 2016.
[46] V. Palamodov. Reconstruction from cone integral transforms. *Inverse Probl.*, 33:104001, 2017.
[47] V. Palamodov. Exact inversion of Funk-Radon transforms with non-algebraic geometries, arxiv math FA 1711.10392, 2017.
[48] G. P. Paternain, M. Salo, and G.Uhlmann. Tensor tomography on surfaces. *Invent. Math.*, 193:229–247, 2013.

[49] M. Quellmalz. A generalization of the Funk-transform. *Inverse Probl.*, 33:036016, 2017.
[50] E. T. Quinto. Radon transforms satisfying the Bolker condition. In: *75 years of Radon transform*, pages 263–270. Publication Press, Boston, 1994.
[51] J. Radon. Über die Bestimmung von Funktionen durch ihre Integralwerte längs gewisser Mannigfaltigkeiten. *Berichte Sächsische Akademie der Wissenschaften, Math-Phys Kl.*, 69:262–267, 1917.
[52] A. Ruhlandt, M. Krenkel, M. Bartels, and T. Salditt. Three-dimensional phase retrieval in propagation-based phase-contrast imaging. *Phys. Rev. A*, 89:033847, 2014.
[53] Y. Salaman Recovering functions defined on the unit sphere by integration on a special family of sub-spheres. *Anal. Math. Phys.*, 7:165–185, 2017.
[54] Y. Salman. An inversion formula for the pherical transform in S^2 for a family of circles of integration. *Anal. Math. Phys.*, 6:43–58, 2016.
[55] D. Schiefeneder and M. Haltmeier. The Radon transform over cones with veriticies on the sphere and orthogonal axes. *SIAM J. Appl. Math.*, 77(N4), 1335–1351, 2017.
[56] V. I. Semjanistyĭ. Some integral transformations and integral geometry in an elliptic space (Russian). *Tr. Semin. Vektorn. Tenzorn. Anal.*, 12:397–441, 1963.
[57] V. I. Semjanistyĭ. Homogeneous functions and some problems of integral geometry in the spaces of constant curvature. *Dokl. Akad. Nauk SSSR*, 136:288–291 (Russian); translated as *Soviet Math. Dokl.*, 2:59–62, 1961.
[58] V. Sharafutdinov. The linearized problem of magneto-photoelasticity. *Inverse Probl. Imaging* 8:247–257, 2014.
[59] B. Smith. Reconstruction methods and completeness conditions for two Compton data models. *J. Opt. Soc. Am. A*, 22:445–459, 2005.
[60] P. Stefanov and G. Uhlmann Thermoacoustic tomography with variable sound speed. *Inverse Probl.*, 25:075011, 2009.
[61] F. Terzioglu. Some inversion formulas for the cone transform. *Inverse Probl.*, 31:115010, 2015.

Gaik Ambartsoumian

7 V-line and conical Radon transforms with applications in imaging

Abstract: This chapter is dedicated to the recent developments and current knowledge on the V-line and conical Radon transforms (VLT and CRT). The primary motivation for the study of such operators has been their relation to several imaging modalities based on physics of scattered particles. We discuss many of these applications and explain why the measured data is modeled through VLT and CRT. The mathematical results presented in the chapter are primarily focused on a special class of operators, which use integration trajectories with vertices inside the image domain. The text is supplemented with an extensive list of references on both theoretical and applied aspects of the subject.

Keywords: V-line, conical, broken ray, single scattering, Compton, Radon transform

MSC 2010: 44-02, 44A12, 92C55

7.1 Introduction

A generalized Radon transform maps a function $f(x)$ on $\mathbb{R}^n$ to a set of its integrals along a family $\mathcal{M}$ of curves or surfaces in $\mathbb{R}^n$. In other words, for each curve/surface $S \in \mathcal{M}$, $\mathcal{R}f : \mathcal{M} \to \mathbb{R}$ is defined by

$$\mathcal{R}f(S) = \int_S f(x)\,dx, \tag{1}$$

where dx is the standard measure on S.

There exists an expansive and well-developed theory studying various properties of such transforms in the case, when $\mathcal{M}$ consists of smooth curves and surfaces [26, 31, 41, 54, 55, 58, 59, 64, 82].

During the last decade or so, an interesting new class of generalized Radon transforms moved to the forefront of research in integral geometry. A common feature of all of these transforms is the presence of a vertex (or a "corner") in their paths of in-

Acknowledgement: This work was partially supported by NSF grant DMS 1616564, for which the author thanks the US National Science Foundation. A part of this chapter was written during the author's visit to the American University of Armenia (AUA). He thanks the administration and staff of AUA for their support and stimulating environment.

Gaik Ambartsoumian, University of Texas at Arlington, Arlington, USA, e-mail: gambarts@uta.edu

https://doi.org/10.1515/9783110560855-007

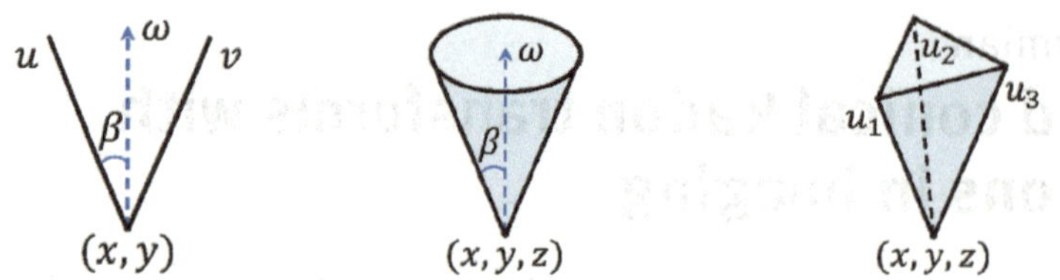

Figure 7.1: Samples of integration trajectories/surfaces with a vertex.

tegration S. For example, the so-called V-line transform (VLT) maps a function in $\mathbb{R}^2$ to its integrals along a family of "broken lines," which consist of two rays emanating from the same vertex (see Figure 7.1(a)). The name of the operator comes from the similarity between its integration trajectories and the letter "V." There is also a closely related broken ray transform (BRT), which maps a function to its integrals along "broken rays," i. e., one of the branches of "V" has a finite length. If these two transforms are used on functions with compact support, then VLT data essentially coincides with a subset of BRT data. Namely, VLT contains the same information as the restriction of BRT to the set of broken rays starting outside (or on the boundary) of the convex hull of the support. Conical Radon transforms (CRT) map a function in $\mathbb{R}^n$ to its integrals along various types of conical surfaces (e. g., see Figures 7.1(b) and 7.1(c)).

The primary motivation for the study of such operators has been their relation to several imaging modalities based on physics of scattered particles. However, as it often happens in mathematics, these applied problems led to a whole slew of others, many of which are of purely mathematical interest.

This chapter discusses the recent developments and current knowledge on the V-line and conical Radon transforms. We start with a fairly detailed description of imaging applications, in which the measured data is modeled by VLT or CRT. We then present the formal definitions of these transformations and list a set of important problems associated with them. In the subsequent sections, we state the significant results on the subject and outline some of the proofs. Our main emphasis here is on the transforms that use V-lines with vertices inside the image domain. The VLTs with vertices on a curve and CRTs are discussed at great length in the section on applications, and we provide a lot of references to appropriate papers there for more details on mathematical aspects. A few other related transformations (e. g., the star transform [85], "lemon" and "apple" transforms [62, 83], etc.) are briefly mentioned in the text, but not thoroughly reviewed. For more information on those operators, as well as details of all results mentioned here, we refer the reader to an extended list of references presented at the end of the chapter.

7.2 Imaging applications

7.2.1 Single scattering optical tomography

In optical tomography, a biological object under investigation is irradiated by a source of light. A portion of that light is absorbed in the body, while the rest is transmitted through or scattered inside it. Detectors placed outside the body measure the outgoing light intensity, and that information is then used to recover the spatially varying functions of light absorption μ_a and/or scattering μ_s [10, 11, 67]. These two functions provide substantial knowledge about the internal structure of the object.

If the object is relatively thick, then vast majority of light photons go through multiple scattering events (diffusive regime). In this case, the process is modeled by the diffusion equation. If the object is very thin, then most photons fly through the object unscattered (ballistic regime). Here, the measured data can be modeled using the (ordinary) Radon transform [67, 80].

In the intermediate case, when the object is of certain moderate thickness, one can assume that majority of photons scatter at most once (mesoscopic regime) (see Figure 7.2). Some pertinent biological applications include imaging of engineered tissues as well as small organisms, such as Zebrafish and Drosophila melanogaster (e. g., see [23, 25, 80, 81]). Image reconstruction from light measurements in such setups is called single scattering optical tomography (SSOT). If μ_s is known, then measured data g here corresponds to the V-line transform integrating the attenuation coefficient of light $\mu = \mu_s + \mu_a$ along broken rays, which coincide with trajectories of scattered

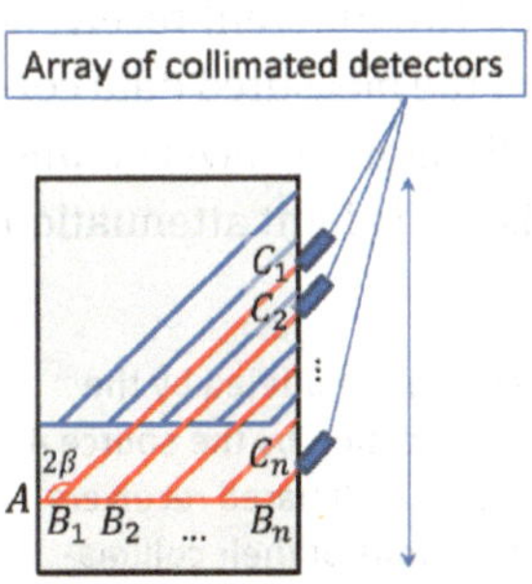

(a) Slab geometry

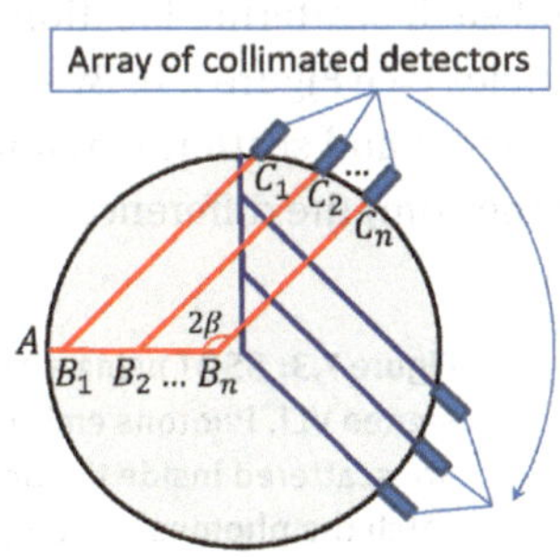

(b) Circular geometry

Figure 7.2: Broken rays with geometric restrictions. The data is measured only for broken rays that enter the domain perpendicular to the boundary and exit in a fixed direction. Here, A denotes the source of light, B_k's are scattering locations, C_k's are detector locations, and 2β is the fixed scattering angle (corresponding to the detector collimation). The array of detectors is shifted (or rotated) to collect data all along (around) the boundary.

photons [23–25, 67]. In other words,

$$g(\mathbf{p}) = \int_{V(\mathbf{p})} \mu \, dl, \tag{2}$$

where the vector of parameters **p** uniquely identifies each V-line corresponding to the available measurements. Naturally, one of the crucial mathematical tasks in SSOT is the inversion of VLT in various geometric setups.

It is easy to notice that the family of all V-lines in a plane is 4-dimensional, while a function defined in that plane depends only on 2 parameters. Similarly, the family of V-lines in $\mathbb{R}^3$ is 7-dimensional, while the image function depends only on 3 variables. Hence, the problem of recovering an image function from its full VLT data is overdetermined. Typically, in integral geometry one considers problems of inverting a generalized Radon transform from limited data, which has the same degrees of freedom as the dimensionality of the image function.

There are numerous ways of restricting the set of V-lines to a subset of appropriate dimension, e. g., constraining the locations of vertices, assigning the opening angles, fixing or limiting the axes of symmetry, etc. The specific choice is usually made based on the application at hand and mathematical considerations, e. g., the possibility and level of difficulty of inverting the transform. For example, in SSOT it is impossible to collect full data, since in order to register the direction of outgoing rays, the *detectors have to be collimated.* As a result, at each point of the boundary of the object one can capture photons coming only from one direction, reducing the dimensionality of measured data by 1 in $\mathbb{R}^2$ and by 2 in $\mathbb{R}^3$. Similar restrictions may be applied to the direction of incoming radiation, e. g., sending the rays along the normals to the boundary (see Figures 7.2(a) and 7.2(b)) [3, 5, 8, 9, 23–25, 28, 68].

Another interesting setup of SSOT uses two sets of collimated detectors so, that for each photon source and each scattering location there are two different receivers catching the scattered particles (see Figure 7.3). By subtracting the measured data corresponding to the same source A and scattering point B, but different receivers C and D, one obtains (instead of the sum) the difference of integrals of the light attenuation

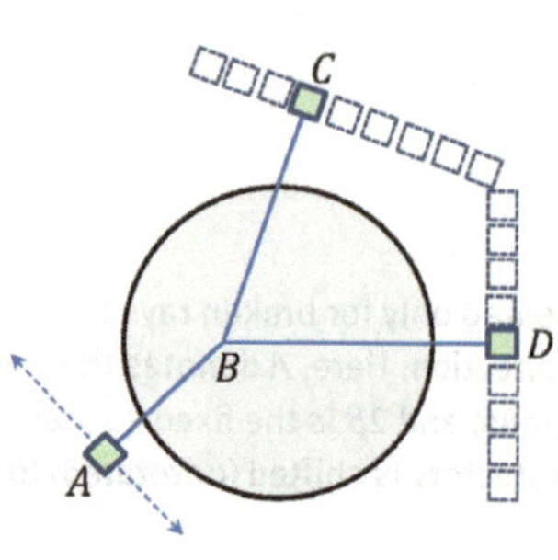

Figure 7.3: SSOT using a two-receiver setup, modeled by the signed VLT. Photons emitted in a fixed direction by the source A are scattered inside the body. Two sets of collimated receivers catch the photons scattered in the directions of their collimation. The data is collected for multiple positions of A along a line perpendicular to the initial direction of photon beams. Knowing the locations of the source A and the collimated receiver C or D, one can uniquely recover the scattering location B. Then, matching the VLT data measured by receivers C and D for the same source A and scattering location B, one can generate the signed VLT data corresponding to the V-line CBD.

coefficient μ along the two branches of V-line *CBD* [5, 23, 24, 68]. This leads to the necessity to study such *signed V-line transforms* (SVLT) and their properties.

It must be mentioned, that in the setup with two sets of receivers, one does not need to know μ_s ahead of time, and both μ_a and μ_s can be reconstructed simultaneously [23, 24]. Also notice, that although here one collects twice more data than in the previous setup, both data sets have the same dimensionality.

7.2.2 Single Scattering X-ray Tomography (SSXT)

A more general, *weighted V-line transform* (WVLT) is used in certain models of X-ray computerized tomography (CT). In a standard CT model, it is assumed that X-rays propagate along straight lines, ignoring any scattering phenomenon. This leads to various types of inaccuracies and artifacts in reconstructed images (e. g., see [47] and the references there). One way to enhance the model and improve the image quality is accounting also for single scattered X-rays (e. g., see [2, 17, 39, 40, 45, 57, 84]). While there are several physically different scattering phenomena for X-rays interacting with matter, at high energy levels the dominant one is the Compton scattering [18].

In Compton scattering, the X-ray photons, scattered by charged particles of the matter, lose energy and change their flight direction by an angle that corresponds to the amount of lost energy. In other words, the path of a single scattered X-ray corresponds to a V-line with a vertex at the scattering location. The opening angle of that V-line depends on the difference of X-ray energy levels before and after scattering. Similar to SSOT, if μ_s is known, the single-scattered X-ray data measured by *collimated detectors* then corresponds to integrals of X-ray attenuation coefficient μ along appropriate V-lines [39, 40, 85].

If μ depends only on the spatial variable, then one obtains exactly the same mathematical model as for the case of SSOT discussed above [39]. However, in reality μ depends also on the energy level of the X-ray, i. e., $\mu = \mu(x, E)$, where E changes after scattering. In the first approximation, one can assume that the dependence of μ on E is linear (at least in some small interval of appropriate energies). In other words, $\mu = c_k f(x)$, $k = 1, 2$, where c_1 and c_2 are different constants corresponding to the energy levels before and after the scattering. As a result, the measured data g corresponds to the weighted V-line transform, mapping f to the sum of its integrals with different weights along each ray of the V-line [5, 40]. In other words, if $R_1(\mathbf{p})$ and $R_2(\mathbf{p})$ are the rays of the V-line $V(\mathbf{p})$, then

$$g(\mathbf{p}) = c_1 \int\limits_{R_1(\mathbf{p})} f\,dl + c_2 \int\limits_{R_2(\mathbf{p})} f\,dl. \tag{3}$$

Notice, that using two sets of detectors (as described for SSOT) and subtracting the corresponding measurements, one may arrive to a WVLT with weights of different algebraic signs [5, 39, 40].

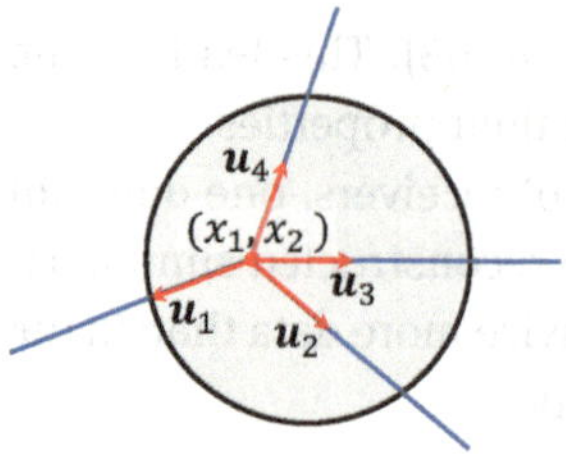

Figure 7.4: The integration set of a star transform.

An interesting generalization of VLT in $\mathbb{R}^2$, called the *star transform*, has been used in SSXT with multiple sets of collimated receivers [85]. Similar to the case with two sets of receivers, the data from the same source and the same scattering location (but different scattering angles) is captured by several receivers, each admitting rays coming only from one direction. A linear combination of VLTs corresponding to these receivers then defines the star transform $\mathfrak{S}$, which maps a function to a sum of its integrals (with different constant weights) along multiple rays emanating from the same vertex. In other words,

$$\mathfrak{S}f\,(x_1, x_2, \mathbf{u_1}, \dots, \mathbf{u_n}) = \sum_{k=1}^{n} c_k \int_0^\infty f(x + l\mathbf{u_k})\,dl, \tag{4}$$

where $\mathbf{u_k}$ is the unit vector pointing in the direction of the kth ray of the "star" (see Figure 7.4).

It has been shown in [85] that there are several advantages in image reconstruction using an inversion of the star transform. First, this method allows using classical Tikhonov regularization. Second, it is flexible with respect to the choice of ray geometry. And finally, it allows robust incorporation of data measured from ballistic photons. Also, similar to the case of VLTs with 2 or more sets of receivers (e. g., [23, 24, 39, 40]), when using the star transform one does not need to know μ_s ahead of time and can recover both μ_s and μ_a from the transform data.

Just like for the V-line transform, inversion of the star transform from full data is an overdetermined problem. One can assign additional constraints on the geometry of the family of "stars" to reduce its dimensionality to 2. For example, in [85] the directions of star rays are fixed, while the location of the vertex is arbitrary.

7.2.3 Compton Camera Imaging (CCI)

In the previous two imaging modalities described above, the data was measured by collimated receivers. The benefit of that, of course, is the knowledge of the direction from which the photons arrive at the detector. But on the flip side, collimation results in a huge loss of data, since particles arriving to the detector from all but one direction

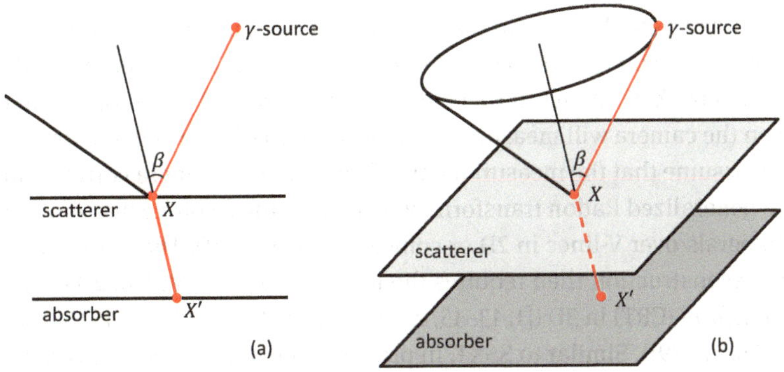

Figure 7.5: Sketches of two Compton cameras. In (a) the detectors consist of 1-dimensional arrays of digital sensors. In (b), the detectors consist of 2-dimensional digital sensor plates.

are discarded. An alternative to collimated receivers, that does not have this shortcoming, is the Compton camera, sometimes (especially in early literature) also called an electronically collimated camera.

The simplest forms of such imaging devices have been proposed as early as in the1970s [22, 76] and 1980s [69, 70]. With the development and advancement of high sensitivity digital detectors, Compton cameras rapidly became one of the primary tools for detection, identification and imaging of sources of γ-radiation. They have a wide range of applications including astrophysics (see [12] and the references there), nuclear medicine (primarily in SPECT) [13–15, 21, 30, 48] and homeland security (nuclear threat detection) [1, 46]. Notice, that in all these modalities the γ-rays are generated by the object of interest itself, and not some external source used to probe the imaging object.

A typical Compton camera consists of two parallel digital sensors called a scatterer and an absorber (see Figure 7.5). When a photon in a γ-ray hits the scatterer at a point X, it changes its flight trajectory and hits the absorber at another point X'. The sensors of the Compton camera register these locations X and X', as well as the energy of the particle at each point. The well-known Compton scattering relation then allows to recover the scattering angle β:

$$\cos\beta = 1 - \frac{mc^2 \triangle E}{(E - \triangle E)E},$$

where E is the original energy of the photon, $\triangle E$ is the energy lost due to scattering, and m is the electron rest mass.

It is easy to notice, that as opposed to the case of collimated receivers, Compton cameras do not discard any data, since all scattering events are registered by the sensors. The trade-off is the loss of directional information. Namely, in the case of 1-dimensional camera (imaging a 2D function), the knowledge of scattering angle β only implies that the γ-source is somewhere on the V-line with vertex at X, opening angle 2β and axis of symmetry $X'X$. In the case of a 2D camera (imaging a 3D function), the γ-source is somewhere on the surface of the cone with vertex at X, opening angle 2β and axis of symmetry $X'X$. In fact, if there are several sources of radiation on that V-line (cone), then the camera will measure the superposition of their signals. Hence, it is reasonable to assume that the measurements of the Compton camera correspond to the values of a generalized Radon transform, which maps the γ-source distribution function to its integrals over V-lines in 2D or conical surfaces in 3D. The mathematical task of image reconstruction then requires the inversion of such VLT in 2D or the *conical Radon transform* (CRT) in 3D ([1, 13–15, 21, 29, 30, 32, 37, 38, 42, 43, 48–53, 56, 61, 66, 71, 72, 74, 75, 77–79]). Similar to SSXT, in practice one may have to use various weighted versions of VLTs and CRTs, e. g., to account for energy loss of the particles, probability distribution of scattering angles, etc. [30, 43, 74, 75, 77].

A few clarifying remarks are in order here.

- There is an important distinction between generalized Radon transforms appearing in SSOT/SSXT and CCI. The former integrate functions along V-lines with vertices *inside the support* of the image function. The latter uses V-lines (or cones) with vertices restricted to a curve (or a surface) corresponding to the scattering detector, with the image function *supported away* and on one side of that curve (surface).
- It is easy to notice that even with such a restriction, the problem of inverting the CRT is overdetermined. The family of all cones with vertices located on a surface is 5-dimensional, while the image function depends only on 3 variables. There are multiple options of restricting the 5-dimensional set of cones to a 3-dimensional family, e. g., by using various combinations of fixing the direction of their axes of symmetry, assigning their opening angle, limiting the vertices to a curve, etc. [21, 38, 56, 79]. The same can be said about the 3-dimensional family of all V-lines with vertex on a curve applied to image functions that depend on 2 variables [13, 77, 78].
- Other geometric shapes of Compton cameras have been proposed recently, e. g., concentric circles in 2D [30], concentric spheres in 3D [66], cylinders with a common symmetry axis in 3D [51, 52] and some others (e. g., see [72]).
- As it was mentioned before, it is customary in integral geometry to consider problems of inverting an integral transform from data sets, that have the same dimensionality as the functions to which that transform is applied. However, in certain applications inversion of transforms from overdetermined data is of utmost importance. Typically, this is due to the scarcity of available measurements, critically low signal to noise ratio, and other similar factors. As a result, the only hope for any meaningful inversion process is based on the appropriate use of extra data. Compton camera imaging is used in several such application, e. g., in nuclear medicine or search and detection of illicit nuclear materials. Multiple interesting results on inversion of VLT and CRT in overdetermined setups have been obtained

using different techniques (see [1, 15, 32, 37, 42, 43, 48, 49, 71, 74] and the review paper [75]).

7.2.4 Compton Scattering Imaging (CSI)

SSOT and SSXT (which use collimated emitters and receivers) are *transmission* tomographic techniques, i. e., both the source and the receiver of radiation are outside the object of investigation. In contrast to that, CCI (with no collimation) described above is an *emission* modality, where the source of radiation is the image function, and hence, by default, it is inside the object of investigation. Compton scattering imaging is a hybrid of the previous two techniques. Namely, it is a transmission modality that uses uncollimated emitters (e. g., a fan beam source) and uncollimated receivers (digital sensors).

As one may expect, due to the absence of directional information and the presence of scattering points inside the image domain, the mathematical models in this modality are quite different from the previously considered ones. The measured data corresponds to weighted integrals of the image function along circular arcs (see [57, 63] and the references there) in 2D, and over toric surfaces in 3D [62, 83]. Depending on the data acquisition geometry, the families of circular arcs and toric surfaces may vary and can include quite exotic sets. For example, in 3D one may have to deal with generalized Radon transforms mapping the image function to its integrals over "lemons" and "apples" (interior and exterior surfaces of spindle tori) [62, 83]. Such operators appear also in mathematical models of positron emission tomography (PET), which account for scattered coincidences [16, 73]. We will not discuss these transforms in this chapter, and instead refer the reader for additional information and details to the papers mentioned above.

7.3 VLT and CRT with vertices inside the image domain

In this section, we discuss the known results about the V-line and conical Radon transforms appearing in transmission tomographic modalities (SSOT and SSXT). The distinguishing feature of these transforms is the fact that they use integration trajectories with vertices inside the image domain. Following the arguments presented in the previous section, we will consider geometric setups with matching dimensions of the data and image function. As with any other generalized Radon transform, the primary questions of interest include inversion formulas and procedures, their numerical implementation, support theorems, range description and microlocal analysis of the operators. We address many of these questions here, and refer the readers to the appropriate literature for the ones that we do not cover.

7.3.1 VLT with fixed axis of symmetry and opening angle

The first inversion formulas for VLT with vertices inside the support of image function were presented in [24]. Here, the authors considered two different sets of V-lines in 2D rectangular slab geometry with a fixed opening angle, fixed axis of symmetry and arbitrary locations of the vertex (see Figure 7.6).

For the first set of V-lines (see Figure 7.6(a)), an inversion formula was derived for the ordinary (unweighted) VLT. Namely, if $f(y,z)$ is the image function supported inside the slab, and its V-line transform $g(y_1,h) \equiv Tf(y_1,h)$ is defined by

$$g(y_1,h) = \int\limits_{L(y_1,h)} f[y(l),z(l)]dl, \tag{5}$$

then

$$\begin{aligned} f(y,z) = \lambda\Bigg\{\left[\frac{\partial}{\partial h} - (1+\kappa)\frac{\partial}{\partial y}\right]g(y,h) + \kappa\frac{\partial}{\partial y}g(y+\lambda z, h_{\max}) \\ - \kappa(1+\kappa)\frac{\partial^2}{\partial y^2}\int\limits_h^{h_{\max}} g(y+\kappa(l-h),l)dl\Bigg\}\Bigg|_{h=h_z}, \end{aligned} \tag{6}$$

where $\lambda = \cot(\theta/2)$, $\kappa = \lambda\cot\theta$, $Z_{\max}$ is the width of the slab, $h_{\max} = Z_{\max}\tan\theta$, and $h_z = (Z_{\max} - z)\tan\theta$.

For the second set of V-lines (see Figure 7.6(b)), an inversion formula was derived for the signed VLT, where the integral of f along the bottom ray $R_2(y,h)$ is subtracted from the integral of f along the upper ray $R_1(y,h)$, and $R_1(y,h) \cup R_2(y,h) = L(y,h)$.

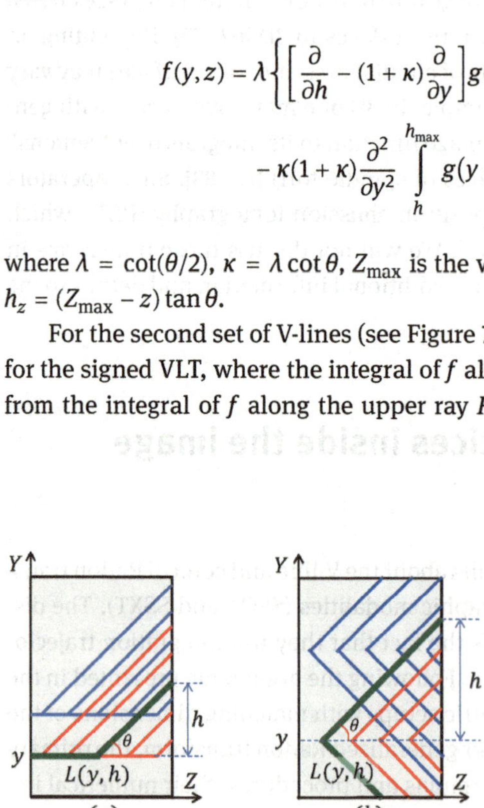

Figure 7.6: V-lines with fixed axis of symmetry and opening angle in a rectangular slab. Here $L(y,h)$ denotes the V-line uniquely identified by parameters y and h. In (a), the V-line enters the slab at the point $(0,y)$ orthogonally to the Y-axis, travels certain distance, breaks at a fixed angle $\pi - \theta$, and exits the slab at the point $(Z_{\max}, y+h)$. In (b), the V-line enters the slab at the point $(Z_{\max}, y+h)$ along the line with slope $\tan\theta$, travels certain distance, breaks at a fixed angle 2θ, and exits the slab at the point $(Z_{\max}, y-h)$.

Namely, if

$$g_s(y_1,h) = \int\limits_{R_1(y_1,h)} f[y(l),z(l)]\,dl - \int\limits_{R_2(y_1,h)} f[y(l),z(l)]\,dl, \tag{7}$$

then

$$f(y,z) = \frac{\sin\theta}{4}\left[\frac{\partial^2}{\partial h^2}\int\limits_{-\infty}^{\infty}\operatorname{sgn}(y-w)\,g_s(w,h)\,dw - 2\frac{\partial}{\partial y}g_s(y,h)\right]\Bigg|_{h=h_z}. \tag{8}$$

Both results were obtained by fairly technical computations using the Fourier transform. It is easy to notice, that formula (6) for VLT is more complicated than formula (8) for SVLT. That is not due to some sort of inherent advantage of SVLT over VLT, but simply a result of data parametrization chosen in [24].

Simpler inversion formulae for the same setup were obtained later in [39] (using PDE techniques), [28] (by Fourier analysis) and [68] (by direct verification). Another work, [5], outlined a general approach for inversion of VLT with various weights, and used it to obtain a simple inversion formula equivalent to the ones mentioned above. Below we present some of the relevant ideas and results of [5].

We start with an introduction of necessary notations and formal definitions. Let $\omega = (\omega_x, \omega_y)$ denote the fixed direction of axes of symmetry of all V-lines, $2\beta \in (0,\pi)$ be their opening angle and unit vectors u, v denote the directions of the rays of the V-lines (see Figure 7.7). For $(x,y) \in \mathbb{R}^2$, we denote by $R_u(x,y) = \{(x,y) + tu : t \geq 0\}$ the ray emanating from (x,y) in the direction of u. Then the unique V-line with a vertex at (x,y) can be represented by the union $L(x,y) = R_u(x,y) \cup R_v(x,y)$.

In the definitions below, we assume that $f \in L^1(\mathbb{R}^2)$ and the transformations are defined for almost every $(x,y) \in \mathbb{R}^2$. With some additional regularity assumption on f (e. g., continuity) the transformations will be defined at every $(x,y) \in \mathbb{R}^2$.

Definition 1. The weighted V-line transform (WVLT) T_w of f is defined by

$$(T_w f)(x,y) = c_v \int\limits_{R_v(x,y)} f\,dl + c_u \int\limits_{R_u(x,y)} f\,dl, \tag{9}$$

where $c_u \neq 0$ and $c_v > 0$ are some constants, and dl is the standard Lebesgue measure on the line.

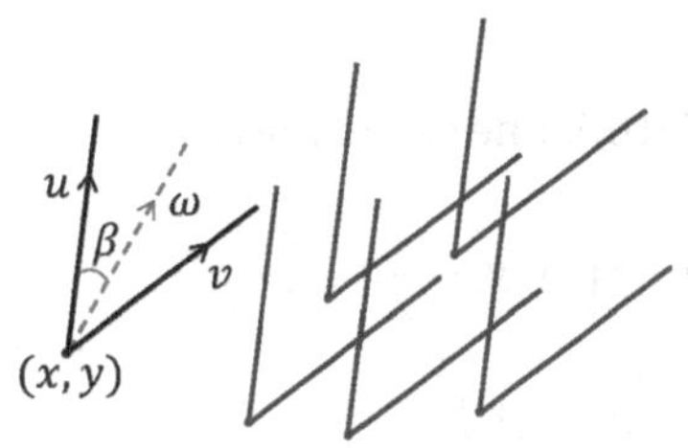

Figure 7.7: V-lines with axes of symmetry along a fixed vector ω and a fixed opening angle 2β.

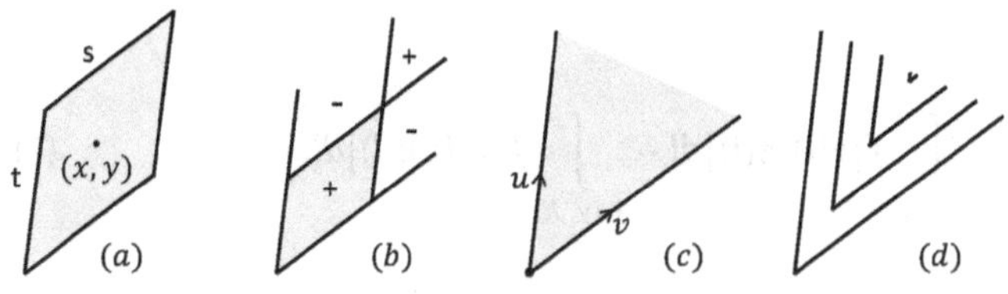

Figure 7.8: A sketch of the main idea behind the inversion of VLTs.

We introduce additional notation for two special cases, $c_u = c_v = 1$ and $c_v = -c_u = 1$.

Definition 2. The (ordinary) V-line transform (VLT) T of f is defined by

$$(Tf)(x,y) = \int_{L(x,y)} f\,dl. \tag{10}$$

Definition 3. The signed V-line transform (SVLT) T_s of f is defined by

$$(T_s f)(x,y) = \int_{R_v(x,y)} f\,dl - \int_{R_u(x,y)} f\,dl. \tag{11}$$

Let $A_{t,s}(x,y)$ be the average of f over the parallelogram P centered at (x,y), with sides of length t, s and directions u, v (see Figure 7.8(a)).

Since the area of the parallelogram made with vectors tu, sv is equal to $|\det(tu,sv)| = ts|\det(u,v)|$, we have

$$A_{s,t}(x,y) = \frac{1}{ts|\det(u,v)|}\int_P f\,d\mu.$$

The main idea behind inversion of VLTs in [5] can be described as follows. *Use the VLT data to get $A_{s,t}(x,y)$ for each point (x,y) and $s,t > 0$. Then take the limit of those averages (when the measure of P goes to 0) to recover $f(x,y)$. The second step is possible due to the well-known Lebesgue differentiation theorem (e. g., see* [65]*).*

Let us now outline how $A_{s,t}(x,y)$ can be obtained from the VLT data.

Denote by $C(x,y)$ the region of the plane bounded by the V-line $L(x,y)$ (see Figure 7.8(c)). Let $F(x,y)$ be the integral of the image function f over the cone $C(x,y)$, i. e.,

$$F(x,y) = \int_{C(x,y)} f d\mu, \tag{12}$$

where μ is the standard Lebesgue measure on $\mathbb{R}^2$.

Using a simple geometric argument (see Figure 7.8(b)), one can express

$$\begin{aligned} A_{t,s}(x,y) = \frac{1}{ts|\det(u,v)|}\Big[&F\Big((x,y)+\frac{t}{2}u+\frac{s}{2}v\Big) - F\Big((x,y)-\frac{t}{2}u+\frac{s}{2}v\Big) \\ &- F\Big((x,y)+\frac{t}{2}u-\frac{s}{2}v\Big) + F\Big((x,y)-\frac{t}{2}u-\frac{s}{2}v\Big)\Big]. \end{aligned} \tag{13}$$

Finally, $F(x,y)$ can be obtained from VLT data by "stacking up" (integrating) the integrals of f along appropriate V-lines to get the integral of f over $C(x,y)$ (see Figure 7.8(d)).

The rigorous formulations of these statements are presented below. For proofs and technical details, we refer the reader to article [5].

Lemma 1. *Consider Tf and F defined by equations* (10) *and* (12). *Then*

$$F(x,y) = \sin\beta \int_0^{\infty} (Tf)(x + t\omega_x, y + t\omega_y)\, dt. \tag{14}$$

Combining the values of F to get $A_{t,s}$ as described above, and then taking an appropriate limit one obtains the following.

Theorem 1 (Inversion of VLT [5]). *Let $f \in C_c(\mathbb{R}^2)$. Then*

$$f(x,y) = \frac{1}{2\cos\beta}\frac{\partial}{\partial u}\frac{\partial}{\partial v}\int_0^{\infty} (Tf)(x + t\omega_x, y + t\omega_y)\, dt, \tag{15}$$

where $\frac{\partial}{\partial u}$ and $\frac{\partial}{\partial v}$ are the directional derivatives in the direction of u and v.

If it is only known that $f \in L^1(\mathbb{R}^2)$, then we have the following inversion formula for a.e. (x,y):

$$f(x,y) = \lim_{t\to 0} A_{t,t}(x,y), \tag{16}$$

where $A_{t,t}(x,y)$ is defined by formula (13).

In the case of WVLT, one has to stack up the data differently to generate $F(x)$. In other words, $T_w f$ has to be integrated in a direction $\widetilde{\omega}$, which is different from ω, unless $c_u = c_v$. Namely, define

$$\widetilde{\omega} = \frac{c_u v + c_v u}{\|c_u v + c_v u\|}. \tag{17}$$

Let us express the fixed opening angle 2β of the V-lines of integration as a sum of two directed angles: $2\beta = \beta_1 + \beta_2 < \pi$ so, that

$$\begin{aligned} \text{angle}\,(v,\widetilde{\omega}) &= \beta_1, \\ \text{angle}\,(\widetilde{\omega},u) &= \beta_2. \end{aligned} \tag{18}$$

It is easy to show that

$$\frac{\sin\beta_1}{\sin\beta_2} = \frac{c_v}{c_u}. \tag{19}$$

Lemma 2. *Consider $T_w f$ and F defined by equations (9) and (12).*

If β_1, β_2, and $\widetilde{\omega}$ satisfy equations (17)–(19), then

$$F(x,y) = \frac{\sin\beta_1}{c_v}\int_0^\infty (T_w f)(x+\widetilde{\omega}_x t, y+\widetilde{\omega}_y t)\,dt. \tag{20}$$

Theorem 2 (Inversion of WVLT [5]). *Let $f \in C_c(\mathbb{R}^2)$. Then*

$$f(x,y) = \frac{1}{\|c_u v + c_v u\|}\frac{\partial}{\partial u}\frac{\partial}{\partial v}\int_0^\infty (T_w f)(x+\widetilde{\omega}_x t, y+\widetilde{\omega}_y t)\,dt. \tag{21}$$

The coefficient in the above formula can be expressed as

$$\frac{1}{\|c_u v + c_v u\|} = \frac{\sin\beta_1}{c_v \sin(2\beta)}. \tag{22}$$

If it is only known that $f \in L^1(\mathbb{R}^2)$, then we have the following inversion formula for a.e. (x,y):

$$f(x,y) = \lim_{t\to 0} A_{t,t}(x,y), \tag{23}$$

where $A_{t,t}(x,y)$ is defined by formula (13).

Notice, that if $c_u > 0$, then $\widetilde{\omega}$ points inside the cone $C(x,y)$, while for $c_u < 0$ the vector $\widetilde{\omega}$ points outside $C(x,y)$.

In the special case of SVLT (i. e., $c_v = -c_u = 1$), we have $\widetilde{\omega} = \frac{u-v}{\|u-v\|} = \omega^\perp$, which is the unit vector perpendicular to the cone direction ω.

Remark 1. If the image function f is supported inside a compact set K and $(x,y) \in K$, then $(T_w f)(x+\widetilde{\omega}_x t, y+\widetilde{\omega}_y t) = 0$ for large values of t. In other words, the formulas (14), (15), (20) and (21) will require integration only along finite intervals.

Moreover, if $f \in C_c(\mathbb{R}^2)$ and its values are known on the boundary of some bounded, open, convex set Ω, then one can recover f inside Ω using just its WVLT data restricted to the V-lines with vertices inside Ω. Namely, we have the following.

Theorem 3 ([5]). *Consider a bounded, open, convex set Ω in $\mathbb{R}^2$ and let $g = T_w f$ be the weighted V-line transform of $f \in C_c(\mathbb{R}^2)$. For each point $p = (x,y) \in \Omega$, let*

$$\begin{aligned} m_p &= \inf\{s \ge 0 : p + s\widetilde{\omega} \in \partial\Omega\},\\ m_n &= m_p - 1/n,\\ (x_n, y_n) &= (x,y) + m_n\widetilde{\omega},\\ (x_0, y_0) &= (x,y) + m_p\widetilde{\omega}. \end{aligned}$$

Then $(x_n, y_n) \to (x_0, y_0)$ *and*

$$f(x,y) = f(x_0,y_0) + \lim_{n\to\infty} \frac{\sin\beta_1}{c_v \sin(2\beta)} \frac{\partial}{\partial u}\frac{\partial}{\partial v} \int_0^{m_n} g(p + s\widetilde{\omega})\, ds. \tag{24}$$

In the case of VLT and SVLT, similar results were obtained for $f \in C_c^2(\mathbb{R}^2)$ in [39], by solving a first-order PDE using the method of characteristics. It was also shown there, that if Ω is convex, then one does not need to require compact support for f. Instead of using $f = 0$ on the boundary, in this case they use internal data to recover f along a diagonal of Ω and then apply the method of characteristics (see [39] for more details).

The inversion results presented above lead to various support theorems for WVLTs. For example, we have the following.

Theorem 4 (Support theorem [5]). *Let* $f \in L^1(\mathbb{R}^2)$ *be continuous and* $T_w f = 0$ *on some* $S \subset \mathbb{R}^2$. *Let* L *be a line parallel to the vector* $\widetilde{\omega}$ *defined in* (17) *such that* $L \cap S \neq \varnothing$. *Then* f *is constant on each connected component of* $L \cap S$.

In particular, if S *is a compact set,* $T_w f = 0$ *in* S, *and* $f = 0$ *on the boundary* ∂S *of* S, *then* $f \equiv 0$ *in* S.

Another interesting property of WVLTs studied in [5] is the description of the range of those operators, when applied to the class of nonnegative integrable functions on the plane. It was shown that a function g is in the range, if and only if its conical integrals $F(x,y)$ satisfy certain conditions on smoothness, growth and asymptotic behavior (see [5] for more details). To the best of our knowledge, this is the only known range description for the WVLTs in the geometric setup considered above.

The results of various numerical implementations of the discussed inversion formulas have been presented in [5, 24, 28, 39, 68]. In Figure 7.9, we present a couple of numerical reconstructions based on equation (16). It must be noted that the numerical implementation of equation (15) leads essentially to the same algorithm (think of the discretized version of the directional derivatives).

All numerical implementations show excellent quality of image reconstruction, despite the fact that the available VLT data corresponds to integrals of f in essentially

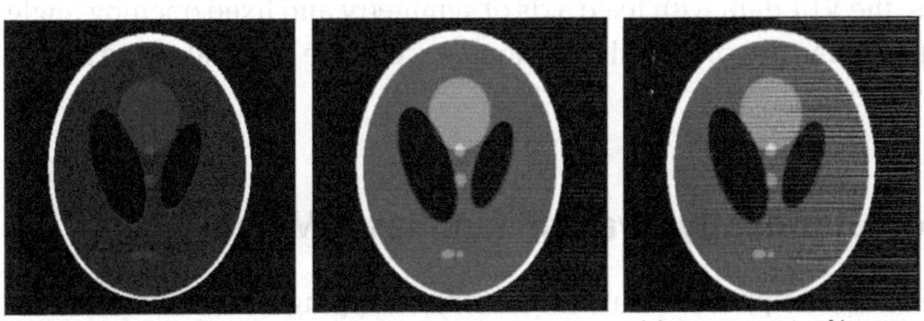

(a) Original phantom (b) Recon. with no noise (c) Recon. with 5% noise

Figure 7.9: Numerical reconstructions courtesy of Mohammad J. Latifi Jebelli. The VLT uses V-lines with branches in the direction of vectors $(-2, 1)$ and $(-2, -1)$.

two fixed directions (if one "forgets" the presence of the vertices of V-lines). In the case of the ordinary Radon transform, any attempt of recovering the function from its integrals along lines with two fixed directions will fail. This is due to the fact that a generalized Radon transform, mapping a function f to its integrals along a family of *smooth* curves, preserves only certain information about the wavefront set of f. Namely, if f has a singularity at point x in the direction ξ, then it can be recovered stably from the Radon data only if the latter includes integration curves passing through x and normal to ξ. (Here, and through the rest of this subsection we assume that x is a point in the plane, and ξ is a vector in the plane.) Hence, in the case of only two available directions, the set of "visible" singularities will have measure zero.

However, the presence of a vertex in the path of integration completely changes the microlocal analysis of the transform. In [68], the VLT was defined and studied in the distributional sense. In particular, a thorough study was done on the propagation of singularities from f to Tf, and from Tf to f (through the inversion formulas) for various geometric setups of V-line transform T. In particular, for the setup discussed in this subsection the following theorem was proved.

Theorem 5 (Propagation of singularities [68]). *Let $f \in \mathcal{E}'(\mathbb{R}^2)$. Then*

$$\begin{aligned}\mathrm{WF}(Tf) \subseteq \mathrm{WF}(f) &\cup \{(x - tu, \xi) \mid (x, \xi) \in \mathrm{WF}(f),\ \xi \in u^{\perp},\ t > 0\} \\ &\cup \{(x - tv, \xi) \mid (x, \xi) \in \mathrm{WF}(f),\ \xi \in v^{\perp},\ t > 0\},\end{aligned} \tag{25}$$

and

$$\begin{aligned}\mathrm{WF}(f) \subseteq\ & \mathrm{WF}(Tf) \\ &\cup \{(x + t(u + v), \xi) \mid (x, \xi) \in \mathrm{WF}(Tf),\ \xi \in (u + v)^{\perp},\ t > 0\},\end{aligned} \tag{26}$$

$$\begin{aligned}\mathrm{WF}(f) \subseteq\ & \mathrm{WF}(Tf) \\ &\cup \{(x - t(u + v), \xi) \mid (x, \xi) \in \mathrm{WF}(Tf),\ \xi \in (u + v)^{\perp},\ t > 0\}.\end{aligned} \tag{27}$$

Moreover, if $(x_0, \xi_0) \in \mathrm{WF}(f) \setminus \mathrm{WF}(Tf)$, then $\xi_0 \in (u + v)^{\perp}$, and x_0 must lie on some line segment $x_0 + Iw$, for which $(x_0 + Iw) \times \{\xi_0\} \subseteq \mathrm{WF}(f)$, and whose endpoints lie in $\mathrm{WF}(Tf)$.

In other words, the VLT data with fixed axis of symmetry and fixed opening angle allows recovery of singularities of f in all directions, except one (normal to the axis of symmetry of the V-lines).

7.3.2 VLTs over rotationally invariant families of V-lines

The VLTs considered in the previous subsection were mapping a function to its integrals over *shift-invariant* families of V-lines. As a result, several inversion formulas were obtained by utilizing Fourier transform techniques. In this subsection, we consider transforms using *rotation-invariant* families of V-lines. The conventional wisdom

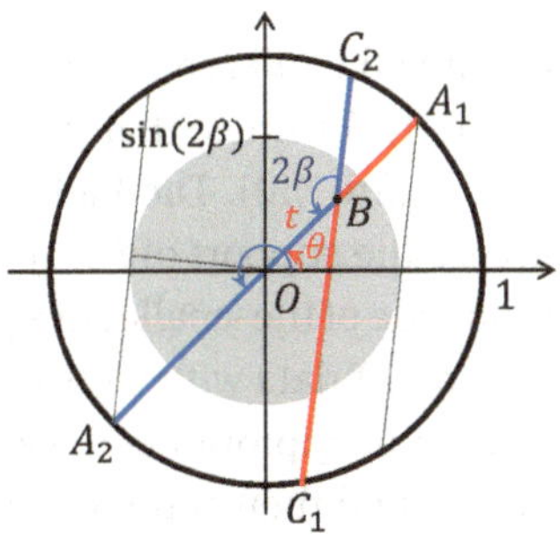

Figure 7.10: Parametrization of V-lines in the unit disc. The two parallel lines tangent to the small circle are lines farthest from the origin in the Radon data set.

in integral geometry suggests that these transforms should be invertible using Fourier series expansions (e. g., see [4, 6, 7, 19, 20]).

We start with a circular setup that was motivated by the experimental ideas of [23, 25] applied in the rectangular slab (see Figure 7.2(b)). Namely, we assume that the image domain is supported inside the unit disc. The V-lines, corresponding to paths of single scattered photons, enter the disc along the normals to its boundary. The receivers are collimated in a way, that all V-lines have the same opening angle 2β. The resulting family of V-lines is clearly rotation-invariant and 2-dimensional.

Let us parameterize each V-line by the signed distance $t \in [-1,1]$ of its vertex to the origin and the polar angle $\theta \in [0,2\pi]$ of the vertex (or equivalently, of the V-line's point of entry into the disc) (see Figure 7.10). For example, in Figure 7.10 the V-line A_1BC_1 has parameters (θ,t), while the V-line A_2BC_2 has parameters $(\theta+\pi,-t)$. We will denote the V-line with parameters (θ,t) by $L(\theta,t)$, and the corresponding value of VLT of f by $Tf(\theta,t)$.

If the VLT data is known for all $t \in [-1,1]$ and $\theta \in [0,2\pi]$, then for a certain restricted class of functions f, one can invert the VLT by recovering the ordinary Radon transform (RT) of the image function, and then applying a standard approach to invert RT [3]. Namely, a simple geometric observation shows that

$$Tf(\theta,t)+Tf(\theta+\pi,-t)=\mathcal{R}f(\theta-\pi/2,0)+\mathcal{R}f(\theta+\pi/2-2\beta,t\sin(2\beta)),$$

where $\mathcal{R}f$ denotes the (ordinary) Radon transform of f:

$$\mathcal{R}f(\theta,t)=\int\limits_{\omega\cdot x=t} f dl,$$

and $\omega=(\cos\theta,\sin\theta)$.

Notice, that $\mathcal{R}f(\theta-\pi/2,0)=Tf(\theta,-1)$. Then one can recover the (ordinary) Radon data from VLT data by the following formula:

$$\begin{aligned}&\mathcal{R}f(\theta,t\sin(2\beta))\\&\quad=Tf(\theta+2\beta-\pi/2,t)+Tf(\theta+2\beta+\pi/2,-t)-Tf(\theta+2\beta-\pi/2,-1).\end{aligned} \tag{28}$$

Unfortunately, there is a snag in this simple approach. Using the full range of values of both θ and t, one obtains the (ordinary) Radon data only for lines, that pass through the disc of radius $\sin(2\beta)$ (see the shaded region in Figure 7.10). The Radon transform with such (interior) data is not injective [54], unless the support of f is restricted to that smaller disc. One may argue, that if f is zero outside of the smaller disc, then (in applications) there will be no scattering, and hence no VLT data with vertices there. That can be remedied by placing the object in a medium of appropriate thickness with known properties before making the measurements, and then subtracting from measured data the portion of VLT corresponding to the layer of known material.

The data set used in the above setup includes two different V-lines with a vertex at each point of the disc. It is natural to ask if one can invert the VLT from half of that data, i. e., using $Tf(\theta,t)$ for $\theta \in [0,2\pi]$ and $t \in [0,1]$. A positive answer to that question was given in [8] using Fourier series techniques.

For brevity, let us denote $g(\theta,t) := Tf(\theta,t)$, and let $f(\phi,r)$ be the image function in polar coordinates. Then the Fourier series of $f(\phi,r)$ and $g(\theta,t)$ with respect to their angular variables can be written as follows:

$$f(\phi,r) = \sum_{n=-\infty}^{\infty} f_n(r)\, e^{in\phi}, \quad g(\theta,t) = \sum_{n=-\infty}^{\infty} g_n(t)\, e^{in\theta},$$

where the Fourier coefficients are given by

$$f_n(r) = \frac{1}{2\pi}\int_0^{2\pi} f(\phi,r)\, e^{-in\phi}\, d\phi, \quad g_n(t) = \frac{1}{2\pi}\int_0^{2\pi} g(\theta,t)\, e^{-in\theta}\, d\theta.$$

Due to rotation invariance of the family of V-lines used in our data, the Fourier series expansion diagonalizes our operator. In other words, the nth Fourier coefficient of Tf depend only on the nth Fourier coefficient of f. The exact formula of this relation was established in [8] and then used to “invert it,” i. e., express f_n in terms of g_n.

Theorem 6 ([8]). *Let f be supported inside the disc $D(0,1)$. Then we have*

$$\mathcal{M}f_n(s) = \frac{\mathcal{M}g_n(s-1)}{1/(s-1) + \mathcal{M}h_n(s-1)}, \quad \Re(s) > 1, \tag{29}$$

where $\mathcal{M}F$ denotes the Mellin transform of function F

$$\mathcal{M}F(s) = \int_0^{\infty} p^{s-1} F(p)\, dp,$$

and $h_n(t)$ is an explicitly defined (albeit cumbersome), elementary function of t. (See [8] *for the exact expression of $h(t)$.)*

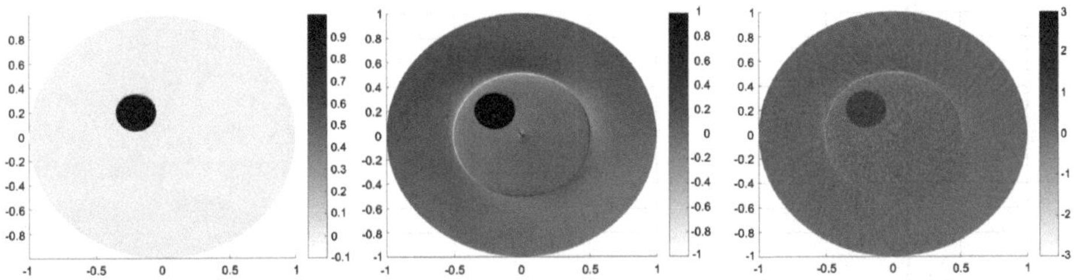

(a) Original phantom (b) Recon. with no noise (c) Recon. with 5% noise

Figure 7.11: Numerical reconstructions courtesy of Souvik Roy. The reconstructed images have streak artifacts at the origin and along the circle of radius $\sin(2\beta) = 0.5$. This is due to the abrupt cut of the (incomplete) VLT data at $t = 0$ and $t = 1$ correspondingly. See [9] for more details.

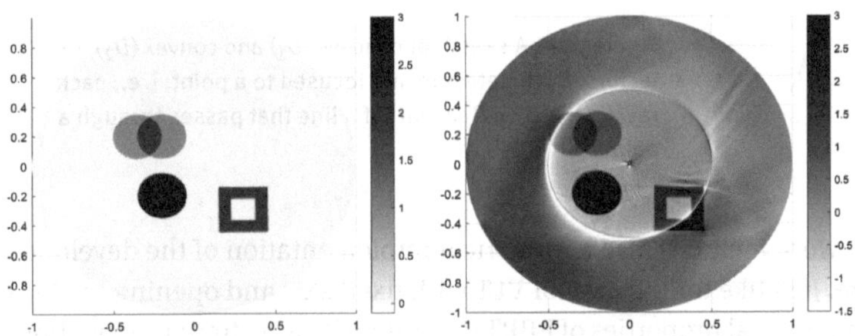

(a) Original phantom (b) Numerical reconstruction

Figure 7.12: Numerical reconstructions courtesy of Souvik Roy. In addition to the streak artifacts visible also in Figure 7.11, the reconstructed image here has some blurring artifacts. Namely, a portion of the boundary of the small disc farthest from the origin is blurred. This is due to the missing parts of the wavefront set in the (incomplete) VLT data. See [9] for more details.

Applying the standard inverse of Mellin transform, one can recover f_n's from g_n's, which finishes the process of inverting VLT in a disc from (radially) half of the data.

A numerical implementation of an inversion procedure based on (29) was presented in [9]. A couple of examples using the algorithm developed in that paper are presented below in Figures 7.11 and 7.12. In both cases, the opening angle of the V-lines is $2\beta = 5\pi/6$, and the VLT data $g(\theta, t) = Tf(\theta, t)$ is known for $\theta \in [0, 2\pi]$ and $t \in [0, 1]$.

Another interesting VLT, using a rotationally invariant family of V-lines, is considered in [68]. This operator, called *polar broken ray transform* (PBRT), maps functions supported inside a closed annulus $A(r_1, r_2) = \overline{D(0, r_2)} \setminus D(0, r_1)$, $0 < r_1 < r_2$ to their integrals along the following trajectories. Each broken ray $L(\theta, t)$ starts at the origin and travels a distance $t > 0$ along the radius vector with polar angle $\theta \in [0, 2\pi]$, before breaking at a fixed angle $2\beta \in (\pi/2, \pi)$ (see Figure 7.13).

A Fourier series approach similar to the one described above is used in [68] to derive an inversion of PBRT. It is shown that the Fourier coefficients f_n and g_n are connected through a Volterra integral equation of the second kind, for which a series solu-

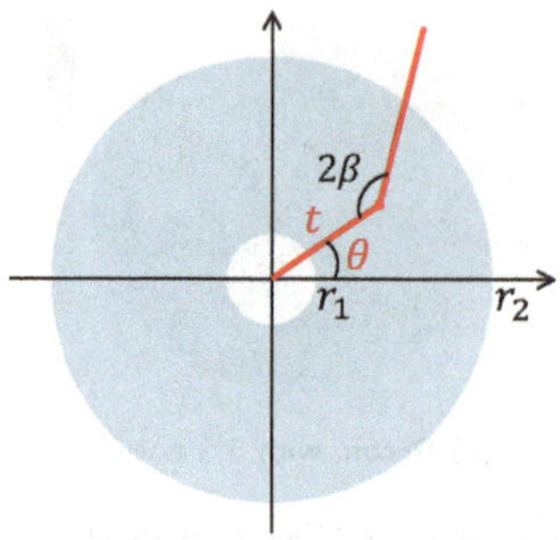

Figure 7.13: Parametrization of polar broken rays in the unit disc.

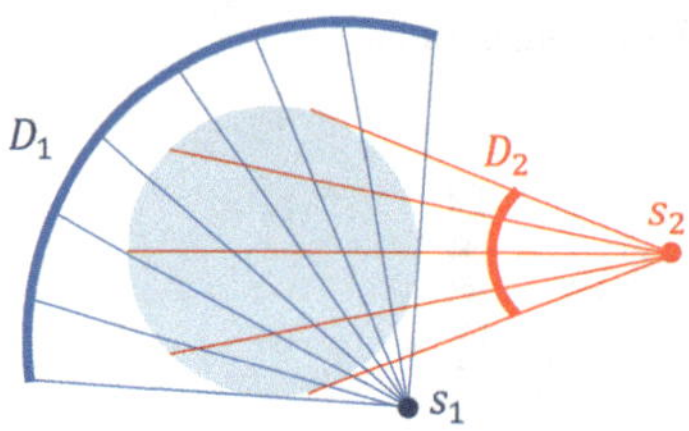

Figure 7.14: A sketch of concave (D_1) and convex (D_2) detectors. Both detectors are focused to a point, i. e., each ray detected by D_i is part of a line that passes through a point s_i.

tion is derived. The work includes also a numerical implementation of the developed inversion procedure. Just like for the case of VLT with fixed axis and opening angle, a thorough study of microlocal properties of PBRT (including its wavefront propagation) is done in [68].

7.3.3 Some Other VLTs on the Plane

Several other types of VLTs on the plane have been considered in [39, 40, 68]. The considered geometries included convex and concave curvilinear detectors, so that each point x of the image domain corresponds to a vertex of exactly one (in the case of 2 detectors) or two (in the case of 3 detectors) V-lines (see Figure 7.14).

In the case of two detectors, the primary object of study is the *signed V-line transform*, which subtracts from the integral of f along one ray of the V-line its integral along the other ray. The authors of [39] invert that transform by solving a first-order partial differential equation for f with nonconstant coefficients, assuming that $f = 0$ on the boundary and integrating along characteristics. The same setup is investigated further in [68], where it is shown, that if both detectors are simultaneously convex or concave, the transform is not injective. Hence, the recovery of f is not possible without additional information, e. g., boundary data. The microlocal properties of the SVLT with curved detectors is also studied in [68].

As we already saw in the case of rotation invariant VLTs in a disc, the presence of additional data substantially simplifies the inversion of the operator. Hence, in the case when the SVLT data includes measurements from three curved detectors, all of the above statements are significantly improved, and some new results are obtained.

Namely, the inversion of SVLT from three detector data requires only the first-order derivatives of the data. Moreover, the formula is purely local and solves the interior problem (see [39]). In the same setup, a range description of the operator on the class of compactly supported and infinitely smooth functions is presented in [39]. The inverse of the weighted version of the SVLT with three or more detectors is given in [40].

7.3.4 Conical Radon transforms with vertices inside the image domain

In this section, we briefly mention some results on inversion of CRTs, which use cones with vertices inside the image domain. It must be noted that none of the imaging modalities discussed in this chapter use these transformations. To the best of our knowledge, these problems are of purely mathematical interest.

A CRT, which maps compactly supported smooth functions in $\mathbb{R}^3$ to their integrals over surfaces of circular cones with fixed axis of symmetry and opening angle is studied in [28]. An exact inversion formula is derived using Fourier analysis. A generalization of the same transform to higher dimensions is studied in [27], and a similar inversion formula is presented there. The weighted version of this transform in $\mathbb{R}^3$ is studied in [60]. New reconstruction formulas are derived using PDE techniques.

A CRT mapping integrable functions in $\mathbb{R}^n$ to their integrals over surfaces of polyhedral cones is studied in [5]. A simple inversion formula is derived, and a possible method for deriving a range description of that transform is outlined (as a generalization of the corresponding strategy in 2D case).

7.4 Additional remarks

1. We did not discuss in detail the mathematical results related to VLT and CRT with vertices on a curve or a surface. However, a lot of references to appropriate papers were provided in Section 7.2.3, where we discussed the relevant applications. For more information on the subject, we refer the reader to those papers and the recent review article [75].
2. One can consider generalizations of the topics discussed in this chapter to manifolds, e. g., broken geodesics and various problems related to them (e. g., see [44]).
3. An interesting area of research in integral geometry is dedicated to the recovery of functions defined inside a compact domain from their integrals along piecewise-linear trajectories that reflect multiple times from the boundary of that domain. As it often happens in mathematics, this transform is also called a broken-ray transform, although it is quite different from the BRT mentioned in this chapter. For more details and interesting results in this field, we refer the reader to [33–36].

Bibliography

[1] M. Allmaras, D. Darrow, Y. Hristova, G. Kanschat, and P. Kuchment. Detecting small low emission radiating sources. *Inverse Probl. Imaging*, 7(1):47–79, 2013.

[2] J. E. Alpuche Aviles, S. Pistorius, R. Gordon, and I. A. Elbakri. A novel hybrid reconstruction algorithm for first generation incoherent scatter CT (ISCT) of large objects with potential medical imaging applications. *J. X-Ray Sci. Technol.*, 19:35–56, 2011.

[3] G. Ambartsoumian. Inversion of the V-line Radon transform in a disc and its applications in imaging. *Comput. Math. Appl.*, 64(3):260–265, 2012.

[4] G. Ambartsoumian, R. Gouia-Zarrad, and M. A. Lewis. Inversion of the circular Radon transform on an annulus. *Inverse Probl.*, 26(10):105015,11, 2010.

[5] G. Ambartsoumian and M. J. L. Jebelli. The V-line transform with some generalizations and cone differentiation. Preprint at arXiv:1612.05113 [math.CA], 2018.

[6] G. Ambartsoumian and V. P. Krishnan. Inversion of a class of circular and elliptical Radon transforms. *Contemp. Math.*, 653:1–12, 2015.

[7] G. Ambartsoumian and L. Kunyansky. Exterior/interior problem for the circular means transform with applications to intravascular imaging. *Inverse Probl. Imaging*, 8(2):339–359, 2014.

[8] G. Ambartsoumian and S. Moon. A series formula for inversion of the V-line Radon transform in a disc. *Comput. Math. Appl.*, 19(66):1567–1572, 2013.

[9] G. Ambartsoumian and S. Roy. Numerical inversion of a broken ray transform arising in single scattering optical tomography. *IEEE Trans. Comput. Imag.*, 2(2):166–173, 2016.

[10] S. R. Arridge and J. C. Schotland. Optical tomography: forward and inverse problems. *Inverse Probl.*, 25(12):123010, 2009.

[11] G. Bal. Inverse transport theory and applications. *Inverse Probl.*, 25(5):053001, 2009.

[12] M. S. Bandstra, E. C. Bellm, S. E. Boggs, D. Perez-Becker, A. Zoglauer, H.-K. Chang, J.-L. Chiu, J.-S. Liang, Y.-H. Chang, et al. Detection and imaging of the Crab Nebula with the nuclear Compton telescope. *Astrophys. J.*, 738(1):8, 2011.

[13] R. Basko, G. L. Zeng, and G. T. Gullberg. Analytical reconstruction formula for one-dimensional Compton camera. *IEEE Trans. Nucl. Sci.*, 44(3):1342–1346, 1997.

[14] R. Basko, G. L. Zeng, and G. T. Gullberg. Fully three dimensional image reconstruction from "V"-projections acquired by Compton camera with three vertex electronic collimation. In *1997 IEEE nuclear science symposium conference record*, volume 2, pages 1077–10812. 1997.

[15] R. Basko, G. L. Zeng, and G. T. Gullberg. Application of spherical harmonics to image reconstruction for the Compton camera. *Phys. Med. Biol.*, 43(4):887, 1998.

[16] Y. Berker and V. Schulz. Scattered PET data for attenuation-map reconstruction in PET/MRI: fundamentals. In *2014 IEEE nuclear science symposium and medical imaging conference (NSS/MIC)*, pages 1–6. 2014.

[17] W. Cong and G. Wang. X-ray scattering tomography for biological applications. *J. X-Ray Sci. Technol.*, 19:219–227, 2011.

[18] M. Cooper, P. Mijnarends, N. Shiotani, N. Sakai, and A. Bansil. *X-ray Compton scattering*. Oxford series on synchrotron radiation. Oxford University Press, Oxford, 2004.

[19] A. M. Cormack. Representation of a function by its line integrals with some radiological applications. *J. Appl. Phys.*, 34:2722–2727, 1963.

[20] A. M. Cormack. Representation of a function by its line integrals with some radiological applications II. *J. Appl. Phys.*, 35:2908–2913, 1964.

[21] M. J. Cree and P. J. Bones. Towards direct reconstruction from a gamma camera based on Compton scattering. *IEEE Trans. Med. Imaging*, 13(2):398–407, 1994.

[22] D. B. Everett, J. S. Fleming, R. W. Todd, and J. M. Nightingale. Gamma-radiation imaging system based on the Compton effect. *Proc. IEEE*, 124(11):995–1000, 1977.

[23] L. Florescu, V. A. Markel, and J. C. Schotland. Single-scattering optical tomography: simultaneous reconstruction of scattering and absorption. *Phys. Rev. E*, 81:016602, 2010.

[24] L. Florescu, V. A. Markel, and J. C. Schotland. Inversion formulas for the broken-ray Radon transform. *Inverse Probl.*, 27(2):025002, 2011.

[25] L. Florescu, J. C. Schotland, and V. A. Markel. Single scattering optical tomography. *Phys. Rev. E*, 79:036607,10, 2009.

[26] I. M. Gelfand, S. G. Gindikin, and M. I. Graev. *Selected topics in integral geometry*. Translations of mathematical monographs. American Mathematical Society, Providence, RI, 2003.

[27] R. Gouia-Zarrad. Analytical reconstruction formula for n-dimensional conical Radon transform. *Comput. Math. Appl.*, 68(9):1016–1023, 2014.

[28] R. Gouia-Zarrad and G. Ambartsoumian. Exact inversion of the conical Radon transform with a fixed opening angle. *Inverse Probl.*, 30(4):045007, 2014.

[29] M. Haltmeier. Exact reconstruction formulas for a Radon transform over cones. *Inverse Probl.*, 30(3):035001, 2014.

[30] M. Haltmeier, S. Moon, and D. Schiefeneder. Inversion of the attenuated V-line transform for SPECT with Compton cameras. *IEEE Trans. Comput. Imag.*, 3(4):853–863, 2017.

[31] S. Helgason. *Integral geometry and radon transforms*. Springer, New York, 2011.

[32] Y. Hristova. Inversion of a V-line transform arising in emission tomography. *J. Coupled Syst. Multiscale Dyn.*, 3(3):272–277, 2015.

[33] M. Hubenthal. The broken ray transform on the square. *J. Fourier Anal. Appl.*, 20(5):1050–1082, 2014.

[34] M. Hubenthal. The broken ray transform in n dimensions with flat reflecting boundary. *Inverse Probl. Imaging*, 9(1):143–161, 2015.

[35] J. Ilmavirta. Broken ray tomography in the disc. *Inverse Probl.*, 29(3):035008, 2013.

[36] J. Ilmavirta and M. Salo. Broken ray transform on a Riemann surface with a convex obstacle. *Commun. Anal. Geom.*, 24(2):379–408, 2016.

[37] C.-Y. Jung and S. Moon. Inversion formulas for cone transforms arising in application of Compton cameras. *Inverse Probl.*, 31(1):015006, 2015.

[38] C.-Y. Jung and S. Moon. Exact inversion of the cone transform arising in an application of a Compton camera consisting of line detectors. *SIAM J. Imaging Sci.*, 9(2):520–536, 2016.

[39] A. Katsevich and R. Krylov. Broken ray transform: inversion and a range condition. *Inverse Probl.*, 29(7):075008, 2013.

[40] R. Krylov and A. Katsevich. Inversion of the broken ray transform in the case of energy-dependent attenuation. *Phys. Med. Biol.*, 60(11):4313, 2015.

[41] P. Kuchment. *The Radon transform and medical imaging*. CBMS-NSF regional conference series in applied mathematics. Society for Industrial and Applied Mathematics, 2014.

[42] P. Kuchment and F. Terzioglu. Three-dimensional image reconstruction from Compton camera data. *SIAM J. Imaging Sci.*, 9(4):1708–1725, 2016.

[43] P. Kuchment and F. Terzioglu. Inversion of weighted divergent beam and cone transforms. *Inverse Probl. Imaging*, 11(6):1071–1090, 2017.

[44] Y. Kurylev, M. Lassas, and G. Uhlmann. Rigidity of broken geodesic flow and inverse problems. *Am. J. Math.*, 132(2):529–562, 2010.

[45] J.-S. Lee and J.-C. Chen. A single scatter model for X-ray CT energy spectrum estimation and polychromatic reconstruction. *IEEE Trans. Med. Imaging*, 34(6):1403–1413, 2015.

[46] W. Lee and T. Lee. A compact Compton camera using scintillators for the investigation of nuclear materials. *Nucl. Instrum. Methods Phys. Res., Sect. A, Accel. Spectrom. Detect. Assoc. Equip..*, 624(1):118–124, 2010.

[47] E. Masahiro, T. Takanori, N. Nobuyuki, and Y. Katsuya. Effect of scattered radiation on image noise in cone beam CT. *Med. Phys.*, 28(4):469–474, 2001.

[48] V. Maxim. Filtered backprojection reconstruction and redundancy in Compton camera imaging. *IEEE Trans. Imag. Proc.*, 23(1):332–341, 2014.

[49] V. Maxim, M. Frandeş, and R. Prost. Analytical inversion of the Compton transform using the full set of available projections. *Inverse Probl.*, 25(9):095001, 2009.

[50] S. Moon. On the determination of a function from its conical Radon transform with fixed central axis. *SIAM J. Math. Anal.*, 48(3):1833–1847, 2016.

[51] S. Moon. Inversion of the conical Radon transform with vertices on a surface of revolution arising in an application of a Compton camera. *Inverse Probl.*, 33(6):065002, 2017.

[52] S. Moon and M. Haltmeier. Analytic inversion of a conical Radon transform arising in application of Compton cameras on the cylinder. *SIAM J. Imaging Sci.*, 10(2):535–557, 2017.

[53] M. Morvidone, T. T. Truong, M. K. Nguyen, and H. Zaidi. On the V-line Radon transform and its imaging applications. *Int. J. Biomed. Imaging*, 2010:208179,6, 2010.

[54] F. Natterer. *The mathematics of computerized tomography*. Classics in mathematics. Society for Industrial and Applied Mathematics, New York, 2001.

[55] F. Natterer and F. Wübbeling. *Mathematical methods in image reconstruction*. Monographs on mathematical modeling and computation. Society for Industrial and Applied Mathematics, New York, 2001.

[56] M. K. Nguyen, T. T. Truong, and P. Grangeat. Radon transforms on a class of cones with fixed axis direction. *J. Phys. A, Math. Gen.*, 38(37):8003, 2005.

[57] S. J. Norton. Compton scattering tomography. *J. Appl. Phys.*, 76(4):2007–2015, 1994.

[58] V. Palamodov. *Reconstructive integral geometry*. Monographs in mathematics. Birkhäuser, 2004.

[59] V. Palamodov. *Reconstruction from integral data*. Monographs and research notes in mathematics. Chapman and Hall/CRC, 2016.

[60] V. Palamodov. Reconstruction from cone integral transforms. *Inverse Probl.*, 33(10):104001, 2017.

[61] L. C. Parra. Reconstruction of cone-beam projections from Compton scattered data. In *1999 IEEE nuclear science symposium. Conference record*, volume 2, pages 1082–1086. 1999.

[62] G. Rigaud and B. N. Hahn. 3D Compton scattering imaging and contour reconstruction for a class of Radon transforms. *Inverse Probl.*, 34(7):075004, 2018.

[63] G. Rigaud, M. K. Nguyen, and A. K. Louis. Novel numerical inversions of two circular-arc Radon transforms in Compton scattering tomography. *Inverse Probl. Sci. Eng.*, 20(6):809–839, 2012.

[64] B. Rubin. *Introduction to Radon transforms: with elements of fractional calculus and harmonic analysis*. Encyclopedia of mathematics and its applications. Cambridge University Press, New York, NY, USA, 2015.

[65] W. Rudin. *Real and complex analysis*, 3rd ed. McGraw-Hill, Inc., New York, NY, USA, 1987.

[66] D. Schiefeneder and M. Haltmeier. The Radon transform over cones with vertices on the sphere and orthogonal axes. *SIAM J. Appl. Math.*, 77(4):1335–1351, 2017.

[67] J. C. Schotland. Direct reconstruction methods in optical tomography. In H. Ammari, editor, *Mathematical modeling in biomedical imaging ii: optical, ultrasound, and opto-acoustic tomographies*, pages 1–29. Springer, Berlin, Heidelberg, 2012.

[68] B. Sherson. Some results in single-scattering tomography. PhD thesis, Oregon State University, 2015. PhD Advisor: David Finch.

[69] M. Singh. An electronically collimated gamma camera for single photon emission computed tomography. Part I: theoretical considerations and design criteria. *Med. Phys.*, 10(4):421–427, 1983.

[70] M. Singh and D. Doria. An electronically collimated gamma camera for single photon emission computed tomography. Part II: image reconstruction and preliminary experimental measurements. *Med. Phys.*, 10(4):428–435, 1983.
[71] B. D. Smith. Reconstruction methods and completeness conditions for two Compton data models. *J. Opt. Soc. Am. A*, 22(3):445–459, 2005.
[72] B. D. Smith. Line-reconstruction from Compton cameras: data sets and a camera design. *Opt. Eng.*, 50(5):053204, 2011.
[73] H. Sun and S. Pistorius. Characterization the annihilation position distribution within a geometrical model associated with scattered coincidences in PET. In *2014 IEEE nuclear science symposium and medical imaging conference (NSS/MIC)*, pages 1–3. 2014.
[74] F. Terzioglu. Some inversion formulas for the cone transform. *Inverse Probl.*, 31(11):115010, 2015.
[75] F. Terzioglu, P. Kuchment, and L. Kunyansky. Compton camera imaging and the cone transform: a brief overview. *Inverse Probl.*, 34(5):054002, 2018.
[76] R. W. Todd, J. M. Nightingale, and D. B. Everett. A proposed γ camera. *Nature*, 251:132–134, 1974. (Awarded the Ambrose Fleming Premium by the IEE in 1977).
[77] T. T. Truong and M. K. Nguyen. On new V-line Radon transforms in R^2 and their inversion. *J. Phys. A, Math. Theor.*, 44(7):075206, 2011.
[78] T. T. Truong and M. K. Nguyen. New properties of the V-line Radon transform and their imaging applications. *J. Phys. A, Math. Theor.*, 48(40):405204, 2015.
[79] T. T. Truong, M. K. Nguyen, and H. Zaidi. The mathematical foundation of 3D Compton scatter emission imaging. *Int. J. Biomed. Imaging*, 2007:92780,11, 2007.
[80] M. C. W. van Rossum and T. M. Nieuwenhuizen. Multiple scattering of classical waves: microscopy, mesoscopy, and diffusion. *Rev. Mod. Phys.*, 71:313–371, 1999.
[81] C. Vinegoni, C. Pitsouli, D. Razansky, N. Perrimon, and V. Ntziachristos. In vivo imaging of Drosophila melanogaster pupae with mesoscopic fluorescence tomography. *Nat. Methods*, 9:45–47, 2008.
[82] V. V. Volchkov. *Integral geometry and convolution equations*. Kluwer Academic Publishers, Dordrecht, 2003.
[83] J. W. Webber and W. R. B. Lionheart. Three dimensional Compton scattering tomography. *Inverse Probl.*, 34(8):084001, 2018.
[84] T. Yuasa, M. Akiba, T. Takeda, M. Kazama, A. Hoshino, Y. Watanabe, K. Hyodo, A. Dilmanian, T. Akatsuka, and Y. Itai. Incoherent-scatter computed tomography with monochromatic synchrotron X-ray: feasibility of multi-CT imaging system for simultaneous measurement-of fluorescent and incoherent scatter X-rays. *IEEE Trans. Nucl. Sci.*, 44(5):1760–1769, 1997.
[85] F. Zhao, J. C. Schotland, and V. A. Markel. Inversion of the star transform. *Inverse Probl.*, 30(10):105001, 2014.

Alfred K. Louis

8 Uncertainty, ghosts, and resolution in Radon problems

Abstract: We study the nonuniqueness problem for Radon transforms for finitely many directions. In the early days of the application of computed tomography, they caused some confusion about the possible information content in the reconstructions from tomographic data. The existence of nontrivial functions in the null space started the analysis of these then so-called ghosts. A result of Logan [25] described properties of the spectrum of those functions. Only with the description of those functions in terms of special functions by Louis [27] a more detailed study and an improvement of earlier results was possible. Here, we describe the essential steps to find those characterizations and the analysis of the spectral properties allowing for resolution results.

Keywords: Radon transform, nonuniqueness for finitely many directions, resolution, ghosts

MSC 2010: 44A12, 65R10, 33C55

8.1 Introduction

After the pioneering work of Hounsfield and Cormack introducing computed tomography the practical consequences of the limitations of the measuring system started moving in the focus of interest. The mathematical model of x-ray computed tomography in two dimensions is given by the two-dimensional Radon transform

$$\mathbf{R}f(\omega, s) = \int_{\mathbb{R}^2} f(x)\delta(s - x^\top \omega)\, dx$$

where f is the searched-for density distribution, ω is the direction orthogonal to the propagation of the x-ray beams, and s is the signed distance of the ray from the origin. If $\mathbf{R}f(\omega, \cdot)$ is given for a direction ω and all $s \in \mathbb{R}$, then those data are called complete projections. A first example of a nontrivial function whose complete projections are 0 was constructed on a square where on the four equally sized subquarters the function has the values +1 or −1 alternating and the two measuring directions are parallel to the sides of the squares. This shows that the Radon transform for those directions

Acknowledgement: The work of the author was supported by Deutsche Forschungsgemeinschaft under grant LO310/13-1 and by the Sino-German Center for Research Promotion under grant GZ 1025.

Alfred K. Louis, Saarland University, Department of Mathematics, D-66041 Saarbrücken, Germany, e-mail: louis@num.uni-sb.de

https://doi.org/10.1515/9783110560855-008

has a non-trivial null space. The argument that typical densities are positive does not prevent from nonuniqueness, because when one adds such a null space function to another density such that the sum is positive, there is still nonuniqueness as the data are the same.

Those considerations lead Katz [21] to study the nonuniqueness problem in the space of piecewise constant functions on a regular pixel basis as used in those days for ART, the algebraic reconstruction technique; see, e. g., Herman [19].

Another approach was followed in the group of K. T. Smith [48] in Corvallis. They described the functions in the null space as the solution of a system of homogeneous hyperbolic equations. The result he reformulated as: *A finite set of radiographs tells nothing at all.* Here, radiographs stands for complete projections. With that remark, Smith tried to stop several projects in the 1970s, but fortunately enough his result was shown too late; already one decade of positive experience with x-ray tomography existed.

In a next step, characterizations of the functions in the null space were addressed. In the meantime, the wording ghosts for those functions came in use. Indeed this is in contrast to the everyday life: there some people see ghosts but everyone knows that they do not exist. Here, one can prove that they exist but they are invisible for certain directions.

The first remarkable characterization result was shown by Logan [25] in a paper with some 500 formulas. Logan studied the frequency distribution of the ghosts and showed that for the two-dimensional case the largest part of the spectrum is contained outside of a ball around 0 with radius equal to the number of complete projections. Based on consistency conditions for the Radon transform, it then was possible to give a constructive description of the functions in the null space, and hence to derive the results of Logan in a more elegant form and to improve the estimates; see Louis [27].

Some related result should be mentioned here. The Radon transform for the exterior problem in $\mathbb{R}^N$ with suitable weight, where planes do not intersect the interior of the unit ball was studies by Quinto [40]. For the x-ray transform, a singular value decomposition and the case of a finite number of data with resolution properties was a groundbreaking work by Maass [35]. Null space results for cone-beam tomography were given by Derevtsov [11]. The singular value decompositions for the cone beam transform (see Kazantsev [22] and Quellmalz, Hielscher, and Louis [39]) have yet to be exploited in this direction. For line integrals on Cormack-type of curves, the nonuniqueness problem was considered by Rigaud [41]. In case of incomplete data problems even for an infinite number of projections, there is nonuniquess for the case of truncated projection, where only those line integrals are given which intersect a ball $V(0, a) \subset V(0, 1)$ with $a < 1$, as shown by Smith, Solmon, and Wagner [48]. Functions in the corresponding null space were constructed by Louis and Rieder [33]. Recently Hahn [16], studied dynamic tomographic problems where the structure of the object moves during the measurements. With a far reaching extension of the methods presented here, she characterized the achievable resolution. For the case of vector tomography, the situation is more complicated. There exist nonuniqueness problems even for an infinite number of data; see Sharafutdinov [45] or Schuster [42]. Also the singular value decomposition for tensor tomography, presented by Derevtsov, Efimov, Louis, and Schuster [12] has yet to be exploited. In the case of discrete tomography, ghosts were used to improve the resolution; see, e. g., Chandra et al. [7]. The list of contributions to this field cannot be complete due to magnificent results achieved in the last 40 years. As recent contributions, we mention [1, 5, 17, 23].

In this paper, we present in the next section the problem of nonuniqueness of the Radon transform for finitely many directions. We derive the main ideas for the two-dimensional Radon transform, where we derive series expansions for the functions in the null space, the so-called ghosts. We then extend the results to arbitrary dimension and elaborate on the main differences. The third section is devoted to the question of resolution, directly for arbitrary dimensions where the spectral properties of the ghosts are determined, allowing for analyzing the resolution. Finally, an outlook on increasing the resolution by introducing additional information including compressive sensing and deep learning.

8.2 The ghosts

8.2.1 Properties of the null space

In the following, we concentrate on the two-dimensional case, peculiarities of the higher dimensional case are mentioned at the end of this section. We consider the Radon transform as continuous mapping between L_2 spaces, say

$$\mathbf{R} : L_2(\Omega_2) \mapsto L_2(C_2) \tag{1}$$

$$\mathbf{R} : L_2(\Omega_2) \mapsto L_2(C_2, w_1^{-1}) \tag{2}$$

where Ω_2 denotes the unit ball in $\mathbb{R}^2$, C_2 is the unit cylinder

$$C_2 = S^1 \times [-1, 1]$$

with S^1 the unit sphere in $\mathbb{R}^2$ and w_1 is the weight

$$w_1(s) = (1 - s^2)^{1/2}.$$

The continuity in (2) follows by straightforward calculation, the other simply by the continuous embedding of the space with weight in the one without weight.

We now consider the null space for finitely many directions. Let

$$\mathscr{A}_p = \{\omega_1, \dots, \omega_p\} \subset S^1 \tag{3}$$

be a set of p distinct directions. The distribution of the directions on the unit circle plays no role for the uniqueness questions, whereas for the stability of the inversion it is crucial.

Next, we consider the null space of the Radon transform for those directions

$$\begin{aligned}\mathcal{N}_p &= \mathcal{N}(\mathcal{A}_p)\\ &= \{f \in L_2(\Omega_2) : \mathbf{R}f(\omega, s) = 0 \text{ for all } \omega \in \mathcal{A}_p \text{ and almost all } s \in [-1, 1]\}.\end{aligned} \tag{4}$$

We note some obvious information.

Lemma 8.2.1. *Let f be in the null space $\mathcal{N}(\mathcal{A}_p)$. Then:*
(i) *if T^a is a translation such that $T^a f(x) = f(x - a)$, then $T^a f \in \mathcal{N}(\mathcal{A}_p)$;*
(ii) *if $t \neq 0$ and D^t is a dilation such that $D^t f(x) = f(x/t)$, then $D^t f \in \mathcal{N}(\mathcal{A}_p)$;*
(iii) *if D^U is a rotation such that $D^U(\mathcal{A}_p) \subset \mathcal{A}_p$ and $D^U f(x) = f(Ux)$, then $D^U f \in \mathcal{N}(\mathcal{A}_p)$.*

Proof. The proof follows from properties of the Radon transform. For (i), we observe that $\mathbf{R}T^a f(\omega, s) = \mathbf{R}f(\omega, s - a^\top \omega)$. Hence, if $\mathbf{R}f(\omega, \cdot) = 0$, then also $\mathbf{R}f(\omega, s - a^\top \omega) = 0$. For the dilation in (ii), we use $\mathbf{R}D^t(\omega, s) = \frac{1}{t}\mathbf{R}f(\omega, \frac{s}{t}) = 0$ with the same argument as above. Finally for the rotation in (iii) we get $\mathbf{R}D^U f(\omega, s) = \mathbf{R}f(D^U \omega, s) = 0$. □

The following presentation follows Louis [27]. Occasionally, this paper is referenced as authored by Louis–Törnig. But the esteemed colleague, Professor Törnig, who accepted the paper as a member of the editorial board of that journal for publication is correctly mentioned as "communicated by W. Törnig" on the printed version. Only much later he was made by some ignoramus, probably a computer, to a coauthor in some lists. The main ingredient for finding a representation of functions in the null space is now the consistency conditions named after Gel'fand–Graev–Vilenkin [14], respectively, Helgason [18] or Ludwig [34] depending on the background of the authors, respectively. It states, among others, that when p_m is a polynomial of degree m in s then

$$\int \mathbf{R}f(\omega, s) p_m(s)\, ds = q_m(\omega)$$

with a polynomial q_m of degree m in the directions. With the weight in the mapping property (2), the use of Chebyshev polynomials of the second kind, U_m, is promising. These polynomials form a complete orthogonal system on $L_2([-1, 1], w_1)$. In order to apply the consistency conditions, we consider

$$p_m = w_1 U_m$$

and the weighted L_2 space from (2) to find

$$\langle \mathbf{R}f(\omega, \cdot), w_1 U_m \rangle_{L_2([-1,1], w_1^{-1})} = \int_{-1}^{1} \mathbf{R}f(\omega, s) U_m(s)\, ds = q_m(\omega). \tag{5}$$

With these considerations, we can decompose the functions in the range of the Radon transform as

$$\mathbf{R}f(\omega, s) = w_1(s) \sum_{m=0}^{\infty} U_m(s) q_m(\omega) \tag{6}$$

where the expansion coefficients are, taking the orthogonality of the Chebyshev polynomials into account,

$$q_m(\omega) = \frac{2}{\pi} \int_{-1}^{1} \mathbf{R}f(\omega, s) U_m(s)\, ds \tag{7}$$

$$= \sum_{\substack{\lambda=-m \\ m+\lambda \text{ even}}}^{m} c_\lambda^m e^{\imath\lambda\varphi} \tag{8}$$

with coefficients $c_\lambda^m \in \mathbb{C}$. where we additionally used that the consistency conditions also state that $\mathbf{R}f(-\omega, -s) = \mathbf{R}f(\omega, s)$, hence the q_m are even (odd) when m is even(odd) as the Chebyshev polynomials.

In the following, we use the abbreviation:

$$\sum_{\lambda=0}^{m*} := \sum_{\substack{\lambda=0 \\ m+\lambda \text{ even}}}^{m} . \tag{9}$$

Now we can derive a presentation of the Radon transform of a function in the null space.

Theorem 8.2.2. *Let $f \in \mathcal{N}(\mathcal{A}_p)$ where the directions are mutually different. Then*

$$\mathbf{R}f(\omega, s) = w_1(s) \sum_{m=p}^{\infty} U_m(s) q_m(\omega) \tag{10}$$

with q_m a polynomial of degree $\leq m$ in ω, even (odd) for m even (odd) and

$$q_m(\omega) = 0 \quad \textit{for all } \omega \in \mathcal{A}_p. \tag{11}$$

Proof. The polynomials q_m in the representation of $\mathbf{R}f$ have p zeroes, hence with the parity condition the polynomials up to order $p - 1$ vanish identically. □

For the representation of the functions in the null space themselves, we use the inverse Radon transform of the Chebyshev polynomials and the trigonometric functions as given in [9, 27, 36]. Notice that when switching from the notion of spherical harmonics Y_m with $m \geq 0$ and for $m > 0$ consisting of two linear independent elements to their representation as functions we either have $e^{\pm\imath m\varphi}$ or $\sin m\varphi$ and $\cos m\varphi$.

Theorem 8.2.3. *Let $Q_{m,\lambda}^{\nu,N}$ be the Zernike polynomials, computable as*

$$Q_{m,\lambda}^{\nu,N}(r) = r^{\lambda} P_{\frac{m-\lambda}{2}}^{(\nu-\frac{N}{2},\lambda+\frac{N}{2}-1)}(2r^2 - 1) \tag{12}$$

with the Jacobi polynomials $P_n^{(\alpha,\beta)}$ for $m + \lambda$ even and $0 \le \lambda \le m$. Then

$$\mathbf{R}Q_{m,\lambda}^{1,2} Y_{\lambda} = \frac{2}{m+1} w_1 U_m Y_{\lambda} \quad \text{for } m \ge 0 \text{ and } 0 \le \lambda \le m \text{ with } m + \lambda \text{ even} \tag{13}$$

where Y_{λ} are the spherical harmonics of order λ, i. e., $Y_{\lambda}(\omega(\varphi)) = \exp(\pm\imath\lambda\varphi)$.

When normalizing both side, then one gets the singular values of the Radon transform as mapping from $\mathbf{R} : L_2(\Omega_2) \mapsto L_2(C_2, w_1^{-1})$ with singular values $\sigma_{m,\lambda} = O(m^{-1/2})$ independent of λ; see, e. g., [9, 32, 38]

Theorem 8.2.4. *The Radon transform*

$$\mathbf{R} : L_2(\Omega_2) \to L_2(C_2, w_1^{-1})$$

has the singular value decomposition

$$\{(v_{m\lambda}, u_{m\lambda}; \sigma_{m\lambda})\}: \quad m \ge 0,\ \lambda \in \mathbb{N}_0,\ m + \lambda \text{ even}$$

with

$$v_{m\lambda}(x) = \begin{cases} \sqrt{\frac{m+1}{\pi}} Q_{m\lambda}^{1,2} Y_{\lambda}(\frac{x}{|x|}), & |x| \le 1 \\ 0, & |x| > 1, \end{cases}$$

$$u_{m\lambda}(\omega, s) = \begin{cases} \frac{1}{\pi} w_1(s) U_m(s) Y_{\lambda}(\omega), & |s| \le 1 \\ 0, & |s| > 1 \end{cases}$$

and

$$\sigma_{m\lambda} = \sigma_m = 2\sqrt{\frac{\pi}{m+1}}.$$

Combining the last theorems, we get a representation of the functions in the null space.

Theorem 8.2.5. *Let $f \in \mathcal{N}(\mathcal{A}_p)$ for mutually different directions. Then*

$$f(r\omega) = \sum_{m=p}^{\infty} \sum_{\lambda=0}^{m^*} c_{\lambda}^m Y_{\lambda}(\omega) Q_{m,\lambda}^{1,2}(r) \tag{14}$$

where

$$\sum_{\lambda=0}^{m^*} c_{\lambda}^m Y_{\lambda}(\omega) = 0$$

for $\omega \in \mathcal{A}_p$.

Consequently, the expansion of a function in the null space $\mathcal{N}(\mathscr{A}_p)$ starts with a polynomial of degree p.

Theorem 8.2.6. *Let $\mathscr{A}_p$ consist of p mutually different directions and let*

$$\Pi_{p-1} = \{\textit{polynomials of total degree} < p\}.$$

Then a function in Π_{p-1} is uniquely determined by the Radon transform on $\mathscr{A}_p$.

Remark 8.2.7. The result easily follows from Theorem 8.2.5. Of course, the restriction of the reconstruction problem to polynomials is quite unnatural but see the discussion in Section 8.3.2.

Remark 8.2.8. For the uniqueness result, the distribution of the directions plays no role, but for the stability it is essential, as the limited angle case shows. See [10] for the proof that the smallest singular values decay exponentially and [31] for the analysis of the dichotomy of the singular values: some of them, where the number depends on the size of the given range, are of the same order as for the full range problem, and only the rest decays exponentially, which means that some parts of the solution can be stably recovered. The artifacts produced by this incomplete data problem are very nicely and convincingly described by Frikel and Quinto [13] using wavefront sets. Here, iterative methods may be an alternative; see Herman [19] or Jiang and Wang [20]. Introducing additional information is also a possibility as Vogelgesang and Schorr [49] show.

8.2.2 Equidistributed directions in two dimensions

As the original situation in computed tomography was modeled by the Radon transform, mostly with equidistributed directions, this case was often studied in the literature. Here, we call the directions $\omega_j = (\cos\varphi_j, \sin\varphi_j)^\top$ equidistributed when

$$\varphi_j = (j-1)\frac{\pi}{p} \quad \text{for } j = 1, \ldots, p. \tag{15}$$

Simple examples for ghosts were given by Shepp–Kruskal [46], here generalized from their case $p = 4$ to arbitrary p, as

$$f(r\omega(\varphi)) = \begin{cases} \operatorname{sign}(\sin(p\varphi)), & r < 1 \\ 0, & r > 1 \end{cases} \tag{16}$$

and by Herman [19] as

$$f(r\omega(\varphi)) = \begin{cases} \sin(\frac{\pi r}{\delta})\sin(p\varphi), & 0 \le r \le \delta \\ 0, & r > \delta. \end{cases} \tag{17}$$

Both functions are depicted in Figure 8.1. Without difficulties, those functions are recognized as ghosts, and hence do not really disturb the reconstructions. In order to construct ghosts that are not as easily recognizable as such, we consider the results from Theorem 8.2.2. The polynomials q_m from (8) with the zeroes φ_j from (15) can be represented as

$$q_m(\omega(\varphi)) = \sin p\varphi \quad \widetilde{q_{m-p}}(\omega(\varphi)) \tag{18}$$

where $\widetilde{q_{m-p}}$ is an even polynomial of degree $m-p$ for $m \geq p$; i. e.,

$$q_m(\omega(\varphi)) = \sin(p\varphi) \sum_{j=0}^{(m-p)/2} (\alpha_{2j}^m \cos(2j\varphi) + \beta_{2j}^m \sin(2j\varphi)). \tag{19}$$

Consequently, the first term in the expansion of the functions in the null space is

$$g(\omega, s) = c \sin p\varphi \cdot w_1(s) U_p(s)$$

for $c \in \mathbb{R}$, and hence the function itself is

$$f(r\omega(\varphi)) = c \sin p\varphi \cdot r^p. \tag{20}$$

This function is depicted in Figure 8.2.

With this representation, we are now able to construct more realistic ghosts. To this end, we project the characteristic function χ_ρ of a ball around 0 with radius ρ on the null space $\mathcal{N}(\mathcal{A}_p)$. As noted in Lemma 8.2.1, this is no limitation as shifted and dilated versions of those functions are also in the null space.

We have to compute the L_2 scalar product of the characteristic function χ with the basis of the null space. Using the normalized functions from the singular value decomposition, the relation

$$v_{m\lambda} = \frac{1}{\sigma_m^2} \mathbf{R}^* \mathbf{R} v_{m\lambda}$$

and that the Radon transform of the characteristic function of the ball around 0 with radius ρ is

$$\mathbf{R}\chi_\rho(\omega, s) = \begin{cases} 2\sqrt{\rho^2 - s^2}, & s < \rho \\ 0, & s > \rho \end{cases}$$

we compute, remembering the weighted scalar product in $L_2(C_2, w_1^{-1})$,

$$c_m = \frac{4}{\pi} \int_0^\rho U_m(s) \sqrt{\rho^2 - s^2}\, ds \int_0^{2\pi} q_m(\omega(\varphi))\, d\varphi$$

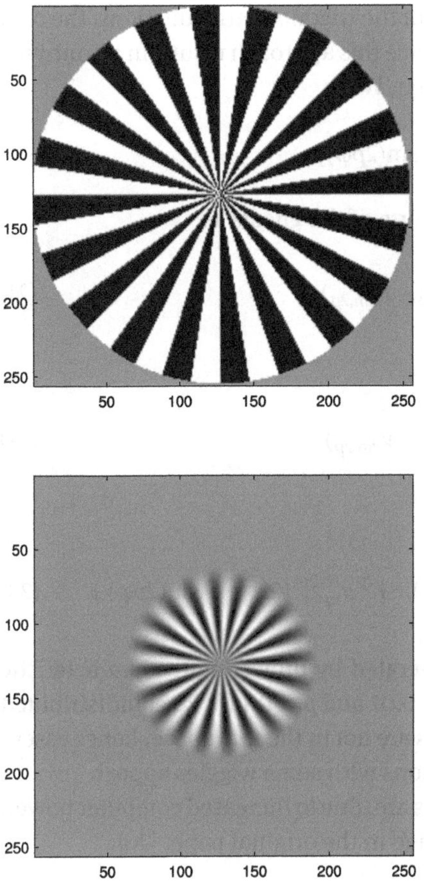

Figure 8.1: Simple examples of functions in the null space of the Radon transform for 20 directions of Shepp–Kruskal [46] (top) and Herman [19] (bottom).

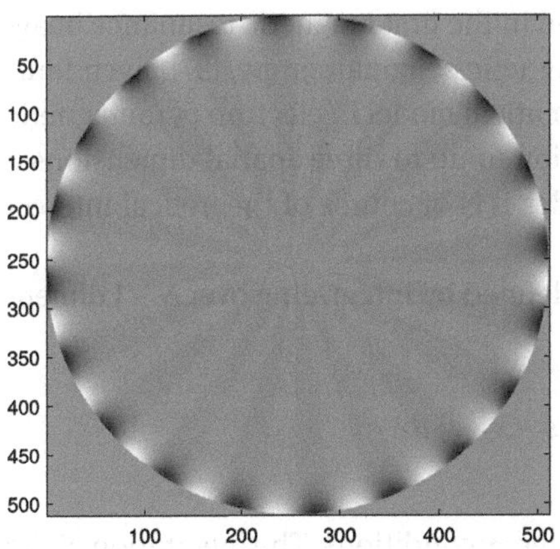

Figure 8.2: Function in the null space with smallest index for 20 equidistributed directions in a ball around 0 with radius 1.

for q_m from (19). Because of the orthogonality of the trigonometric functions, the only nonzero term appears for $2j = p$ and $\sin p\varphi$. Hence this approach results in a nontrivial example only if the number of directions is even. With

$$\sin^2(p\varphi) = \frac{1}{2}(1 - \sin(2p\varphi))$$

we get the representation for the Radon transform of the ghosts as

$$\mathbf{R}f = \sum_{m=p}^{\infty} c_m(u_{2m,0} - u_{2m,2p}), \tag{21}$$

and hence for the function itself

$$f = \sum_{m=p}^{\infty} \frac{c_m}{\sigma_m}(v_{2m,0} - v_{2m,2p}) \tag{22}$$

or explicitly

$$f(r\omega(\varphi)) = \sum_{m=p}^{\infty} \frac{c_m(m+1)}{2\pi^{3/2}}(P_m^{(0,0)}(2r^2-1) - r^{2p}P_{m-p}^{(0,2p)}(2r^2-1)\sin(2p\varphi)). \tag{23}$$

In Figures 8.3 and 8.4, we present ghosts generated by projecting the characteristic functions of the ball around 0 with radius $\rho = 0.01$ and $\rho = 0.1$ for 20 equidistributed directions. Observe that the constant functions are not in the null space, hence essentially the boundary of the characteristic functions with some wiggles appear.

We observe that the quality of the examples are, due to increased computer power, much superior to those from almost 40 years ago in the original paper [26].

8.2.3 Higher dimensional case

Three-dimensional Radon transforms appeared in the first magnetic resonance imaging scanners, in those days called nuclear magnetic zeugmatograpy, [37]. Even four-dimensional Radon transforms are the mathematical model in electron paramagnetic resonance imaging (EPRI) [24] where in addition to up to three spatial dimensions a further spectral dimension is measured. Besides it is of course of theoretical interest to analyze the null space.

In N dimensions, the Radon transform is defined by integrating over $N-1$ dimensional hyperplanes

$$\mathbf{R}f(\omega, s) = \int_{\mathbb{R}^N} f(x)\delta(s - x^\top\omega)\,dx.$$

In principle, we can start again from the consistency conditions. The integration of the Radon transform of a function against a polynomial in s results in a polynomial in the

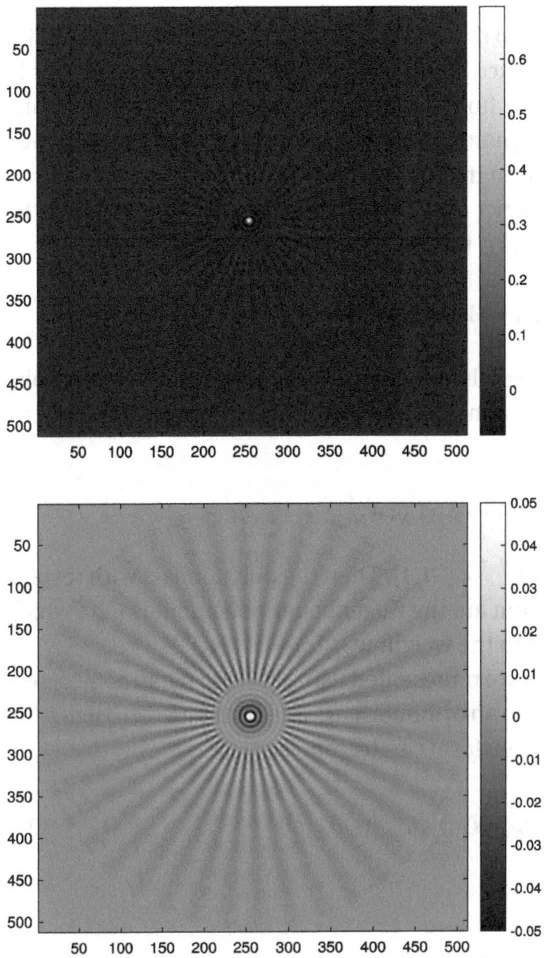

Figure 8.3: Projection of the characteristic function of a ball around 0 with radius 0.01 on the null space for 20 equidistributed directions. Top: Full display window such that the largest value (0.695) presented as white, Bottom: display window truncated such that everything larger than 0.05 is white in order to show the wiggles outside the center.

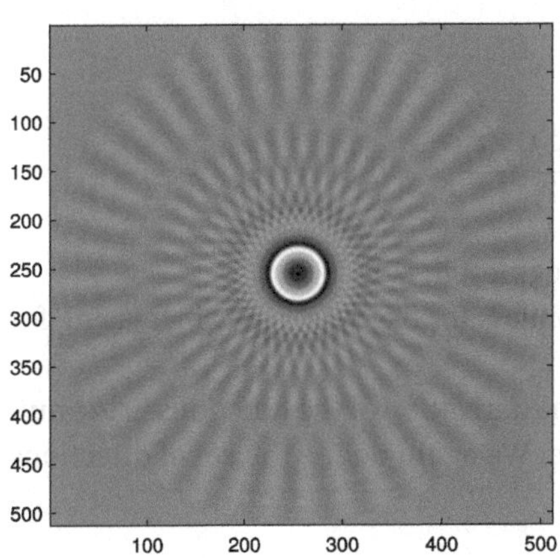

Figure 8.4: Projection of the characteristic function of a ball around 0 with radius 0.1 on the null space, full display window. Compare to Figure 8.3 TOP.

directions. The first selection we have to make is the polynomials. When we start with a radial symmetric weight W on the space of the searched for functions, the weight in the image space is $w = \mathbf{R}W$ which is a function of the distance s. For different pairs of weights and the corresponding polynomials inversion formulas, and consequently singular value decompositions are given in [30].

For the sake of simplicity, we restrict the considerations in the following on the weight $W \equiv 1$ resulting in the weight $w_{N/2}$ in N dimensions where

$$w_{N/2}(s) = (1 - s^2)^{(N-1)/2}.$$

The corresponding orthogonal polynomials are the Gegenbauer or ultraspherical polynomials, denoted by $C_m^{N/2}$. Hence we consider the Radon transform as continuous mapping

$$\mathbf{R} : L_2(\Omega_N) \to L_2(C_N, w_{N/2}^{-1})$$

with Ω_N the unit ball in $\mathbb{R}^N$ and $C_N = S^{N-1} \times [-1, 1]$. Then the polynomials with respect to s in the range of the Radon transform are the Gegenbauer polynomials $C_m^{N/2}$ which are orthogonal on $[-1, 1]$ with respect to the weight $w_{N/2}(s) = (1 - s^2)^{(N-1)/2}$.

The polynomials in the directions are represented using spherical harmonics; see, e. g., [43]. The spherical harmonics form an orthonormal system on $L_2(S^{N-1})$. Hence the functions in the range of the Radon transform can be represented as

$$\mathbf{R}f(\omega, s) = w_{N/2}(s) \sum_{m=0}^{\infty} C_m^{N/2}(s) q_m(\omega) \tag{24}$$

where again

$$q_m(\omega) = \sum_{\lambda=0}^{m^*} \sum_{\mu=1}^{M(N,\lambda)} c_{\lambda,\mu}^m Y_{\lambda,\mu}(\omega) \tag{25}$$

and

$$M(N, \lambda) = \frac{(2\lambda + N - 2)(N + \lambda - 3)!}{\lambda!(N - 2)!} = O(\lambda^{N-2})$$

the dimension of the spherical harmonics of degree λ in $\mathbb{R}^N$.

The number of linear independent polynomials for fixed degree is increasing for increasing dimension.

With this representation, we can transfer the derivation of the functions in the null space from the two to higher dimensional case. We again define the set of directions as

$$\mathscr{A}_p = \{\omega_1, \ldots, \omega_p\} \subset S^{N-1}, \tag{26}$$

a set of p distinct directions. The distribution of the directions on the unit sphere again plays no role for the uniqueness questions, whereas for the stability of the inversion it is crucial.

Next, we consider the null space of the Radon transform for those directions:

$$\begin{aligned}\mathcal{N}_p &= \mathcal{N}(\mathcal{A}_p)\\ &= \{f \in L_2(\Omega_N) : \mathbf{R}f(\omega, s) = 0 \text{ for all } \omega \in \mathcal{A}_p \text{ and almost all } s \in [-1,1]\}.\end{aligned} \tag{27}$$

In order to be in the null space, the expansion has to fulfill the condition that

$$q_m(\omega) = 0 \quad \text{for all } m \geq 0 \text{ and for all } \omega \in \mathcal{A}_p \text{ and almost all } s \in [-1,1].$$

But there is an essential difference in the higher dimensional case. In the 2D case, it suffices that the directions are mutually distinct in order to conclude that for p directions the polynomials up to order $p-1$ vanish identically. As already known from the algebraic polynomials, this does not suffice in higher dimensions. For example, a polynomial of degree 1 in two dimensions has three coefficients, namely for 1, x_1, and x_2. If such a polynomial has 3 zeroes, this does not imply that it vanishes identically. Only if the zeroes do not lie on a line one can conclude that the polynomial vanishes identically. To express this rigorously, we define, following [30],

$$\mathcal{P}_m = \operatorname{span}\{Y_{\lambda\mu} : 0 \leq \lambda \leq m, \lambda + m \text{ even}, 1 \leq \mu \leq M(N,\lambda)\} \tag{28}$$

with

$$\dim \mathcal{P}_m = \binom{m+N-1}{N-1}.$$

Theorem 8.2.9 ([30, Theorem 5.2]). *Let $n > 0$ and $p \geq \dim \mathcal{P}_{n-1}$. Let $\mathcal{A}_p$ not be contained in an algebraic variety of degree $< n$; i. e., there is no $q \in \mathcal{P}_{n-1}$, $q \not\equiv 0$ with $q(\omega) = 0$ for all $\omega \in \mathcal{A}_p$. Let $f \in \mathcal{N}_{A_p}$. Then*

$$\mathbf{R}f = w_{N/2} \sum_{m=n}^{\infty} C_m^{N/2} q_m(\omega) \tag{29}$$

where $q_m \in \mathcal{P}_m$ with $q_m(\omega) = 0$ for all $\omega \in \mathcal{A}_p$.

Remark 8.2.10. The condition in the above theorem concerning the distribution of the directions not lying in an algebraic variety of degree $< n$ was later called by Natterer [38] *n-resolving*.

The reasoning for this expression becomes clear in the next section. It also gives conditions for optimally distributing the directions; see [28]. Again, we can conclude a uniqueness result.

Theorem 8.2.11. *Let the conditions from Theorem 8.2.9 be fulfilled. Then the reconstruction problem in Π_{n-1} is unique.*

Finally, we mention the functions in the null space themselves. For that, we need the inverse Radon transform of the $w_{N/2}C_m^{N/2}Y_{\lambda\mu}$ (see Corollary 3.2 in [30]),

$$\mathbf{R}^{-1}(w_{N/2}C_m^{N/2}Y_{\lambda\mu})(r\omega) = c(N,m)r^\lambda P_{(m-\lambda)/2}^{(0,\lambda+N/2-1)}(2r^2-1)Y_{\lambda\mu}(\omega) \tag{30}$$

with the constant

$$c(N,m) = \pi^{1-N/2}2^{1-n}\Gamma(m+N)/[\Gamma(m+1)\Gamma(N/2)].$$

The result for arbitrary weights W_ν is contained in Theorem 3.1 in [30]. The proof uses the projection theorem for the Radon transform and special functions leading to Bessel functions; see the next section and the Funk–Hecke theorem [43], and integrals over Bessel functions of the Weber–Schafheitlin type, for the Hankel transform.

8.3 Resolution

The oscillatory character of the functions in the null space advise to analyze their spectrum. This is of importance for the study of the inversion algorithms where mostly high frequency components of the data are damped or even eliminated. Consequently, we start with the Fourier transform of the functions in the following. It is defined as

$$\mathscr{F}_N f(\xi) = \hat{f}(\xi) = (2\pi)^{-N/2}\int_{\mathbb{R}^N} f(x)\exp(-\imath x^\top\xi)\,dx. \tag{31}$$

A very helpful relation is the so-called following theorem.

Theorem 8.3.1 (Projection theorem). *For fixed direction $\omega \in S^{N-1}$, we have*

$$\mathscr{F}_1\mathbf{R}f(\omega,\sigma) = (2\pi)^{(N-1)/2}\mathscr{F}_N f(\sigma\omega), \quad \sigma\in\mathbb{R}. \tag{32}$$

This relation is also known as central slice theorem. For compactly supported functions f, it simply can be interpreted as applying the dual operator of the Radon transform to the suitably truncated exponential function. Otherwise, it is a mere change of coordinates. The Paley–Wiener theorem states that the Fourier transform of a compactly supported L_2 function is analytic.

Now we can determine the Fourier transform of the ghosts. Due to the projection theorem, it is sufficient to know the Fourier transform of $w_{N/2}C_m^{N/2}$.

Theorem 8.3.2 ([29, Theorem 2.1]). *Let the conditions of Theorem 8.2.9 be fulfilled and let $f \in \mathscr{N}(\mathscr{A}_p)$. Then*

$$\mathscr{F}_N f(\sigma\omega) = \sigma^{-N/2}\sum_{m=n}^{\infty}\imath^m J_{m+N/2}(\sigma)q_m(\omega) \tag{33}$$

where J_k is the Bessel function of the first kind and order k and $q_m(\omega) = 0$ for all $\omega \in \mathscr{A}_p$.

This means that the expansion of the Fourier transform of the ghosts starts with the Bessel function of order $m + N/2$. Following Debye's formula, the Bessel functions are small when the argument is smaller than the index (see, e. g., [2, 9.3.7]). We interpret this is that the Fourier transform of the ghosts is "small" when the argument is smaller than $n + N/2$. Of course "small" is not an adequate expression here, because multiplying a ghost by any number still remains a ghost. So, in the following we study the relative size of the ghost inside that mentioned ball compared to the whole energy.

Definition 8.3.3. If the Fourier transform of a function is compactly supported, say in $[-b, b]^N$ or in $V(0, b)$, the ball around 0 with radius b, then we call the function b band limited.

The above statement about the analyticity of the Fourier transform of a compactly supported function implies that there is no function, besides 0, that is both compactly supported and band limited. As the functions we consider in tomography are always compactly supported, they cannot be band limited. A way out is the concept of essentially band limited functions introduced by Natterer [38].

A strong result about band limited function is the sampling theorem; see Shannon [44].

Theorem 8.3.4 (Sampling theorem). *Let f be b band limited, and let $h \leq \pi/b$. Then f is uniquely determined by the values $f(hk)$, $k \in \mathbb{Z}^N$. The size π/b is called the Nyquist rate.*

In L_2, the function can be represented by the so-called sinc-series. For a simple and elegant proof, see [38, page 56].

8.3.1 The general case

Logan [25] proved in two dimensions a theorem that very roughly states that the knowledge of the full projections in each of p directions is sufficient to reconstruct the searched-for function f up to but not beyond bandwidth p. A little more precisely, Logan shows that a function in $L_2(\Omega)$ is of essential bandwidth $p(1 - \varepsilon)$, it can be essentially reconstructed from any p views. On the other band, he proves that there exist functions of essential bandwidth $p(1 + \varepsilon)$ which project to zero in any p given directions, as reformulated by Shepp–Kruskal [46].

Correct filtering of the functions in the reconstruction process can also reduce the influence of the ghosts. As an incentive for the following considerations, we consider a substantial simplification of the proof of Logan [25] which sheds light upon the essential properties of the ghosts and both improves the results and generalizes them to higher dimensions. The presentation follows Louis [29].

We study the behavior of the spectrum of the ghosts and show that they are high-frequency functions. To this end, we compare the energy of their Fourier transform lying inside a ball around 0 of radius c with their total energy. Remembering the condition from Theorem 8.2.9, where the number of given projections, p, is related to the integer n by $p = \dim \mathscr{P}_{n-1}$ we define the following.

Definition 8.3.5. The quotient between the energy of a function inside a ball $V(0,c)$ around 0 with radius c and the total energy is defined as

$$\mu_n(c) = \sup\left\{\frac{\|\hat{f}\|^2_{L_2((V(0,c))}}{\|f\|^2_{L_2(\mathbb{R}^N)}} : f \in \mathscr{N}(\mathscr{A}_p)\right\} \tag{34}$$

where we have used Parseval's relation that $\|f\|_{L_2(\mathbb{R}^N)} = \|\hat{f}\|_{L_2(\mathbb{R}^N)}$.

If this size is small, then most of the power of the ghosts is lying outside $V(O,c)$ and a cut-off of the frequencies larger than c prevents these functions form disturbing the reconstructions.

In order to determine $\mu_n(c)$, we perform the following steps, always assuming that the conditions from Theorem 8.2.9 are fulfilled. For details, see [29] and for helpful tables of integrals see [2, 15].

We use the *Hankel* transform defined as

$$\mathscr{H}_\nu^N u(s) = s^{1-N/2} \int_0^\infty u(\sigma)\sigma^{N/2} J_{\nu+N/2-1}(s\sigma)\, d\sigma$$

where J_ν is the Bessel function of the first kind of order ν. The Fourier transform of a function $f(r\omega) = u(r)Y_{\lambda\mu}(\omega)$ can be computed with the help of the Hankel transform to

$$\mathscr{F}_N f(\sigma\omega) = \mathscr{H}_\lambda^N u(\sigma) Y_{\lambda\mu}(\omega).$$

In a first step, we connect $\mu_n(c)$ with the largest eigenvalue in modulus $\gamma_{n+N/2-1,0}(c)$ of an integral operator with kernel

$$K(s,\sigma) = (s\sigma)^{1/2} J_{n+N/2-1}(cs\sigma), \quad 0 \le s, \sigma \le 1$$

and prove

$$\mu_n(c) = c^2\gamma^2_{n+N/2-1,0}(c).$$

Classical results for the eigenvalues of integral operators, for special functions and a result of Slepian [47] lead to the mentioned equality.

It remains to calculate lower and upper bounds for $\mu_n(c)$.

Theorem 8.3.6. *Let the conditions from Theorem 8.2.9 be fulfilled. Let c_n be a sequence such that*

$$c_n \le (n + N/2 - 1) - \alpha(n + N/2 - 1)^\beta \tag{35}$$

with $\alpha > 0$ and $\frac{1}{3} < \beta < 1$. Then

$$\lim_{n\to\infty} \mu_n(c_n) = 0.$$

For a lower bound, we find the following.

Theorem 8.3.7. *Let the conditions from Theorem 8.2.9 be fulfilled. Let c_n be a sequence such that*

$$c_n \geq (n + N/2 - 1) + \alpha(n + N/2 - 1)^{\beta} \tag{36}$$

with $\alpha > 0$ and $\beta > 1/2$. Then

$$\lim_{n\to\infty} \mu_n(c_n) = 1.$$

Following the sampling theorem objects of the size $(n + N/2 - 1)^{-1}$ or larger can reliably be recovered in the reconstruction if p directions are used such that condition from Theorem 8.2.9 is fulfilled. Smaller details may be disturbed by functions in the null space.

Shannon's sampling theorem also states that a sampling distance $\Delta s = \pi(n+N/2-1)^{-1}$ is sufficient if the frequencies larger than $n + N/2 - 1$ are filtered. Therefore, we can replace the demand for complete projections by the condition that the data are sampled with a rate less than or equal to Δs.

Note that the cut-off frequency in $\mathbb{R}^N$ is $n+N/2-1$ which in the case $N = 2$ coincides with the number p of given directions. In general, it is $O(p^{1/(N-1)})$. The example in Figure 8.3 shows what happens if one tries to reconstruct disregarding these resolution limits: the ghost is too small to fulfill the limits of the above theory.

8.3.2 Restrictions on the functions space for improving the resolution, compressive sensing and deep learning

Already in 1974, Marr [36] published a uniqueness result when assuming the searched-for function is a polynomial of highest degree depending on the number of data. He even used a finite number of rays in a fan-beam geometry. Shepp–Kruskal [46] honorably mentioned that result, but criticized at the same time that polynomials are quite unnatural as space for the images to be reconstructed. Especially, the pictures in tomography are far from being smooth; they often are considered at most to be piecewise smooth. Hence, the same criticism includes the Theorems 8.2.6 and 8.2.11. Of course, everyone knows polynomials and it is clear that they are not the natural candidates.

More success was reached when the well-known polynomials are replaced by something less precise: the condition that was studied is sparsity, sparsity in a suitable basis. So it is up to the user to find the right basis or dictionary. When the user fails, it is then not the problem of the method. The method was named compressive sensing; see e. g., [6] or [8]. A statement that one can beat Shannon or Nyquist can be considered as dubious because the success is based on the restriction of the set of functions to be reconstructed. For a constant function, one measurement suffices and that cannot be improved, if exact data are considered.

In recent times, deep learning is the magic method. Neural networks are trained for special tasks. Neural networks were in use decades ago, but they were criticized that the layers were not physical. Now one increases largely the number of layers with simple operations, like convolutions, which leads to the expression deep, and one adds a miraculous nonlinear map in each step to be found by the user. With all the effort going in this direction by highly qualified colleagues, success is to be expected; see, e. g., [3, 4].

Bibliography

[1] I. Abraham, R. Abraham, M. Bergounioux, and G. Carlier. Tomographic reconstruction from a few views: a multi-marginal optimal transport approach. *Appl. Math. Optim.*, 75(1):55–73, 2017.

[2] M. Abramowitz and I. A. Stegun. *Handbook of Mathematical Functions*. Dover, New York, 1972.

[3] J. Adler and O. Öktem. Solving ill-posed inverse problems using iterative deep neural networks. *Inverse Probl.*, 33(12):124007, 2017.

[4] J. Behrmann, C. Etmann, T. Boskamp, R. Casadonte, J. Kriegsmann, and P. Maaß. Deep learning for tumor classification in imaging mass spectrometry. *Bioinformatics*, 34(7):1215–1223, 2018.

[5] L. Borg, J. Frikel, J. Jørgensen, and E. T. Quinto. Analyzing reconstruction artifacts from arbitrary incomplete X-ray CT data, 2018. arXiv:1707.03055.

[6] E. Candès and J. Romberg. Sparsity and incoherence in compressive sampling. *Inverse Probl.*, 23(3):969, 2007.

[7] S. Chandra, I. Svalbe, J.-P. Guedon, A. Kingston, and N. Normand. Recovering missing slices of the discrete fourier transform using ghosts. *IEEE Trans. Image Process.*, 21(10):4431–4441, 2012.

[8] K. Choi, J. Wang, L. Zhu, T. Suh, S. Boyd, and L. Xing. (Compressed sensing based cone-beam computed tomography reconstruction with a first-order method). *Med. Phys.*, 37(9):5113–5125, 2010.

[9] A. M. Cormack. Representation of a function by its line integrals, with some radiological applications. II. *J. Appl. Phys.*, 35(10):2908–2913, 1964.

[10] M. E. Davison. The ill-conditioned nature of the limited angle tomography problem. *SIAM J. Appl. Math.*, 43(2):428–448, 1983.

[11] E. Y. Derevtsov. Ghost distributions in the cone-beam tomography. *J. Inverse Ill-Posed Probl.*, 5(5):411–426, 1997.

[12] E. Y. Derevtsov, A. V. Efimov, A. K. Louis, and T. Schuster. Singular value decomposition and its application to numerical inversion for ray transforms in 2D vector tomography. *J. Inverse Ill-Posed Probl.*, 19(4–5):689–715, 2011.

[13] J. Frikel and E. T. Quinto. Characterization and reduction of artifacts in limited angle tomography. Inverse Probl., 29(12):125007, 21, 2013.

[14] I. M. Gel'fand, M. I. Graev, and N. Y. Vilenkin. *Generalized Functions*, volume 5. Academic Press, Boston, MA, 1966.

[15] I. S. Gradshteyn and I. M. Ryzhik. *Table of Integrals, Series, and Products*. Elsevier/Academic Press, Amsterdam, 7th edition, 2007. Translated from the Russian, Translation edited and with a preface by Alan Jeffrey and Daniel Zwillinger.

[16] B. N. Hahn. Null space and resolution in dynamic computerized tomography. *Inverse Probl.*, 32(2):025006, 21, 2016.
[17] K. M. Hanson, G. S. Cunningham, and S. S. Saquib. Inversion based on computational simulations. In *Maximum Entropy and Bayesian Methods (Boise, ID, 1997)*. Fund. Theories Phys., volume 98, pages 121–135. Kluwer Acad. Publ., Dordrecht, 1998.
[18] S. Helgason. The Radon transform on Euclidean spaces, compact two-point homogeneous spaces and Grassmann manifolds. *Acta Math.*, 113:153–180, 1965.
[19] G. T. Herman. *Image Reconstruction from Projections: The Fundamentals of Computerized Tomography*. Academic Press, New York, 1980.
[20] M. Jiang and G. Wang. Convergence of the simultaneous algebraic reconstruction technique (SART). *IEEE Trans. Image Process.*, 12(8):957–961, 2003.
[21] M. Katz. *Questions of Uniqueness and Resolution in Reconstruction from Projections*. Springer-Verlag, Berlin, New York, 1978.
[22] S. Kazantsev. Singular value decomposition for the cone-beam transform in the ball. *J. Inverse Ill-Posed Probl.*, 23(2):173–185, 2014.
[23] V. P. Krishnan and E. T. Quinto. Microlocal analysis in tomography. In *Handbook of Mathematical Methods in Imaging*, volumes 1, 2, 3, pages 847–902. Springer, New York, 2015.
[24] P. Kuppusamy, M. Chzhan, A. Samouilov, P. Wang, and J. Zweier. Mapping the spin-density and lineshape distribution of free radicals using 4D spectral-spatial EPR imaging. *J. Magn. Reson., Ser. B*, 107:116–125, 1995.
[25] B. F. Logan. The uncertainty principle in reconstructing functions from projections. *Duke Math. J.*, 42(4):661–706, 1975.
[26] A. K. Louis. Picture reconstruction from projections in restricted range. *Math. Methods Appl. Sci.*, 2(2):209–220, 1980.
[27] A. K. Louis. Ghosts in tomography—the null space of the radon transform. *Math. Methods Appl. Sci.*, 3(1):1–10, 1981.
[28] A. K. Louis. Optimal sampling in nuclear magnetic resonance (NMR) tomography. *J. Comput. Assist. Tomogr.*, 6:334–340, 1982.
[29] A. K. Louis. Nonuniqueness in inverse Radon problems: the frequency distribution of the ghosts. *Math. Z.*, 185(3):429–440, 1984.
[30] A. K. Louis. Orthogonal function series expansions and the null space of the radon transform. *SIAM J. Math. Anal.*, 15:621–633, 1984.
[31] A. K. Louis. Incomplete data problems in x-ray computerized tomography; I: singular value decomposition of the limited angle transform. *Numer. Math.*, 48(3):251–262, 1986.
[32] A. K. Louis. *Inverse und schlecht gestellte Probleme*. Teubner, Stuttgart, 1989.
[33] A. K. Louis and A. Rieder. Incomplete data problems in x-ray computerized tomography. II: truncated projections and region-of-interest tomography. *Numer. Math.*, 56:371–383, 1989.
[34] D. Ludwig. The Radon transform on Euclidean space. *Commun. Pure Appl. Math.*, 19:49–81, 1966.
[35] P. Maass. The x-ray transform: singular value decomposition and resolution. *Inverse Probl.*, 3(4):729, 1987.
[36] R. B. Marr. On the reconstruction of a function on a circular domain from a sampling of its line integrals. *J. Math. Anal. Appl.*, 45:357–374, 1974.
[37] R. B. Marr, C.-N. Chen, and P. C. Lauterbur. On two approaches to 3D reconstruction in NMR zeugmatography. In G. T. Herman and F. Natterer, editors, *Mathematical Aspects of Computerized Tomography*, pages 225–240. Springer, Berlin, Heidelberg, 1981.
[38] F. Natterer. *The Mathematics of Computerized Tomography*. Classics in Applied Mathematics. SIAM, Philadelphia, 2001.

[39] M. Quellmalz, R. Hielscher, and A. K. Louis. The cone-beam transform and spherical convolution operators. *Inverse Probl.*, 34(10):105006, 2018.

[40] E. T. Quinto. Null spaces and ranges for the classical and spherical Radon transforms. *J. Math. Anal. Appl.*, 90(2):408–420, 1982.

[41] G. Rigaud. On the inversion of the Radon transform on a generalized Cormack-type class of curves. *Inverse Probl.*, 29(11):115010, 21, 2013.

[42] T. Schuster. The 3D Doppler transform: elementary properties and computation of reconstruction kernels. *Inverse Probl.*, 16(3):701–722, 2000.

[43] R. T. Seeley. Spherical harmonics. *Am. Math. Mon.*, 73(4):115–121, 1966.

[44] C. E. Shannon. A mathematical theory of communication. *Bell Syst. Tech. J.*, 27:379–423, 623–656, 1948.

[45] V. A. Sharafutdinov. *Integral Geometry of Tensor Fields*. Inverse and Ill-Posed Problems Series. VSP, Utrecht, 1994.

[46] L. A. Shepp and J. B. Kruskal. Computerized tomography: the new medical x-ray technology. *Am. Math. Mon.*, 85(6):420–439, 1978.

[47] D. Slepian. Prolate spheroidal wave functions, Fourier analysis and uncertainty. IV. Extensions to many dimensions; generalized prolate spheroidal functions. *Bell Syst. Tech. J.*, 43:3009–3057, 1964.

[48] K. T. Smith, D. C. Solmon, and S. L. Wagner. Practical and mathematical aspects of the problem of reconstructing objects from radiographs. *Bull. Am. Math. Soc.*, 83(6):1227–1270, 1977.

[49] J. Vogelgesang and C. Schorr. Iterative region-of-interest reconstruction from limited data using prior information. *Sens. Imag.*, 18(1):16, 2017.

T. Schuster

9 The importance of the Radon transform in vector field tomography

Abstract: Vector tomography is the inverse problem of reconstructing a vector field from scalar data that are usually given as line integrals of the unknown field and are acquired, e. g., by ultrasound measurements. It has various applications ranging from medical imaging, oceanography to atmospheric sciences and even plasma physics. The mathematical model of vector field tomography is the longitudinal ray transform which is related to the Radon transform in many ways. This is the reason, why the development of inversion formulas as well as the study of mathematical properties for the ray transforms are tightly connected to properties of the Radon transform. This article gives an overview on the importance of the Radon transform for the achievement of inversion tools in 2D and 3D vector field tomography, where we especially focus to the parallel and cone beam measurement geometry.

Keywords: Radon transform, vector field tomography, parallel geometry, cone beam geometry, longitudinal ray transform

MSC 2010: 44A12, 65R10, 65R32

9.1 Introduction

Over the last four decades, mathematics in computerized tomography (CT) mainly was concerned with scalar X-ray tomography in different measurement settings and geometries. The mathematics of vector tomography started with Norton's famous article from 1988 [28] and is currently intensively developing. The application areas are widely spread. They range from the reconstruction of velocity fields of liquids and gas flows [14, 42], oceanography [34], electron transmission tomography [32], atmospheric sciences [3, 4, 13, 50], anisotropy properties of materials [20], plasma physics [9], geophysics, and medical diagnosis, to name only a few. The aim of vector field tomography consists of recovering a field $\mathbf{f}$ from integral measurements

$$y(L) = \int_L \mathbf{f} \cdot \tau_L \, \mathrm{d}\ell. \tag{1}$$

Note: Dedicated to Professor Dr. Armin Lechleiter (1982–2018). We miss you.

Acknowledgement: Part of the research of this article was funded by German Science Foundation (Deutsche Forschungsgesellschaft, DFG) under Schu 1978/12-1.

T. Schuster, Department of Mathematics, Saarland University, 66041 Saarbrücken, Germany, e-mail: thomas.schuster@num.uni-sb.de

https://doi.org/10.1515/9783110560855-009

Here, L is a straight line and τ_L its vector of direction. The data are, e. g., acquired by ultrasound time-of-flight or Doppler-shift measurements where $\mathbf{f}$ represents a flow field and L is the propagation path of the ultrasound beam between a transducer and a detector. Though currently propagation along more general rays such as geodesic curves of Riemannian metrics are intensively investigated (cf. [25, 30, 31, 35, 45, 46, 49]), this survey is confined to line integrals, since in this way the connection to the Radon transform is more obvious and directly linked to inversion and reconstruction techniques. It is quite obvious that the reconstruction problem for $\mathbf{f}$ from data (1) is underdetermined: we have a 2D or 3D vector field $\mathbf{f}$ and scalar data y, a fact that already has been addressed in Norton's pioneering article. We will specify this for the different measurement settings. There exists a vast literature for vector field tomography. Main standard results are found in Juhlin [14], Sparr and Stråhlén [43, 44], Defrise and Gullberg [6], Sharafutdinov [40, 41], Braun and Hauck [5], Balandin and Ono [2], Osman and Prince [29], or Schuster [38]. Singular value decompositions are used in [7]. We refer to Svetov et al. [47] as an article for vector field tomography in a refracting medium. The aim of this survey is not to present novel results, but to highlight in so many respects the great importance of the Radon transform for deriving inversion formulas and reconstruction methods as well as for investigating the mathematical properties of the integral operators that represent the mathematical model of vector field tomography.

Outline. This survey is subdivided in two parts: the parallel geometry in 2D and 3D vector field tomography and 3D cone beam vector tomography. The first part (Section 9.2) contains the classical results on vector tomography. We outline how the longitudinal ray transform is obtained as a mathematical model for vector tomography when using time-of-flight measurements. We present important properties of this transform and inversion techniques, all of which make wide use of the 2D Radon transform (Section 9.2.1). Section 9.2.2 deals briefly with 3D Doppler tomography. Again, we derive the longitudinal ray transform as a mathematical model and amazingly also there the 2D Radon transform has a deep impact to prove mathematical properties and to construct numerical solvers. Section 9.3 subsumes the research on 3D vector tomography in the practically relevant cone beam geometry. We present an asymptotic inversion formula in the form of an infinite series (Section 9.3.1) using vector spherical harmonics as well as an exact inversion formula that uses the asymptotic formula as a starting point (Section 9.3.2). Two key ingredients of these inversion techniques are a splitting of the 3D Radon transform on a normal and tangential part on the sphere as well as an identity that was achieved by Kazantsev and Schuster (Theorem 9.3.1) and resembles the famous formula of Grangeat. All these inversion methods are valid if the domain under consideration is the unit ball in $\mathbb{R}^3$. Finally, we sketch how the exact formula could be extended to general convex domains. The article is completed by a concluding section.

9.2 The parallel geometry

In this first section, we consider the *parallel geometry*, that means that all rays are parallel to each other (2D) or to one of the coordinate planes (3D). The first situation is described in Norton's fundamental article [28], the latter one in Juhlin's report [14]. In both situations, the Radon transform plays a crucial role for the deduction of inversion techniques. In Section 9.2.1, we briefly sketch the data acquisition by time-of-flight measurements leading to the 2D ray transform for vector fields as a mathematical model. We show the tight connection to the 2D Radon transform and how we obtain an inversion method of filtered backprojection type from this relation. Section 9.2.2 describes then Norton's theory to reconstruct a vector field from Doppler data. Also, there the 2D Radon transform plays a key role.

9.2.1 2D vector field tomography from time-of-flight measurements

Let $\Omega \subset \mathbb{R}^2$ be a convex, open, bounded domain with piecewise smooth boundary $\partial\Omega$ containing a moving fluid or gas. This situation can be interpreted as a projection of a 3D flow on the two-dimensional domain Ω. The flow field is given by

$$\mathbf{u}(x) = \mathbf{u}(x_1, x_2) \in \mathbb{R}^2, \quad x = (x_1, x_2) \in \Omega,$$

having magnitude $|\mathbf{u}(x_1, x_2)|$. The aim of 2D flow imaging is to recover $\mathbf{u}$ from ultrasonic time-of-flight (TOF) measurements which are acquired between two positions $\mathbf{a}$, $\mathbf{b}$ located at the boundary $\partial\Omega$. The local speed of sound is denoted by $c(x_1, x_2)$ where we assume $c = \text{const}$ in $\mathbb{R}^2 \backslash \overline{\Omega}$. We may assume that the ultrasound beam is traveling from $\mathbf{a}$ to $\mathbf{b}$ along a straight line L which is justified provided that the variations in c are small and the path length $|L|$ is rather short. The line L then just corresponds to the propagation of particles with velocity fields orthogonal to the wavefronts. The travel time $t(\mathbf{a}, \mathbf{b})$ which the signal needs for traveling from $\mathbf{a}$ to $\mathbf{b}$ is given by

$$t(\mathbf{a}, \mathbf{b}) = \int\limits_{L(\mathbf{a},\mathbf{b})} \frac{\mathrm{d}\ell(x)}{c(x) + \mathbf{u}(x) \cdot \tau_L},$$

where $L(\mathbf{a}, \mathbf{b})$ is the line connecting $\mathbf{a}$ and $\mathbf{b}$, τ_L means the vector of direction of $L(\mathbf{a}, \mathbf{b})$ with $|\tau_L| = 1$, and $\mathrm{d}\ell$ denotes the element of length. Writing

$$\frac{1}{c(x) + \mathbf{u}(x) \cdot \tau_L} = \frac{1}{c(x)\,(1 + (\mathbf{u}(x)/c(x)) \cdot \tau_L)}$$

and considering the right-hand side as function of $\mathbf{u}/c$, a first-order Taylor approximation gives

$$t(\mathbf{a}, \mathbf{b}) \approx \int\limits_{L(\mathbf{a},\mathbf{b})} \left(\frac{1}{c(x)} - \frac{\mathbf{u}(x) \cdot \tau_L}{c(x)^2} \right) \mathrm{d}\ell(x). \tag{2}$$

This is a fairly good approximation whenever $|\mathbf{u}|/c \ll 1$, which is the case in most of the relevant applications. Transmitting the ultrasound signal in the opposite direction from **b** to **a**, we get (2) with τ_L replaced by $-\tau_L$ and compute

$$t(\mathbf{a},\mathbf{b}) + t(\mathbf{b},\mathbf{a}) = 2 \int_{L(\mathbf{a},\mathbf{b})} \frac{\mathrm{d}\ell(x)}{c(x)}, \tag{3}$$

$$t(\mathbf{a},\mathbf{b}) - t(\mathbf{b},\mathbf{a}) = -2 \int_{L(\mathbf{a},\mathbf{b})} \frac{\mathbf{u}(x)\cdot\tau_L}{c(x)^2}\,\mathrm{d}\ell(x). \tag{4}$$

Norton suggests to compute $c(x)$ from (3) by applying any inversion scheme for scalar 2D computerized tomography, e. g., the filtered backprojection algorithm. But if $c(x)$ is not constant in Ω, then the ultrasound signals do not propagate along straight lines, but rather along geodesic curves of the Riemannian metric $g_{ij} = c^{-2}\delta_{ij}$ according to Fermat's principle. Computing $c(x)$ from travel times is a famous and very old mathematical problem with important applications, e. g., in geoscience. In [25, 35], the authors suggest numerical inversion schemes and refer to many articles in this field. Since we do not want to consider inhomogeneous media in this survey, we assume that $c = c_0$ in Ω, where c_0 is the sound speed of the background medium. This is even then a reasonable assumption, if the deviations of $c(x)$ from c_0 inside Ω are small. Hence we can define

$$\mathbf{f}(x) := -\frac{2\mathbf{u}(x)}{c_0^2}$$

and finally get for every $\mathbf{a},\mathbf{b} \in \partial\Omega$ the measurement data

$$T(\mathbf{a},\mathbf{b}) := t(\mathbf{a},\mathbf{b}) - t(\mathbf{b},\mathbf{a}) = \int_{L(\mathbf{a},\mathbf{b})} \mathbf{f}(x)\cdot\tau_L\,\mathrm{d}\ell(x). \tag{5}$$

The aim is to calculate **f** from $T(\mathbf{a},\mathbf{b})$. To develop an inversion procedure, Norton introduces a specific parametrization for lines $L(\mathbf{a},\mathbf{b})$. We define $B_R(0) := \{x \in \mathbb{R}^2 : |x| < R\}$ for $R > 0$. A line $L(\mathbf{a},\mathbf{b})$ is uniquely determined by its distance from the origin $s \in \mathbb{R}$ and a normal vector $\omega(\varphi) = (\cos\varphi, \sin\varphi)^t \in S^1$, where $S^1 := \{x \in \mathbb{R}^2 : |x| = 1\} = \partial B_1(0)$ denotes the 2D unit sphere, by

$$L(\mathbf{a},\mathbf{b}) := L(\varphi,s) := \{x \in \mathbb{R}^2 : x\cdot\omega(\varphi) = s\}, \quad s \in \mathbb{R},\ \varphi \in [0,\pi]; \tag{6}$$

see Figure 9.1. The parallel measurement geometry consists of the data acquisition using finitely many parallel lines $L(\mathbf{a},\mathbf{b})$ from different rotating ultrasound sources. The scheme is plotted in Figure 9.2. Parametrization (6) is quite usual in the mathematics of 2D computerized tomography; see, e. g., the books of Natterer [26] and Natterer and Wübbeling [27]. For convenience, we set

$$\mathbf{f}(x) = 0 \quad \text{in } \mathbb{R}^2\backslash\Omega. \tag{7}$$

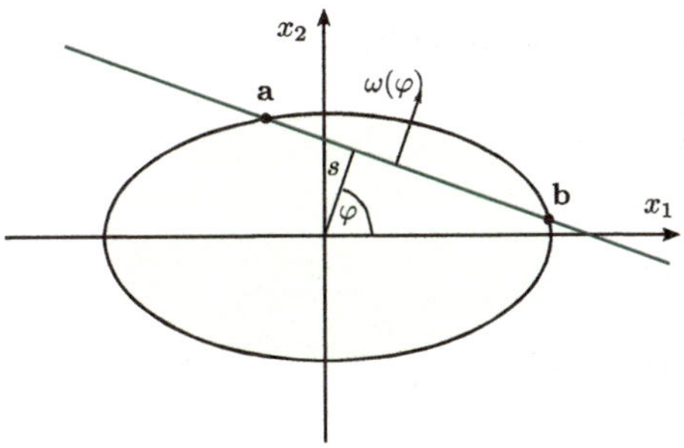

Figure 9.1: Parametrization of a line $L(\mathbf{a}, \mathbf{b})$.

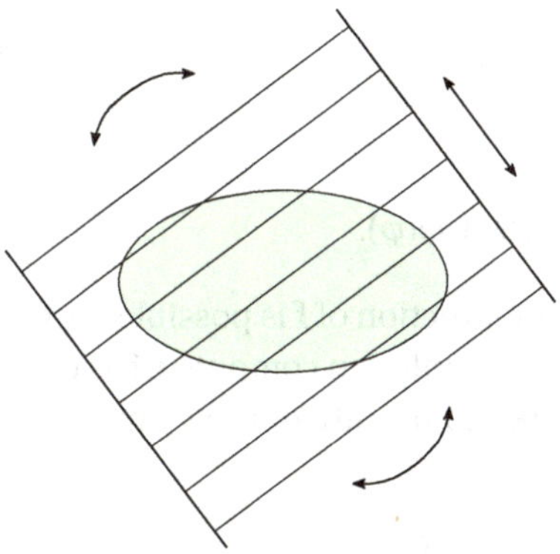

Figure 9.2: Sketch of the parallel measurement geometry.

This together with parametrization (6) leads to the definition of the longitudinal 2D ray transform for vector fields.

Definition 9.2.1 (Longitudinal and transversal 2D ray transform). The *longitudinal 2D ray transform* **D** maps a 2D vector field $\mathbf{f}(x)$ to its line integrals over the projections of **f** along the line

$$(\mathbf{Df})(\varphi, s) = \int_{L(\varphi,s)} \mathbf{f}(x) \cdot \omega(\varphi)^{\perp} \, \mathrm{d}\ell(x)$$
$$= \int_{\mathbb{R}} \mathbf{f}(s\omega(\varphi) + t\omega(\varphi)^{\perp}) \cdot \omega(\varphi)^{\perp} \, \mathrm{d}t,$$

where $\omega(\varphi)^{\perp} = (-\sin\varphi, \cos\varphi)^t$ is perpendicular to $\omega(\varphi)$ and denotes the vector of direction of $L(\varphi, s)$. The *transversal 2D ray transform* $\mathbf{D}^{\perp}$ maps a 2D vector field $\mathbf{f}(x)$ to

$$(\mathbf{D}^{\perp}\mathbf{f})(\varphi, s) = \int_{L(\varphi,s)} \mathbf{f}(x) \cdot \omega(\varphi) \, \mathrm{d}\ell(x)$$
$$= \int_{\mathbb{R}} \mathbf{f}(s\omega(\varphi) + t\omega(\varphi)^{\perp}) \cdot \omega(\varphi) \, \mathrm{d}t.$$

We define the 2D Radon transform for scalar functions f by

$$(Rf)(\varphi, s) = \int_{L(\varphi,s)} f(x)\, \mathrm{d}\ell(x).$$

It has a natural extension for vector fields $\mathbf{f}(x)$ as

$$(\mathbf{Rf})(\varphi, s) = \int_{L(\varphi,s)} \mathbf{f}(x)\, \mathrm{d}\ell(x) = (R\mathbf{f}_1, R\mathbf{f}_2)(\varphi, s).$$

There is a close connection between $\mathbf{D}$ and $\mathbf{R}$. Obviously,

$$(\mathbf{Df})(\varphi, s) = (\mathbf{Rf})(\varphi, s) \cdot \omega(\varphi)^{\perp} \tag{8}$$

and

$$(\mathbf{Rf})(\varphi, s) = (\mathbf{Df})(\varphi, s)\omega(\varphi)^{\perp} + (\mathbf{D}^{\perp}\mathbf{f})\omega(\varphi).$$

Since $\mathbf{R}$ is invertible, this means that a complete reconstruction of $\mathbf{f}$ is possible if $\mathbf{Df}$ as well as $\mathbf{D}^{\perp}\mathbf{f}$ are known. But from TOF measurements, we only can compute $\mathbf{Df}$ as data. Nevertheless, many mathematical properties of $\mathbf{D}$ like continuity or the central slice theorem (Theorem 9.2.2) are inherited from $\mathbf{R}$. Denoting by

$$(Ff)(\sigma) = \hat{f}(\sigma) = (2\pi)^{-d/2} \int_{\mathbb{R}^d} f(x)\, \mathrm{e}^{-\imath\sigma\cdot x}\, \mathrm{d}x$$

the d-dimensional *Fourier transform*, we can easily prove the central slice theorem for vector fields.

Theorem 9.2.2 (Central-slice theorem for 2D vector fields). *Let* $\mathbf{f} \in C_0^{\infty}(\Omega, \mathbb{R}^2)$. *Then*

$$(F\mathbf{Df})(\varphi, \sigma) = (2\pi)^{1/2}\, \hat{\mathbf{f}}(\sigma\omega(\varphi)) \cdot \omega(\varphi)^{\perp}, \tag{9}$$

where we have a 1D Fourier transform with respect to s on the left-hand side and the 2D Fourier transform with respect to x on the right-hand side.

With the help of the central slice theorem and the *Helmholtz decomposition* for vector fields $\mathbf{f} \in C_0^{\infty}(\Omega, \mathbb{R}^2)$,

$$\mathbf{f} = \nabla p + \nabla \times q, \tag{10}$$

where $p, q \in C_0^{\infty}(\Omega)$ and the 2D curl operator is defined by

$$(\nabla \times q)(x) = \left(\frac{\partial q}{\partial x_2}(x), -\frac{\partial q}{\partial x_1}(x)\right)^{\mathsf{T}}$$

one obtains the following reconstruction theorem for 2D flow fields from TOF measurements.

Theorem 9.2.3 (Reconstruction theorem for 2D flow fields). *Let $\Omega \subset \mathbb{R}^2$ be an open, bounded domain with piecewise smooth boundary $\partial\Omega$ which contains a fluid with flow field $\mathbf{u}$ and define $\mathbf{f} = -2\mathbf{u}/c_0^2$. We assume furthermore that the flow is incompressible, $\nabla \cdot \mathbf{f} = 0$, in Ω and that we know the boundary values of $\mathbf{f}$ at $\partial\Omega$. Then the field $\mathbf{f}$ can be completely reconstructed, if we have TOF measurements*

$$T(\mathbf{a},\mathbf{b}) = \int_{L(\mathbf{a},\mathbf{b})} \mathbf{f}(x) \cdot \tau_L \, \mathrm{d}\ell(x) = (\mathbf{Df})(\varphi, s)$$

for all $\mathbf{a}, \mathbf{b} \in \partial\Omega$ and $(\varphi, s) \in [0,\pi] \times \mathbb{R}$ such that $L(\mathbf{a},\mathbf{b}) = L(\varphi, s)$ according to the parametrization (6). More explicitly, the solenoidal part $\nabla \times q$ of $\mathbf{f}$ can be calculated from the data $T(\mathbf{a},\mathbf{b})$ by

$$\begin{aligned}\nabla \times q(r\omega(\theta)) &= (2\pi)^{-1/2} \int_0^\infty \sigma \int_0^{2\pi} \imath\sigma\hat{q}(\sigma\omega(\varphi))\, (\sin\varphi, -\cos\varphi)^t \, \mathrm{e}^{\imath\sigma r \cos(\theta-\varphi)} \, \mathrm{d}\varphi \, \mathrm{d}\sigma \\ &= -(2\pi)^{-1} \int_0^\infty \sigma \int_0^{2\pi} (\mathbf{FDf})(\varphi,\sigma)\, (\sin\varphi, -\cos\varphi)^t \, \mathrm{e}^{\imath\sigma r \cos(\theta-\varphi)} \, \mathrm{d}\varphi \, \mathrm{d}\sigma. \end{aligned} \quad (11)$$

The potential part ∇p of $\mathbf{f}$ is computed as

$$\nabla p(x) = -\frac{1}{4\pi} \int_{\partial\Omega} \frac{x - x'}{|x - x'|^2} \mathbf{f}(x') \cdot \mathbf{n}_{x'} \, \mathrm{d}s_{x'} \quad (12)$$

and $\mathbf{f}(x) = \nabla p(x) + \nabla \times q(x)$. Here, $\mathbf{n}_x$ denotes the outer unit normal field to $\partial\Omega$.

The identity (11) follows from the central slice theorem, whereas the representation (12) uses the fact that $\nabla \cdot \mathbf{f} = \Delta p = 0$. For detailed proofs, we refer to [28, 38].

Remark 9.2.4. A view on (12) tells us that it is sufficient to know the boundary values of the normal component $\mathbf{f} \cdot \mathbf{n}_x$ on $\partial\Omega$. If there is no flux through the boundary, we may assume

$$\mathbf{f}_{|\partial\Omega} \cdot \mathbf{n}_x = 0$$

yielding $\nabla p = 0$. The divergence-free field $\mathbf{f}$ then consists only of the solenoidal part $\nabla \times q$ which can be computed from TOF data $T(\mathbf{a},\mathbf{b})$.

Formula (11) is the basis for methods using Fourier techniques. A drawback is the fact that $\nabla \times q$ is computed on a polar grid, and thus a further interpolation step has to be included for visualizing the field on a rectangular mesh grid. The equality (12) is also problematic from a numerical point of view because of the singularity in the integrand. One would rather first compute $\nabla \times q$ by (11) and then the potential p as solution of a Poisson equation, e. g., if $\nabla \cdot \mathbf{f} = 0$ and $\mathbf{f} \cdot \mathbf{n}_x = g$ on $\partial\Omega$, then p solves the Neumann problem

$$\Delta p = 0 \quad \text{in } \Omega, \qquad \frac{\partial p}{\partial \mathbf{n}_x} = g - \mathbf{n}_x \cdot (\nabla \times q) \quad \text{on } \partial\Omega, \tag{13}$$

which comes from first applying the operator of divergence to both sides of (10) followed by taking the normal component of the decomposition.

TOF measurements to recover 2D fluid flow are also considered by Braun and Hauck [5]. They used the inversion formula for the 2D Radon transform $\mathbf{R}$ as starting point and obtained potentials p, q by

$$p(x) = \frac{1}{2\pi} \int_0^\pi \int_{-R}^R \frac{1}{x \cdot \omega(\varphi) - s} (\mathbf{D}^\perp \mathbf{f})(\varphi, s)\, \mathrm{d}s\, \mathrm{d}\varphi,$$

$$q(x) = -\frac{1}{2\pi^2} \int_0^\pi \int_{-R}^R \frac{1}{x \cdot \omega(\varphi) - s} (\mathbf{D}\mathbf{f})(\varphi, s)\, \mathrm{d}s\, \mathrm{d}\varphi.$$

It is remarkable that there is no differentiation of the data with respect to s as in Radon's inversion formula (see, e. g., [26, Formula (2.5)]), since this differentiation is already contained in the data $\mathbf{Df}$ taking the Helmholtz decomposition into account. An experimental setup for flow investigations is also described in [5].

The following lemma is important for deriving inversion formulas of filtered backprojection type and emphasizes once more the tight connection between the ray transforms for vector fields and the Radon transform.

Lemma 9.2.5. *Let* $\mathbf{f} \in C_0^\infty(\Omega, \mathbb{R}^2)$. *Then*

$$(\partial_s \mathbf{D}\mathbf{f})(\varphi, s) = \mathbf{R}(\nabla \times_\perp \mathbf{f})(\varphi, s), \tag{14}$$

$$(\partial_s \mathbf{D}^\perp \mathbf{f})(\varphi, s) = \mathbf{R}(\nabla \cdot \mathbf{f})(\varphi, s). \tag{15}$$

Here, ∂_s *denotes differentiation with respect to* s *and*

$$\nabla \times_\perp \mathbf{f} = \frac{\partial \mathbf{f}_2}{\partial x_1} - \frac{\partial \mathbf{f}_1}{\partial x_2}.$$

The proof uses differentiation rules for the Radon transform and is found in [38]. Lemma 9.2.5 shows insights of the information about $\mathbf{f}$ that is contained in data $\mathbf{Df}$, respectively, $\mathbf{D}^\perp\mathbf{f}$. Without further information on $\mathbf{f}$, we can only compute the curl $\nabla \times_\perp \mathbf{f}$ from $\mathbf{Df}$, whereas the divergence can be recovered from $\mathbf{D}^\perp\mathbf{f}$. Lemma 9.2.5 is furthermore fundamental for using filtered backprojection to obtain $\nabla \times_\perp \mathbf{f}$, respectively, $\nabla \cdot \mathbf{f}$ as well as for the computation of reconstruction kernels according to the *method of approximate inverse* [21, 22, 24]. We subsume these results that are taken from [38].

Theorem 9.2.6. *Let* $\Omega = B_1(0)$, $\mathbf{f} \in C_0^\infty(B_1(0), \mathbb{R}^2)$, $\phi \in L^2(\mathbb{R})$ *such that* $0 \le \hat{\phi} \le 1$, $\operatorname{supp}(\hat{\phi}) \subset [0,1]$, *and define*

$$\hat{V}_b(\xi) = \frac{1}{2\pi}\,\hat{\phi}\Big(\frac{|\xi|}{b}\Big), \quad \hat{v}_b(\sigma) = \frac{1}{2}(2\pi)^{-3/2}\,|\sigma|\,\hat{\phi}\Big(\frac{|\sigma|}{b}\Big). \tag{16}$$

Then

$$V_b * (\nabla \times_\perp \mathbf{f})(x) = \int_0^{2\pi}\int_{-\infty}^{\infty} v_b'(x\cdot\omega(\varphi) - s)(\mathbf{D}\mathbf{f})(\varphi, s)\,\mathrm{d}s\,\mathrm{d}\varphi,$$

$$V_b * (\nabla \cdot \mathbf{f})(x) = \int_0^{2\pi}\int_{-\infty}^{\infty} v_b'(x\cdot\omega(\varphi) - s)(\mathbf{D}^\perp\mathbf{f})(\varphi, s)\,\mathrm{d}s\,\mathrm{d}\varphi.$$

The recipe to construct the low-pass filter v_b follows exactly the lines of standard filtered backprojection for the 2D Radon transform; we refer to [26, 27]. A smoothed version of $\mathbf{f}$ can be obtained by applying the method of approximate inverse. Using translation and dilation invariance of $\mathbf{D}$, we again end with a formula of filtered backprojection type,

$$\mathbf{f}_\gamma^j(x) = \frac{1}{\gamma^2}\int_0^{2\pi}\int_{-1}^{1} (\mathbf{D}\mathbf{f})(\varphi, s)\,\Phi^j\Big(\varphi, \frac{s - x\cdot\omega(\varphi)}{\gamma}\Big)\,\mathrm{d}s\,\mathrm{d}\varphi, \quad j = 1,2. \tag{17}$$

Here, Φ^j are reconstruction kernels satisfying

$$(\mathbf{D}\mathbf{D}^*\Phi^j)(x) = \mathbf{D}\mathbf{e}^j(\cdot, x), \quad j = 1,2$$

for $\mathbf{e}^j(x) = e(x)\cdot e^j$, where e^j denote the standard unit vectors, e is a smooth function on $B_1(0)$ with mean value one, a so-called *mollifier*, and

$$(\mathbf{D}^* g)(x) = \int_0^{2\pi} g(\varphi, x\cdot\omega(\varphi))\,\omega^\perp(\varphi)\,\mathrm{d}\varphi$$

is the *backprojection operator* of $\mathbf{D}$ which is identical with the L^2-adjoint of $\mathbf{D}$. One can prove that

$$\lim_{\gamma\to 0} \mathbf{f}_\gamma = \mathbf{P}_{N(\mathbf{D})^\perp}(\mathbf{f})$$

in $L^2(B_1(0), \mathbb{R}^2)$, where $\mathbf{f}_\gamma = (\mathbf{f}_\gamma^1, \mathbf{f}_\gamma^2)$, $\mathbf{P}_{N(\mathbf{D})^\perp}$ denotes the orthogonal projection onto $N(\mathbf{D})^\perp$ and $N(\mathbf{D})$ is the null space of $\mathbf{D}$. Since $\mathbf{D}$, in contrast to $\mathbf{R}$, is not injective, this is the best possible approximation result we can achieve. It holds that $N(\mathbf{D}) = \nabla H_0^1(B_1(0))$. Hence $\mathbf{P}_{N(\mathbf{D})^\perp}\mathbf{f}$ corresponds to the solenoidal part of $\mathbf{f}$ according to the Helmholtz decomposition (10).

9.2.2 3D Doppler tomography

Juhlin [14] described in 1992 an experimental setup for three-dimensional vector field tomography and proved that the solenoidal part of a flow field can be completely recovered from acoustic Doppler measurements if specific boundary values of $\mathbf{u}$ are available. In that sense, the reconstruction Theorem 9.2.9 is a 3D analog of Theorem 9.2.3.

We continue with the description of data acquisition by acoustic Doppler measurements and Juhlin's measurement setting, which is a kind of slice-by-slice parallel geometry; cf. also [19, 41].

If we emit a signal $\tau(t) = e^{\imath f_0 t}$ with frequency f_0 along a line L which hits a particle of the fluid with velocity v, then the frequency of the signal will be increased or decreased by the *Doppler shift*

$$\triangle = \frac{2 c f_0 v}{c^2 - v^2},$$

where c again denotes the velocity of sound within the medium. Note that it is necessary that the fluid contains particles causing a Doppler shift for this measurement procedure. Since usually $v \ll c$ the Doppler shift may be approximated by $\triangle \approx \kappa v$ with $\kappa = 2f_0/c$, which means that $\triangle$ is approximately proportional to the velocity v of the particle causing the Doppler shift. The signal $\sigma(t)$ which is received at the detector consists then of a superposition of reflections from particles along L contributing to the signal,

$$\sigma(t) = \frac{1}{2\pi} \int_{\mathbb{R}} e^{\imath (f_0 + \kappa v) t} \, \mathrm{d}S(\mathbf{u}, L, v), \quad \triangle = \kappa v. \tag{18}$$

Here, $\mathrm{d}S$ is a positive Radon measure characterized by

$$\mathrm{d}S(\mathbf{u}, L, v) = \operatorname{meas}\{x \in L : v \le \mathbf{u} \cdot \theta_L(x) < v + \mathrm{d}v\}, \tag{19}$$

where $\theta_L \in S^2 = \{x \in \mathbb{R}^3 : |x| = 1\} = \partial B_1(0)$ is the vector of direction of L. The measure $\mathrm{d}S$ is called the *velocity spectrum* of $\mathbf{u}$ and can be interpreted as the number of particles with velocity v along L. Defining

$$S(\mathbf{u}, L, v) = \operatorname{meas}\{x \in L : \mathbf{u} \cdot \theta_L(x) < v\}$$

the velocity spectrum $\mathrm{d}S$ has the representation $\mathrm{d}S = S' \, \mathrm{d}v$, where S' is the derivative of S with respect to v. Hence (18) essentially is the inverse Fourier transform of S' and we obtain the first moment of the velocity spectrum using Fourier techniques

$$\int_{\mathbb{R}} v \, \mathrm{d}S(\mathbf{u}, L, v) = \int_L \mathbf{u} \cdot \theta_L(x) \, \mathrm{d}\ell(x) =: \mathbf{Du}(L). \tag{20}$$

We call the mapping $\mathbf{u} \mapsto \mathbf{Du}(L)$ the *Doppler transform*.

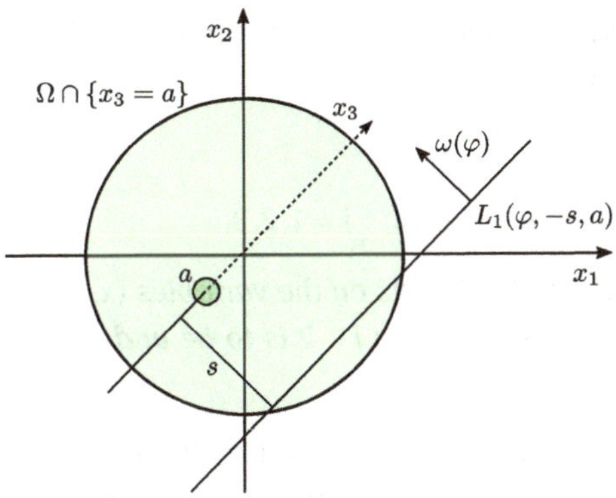

Figure 9.3: Parametrization of a line $L_1(\varphi, s, a)$.

Definition 9.2.7 (Longitudinal ray transform for 3D vector fields). Let $\Omega \subset \mathbb{R}^3$ be an open, bounded domain with piecewise C^1-boundary $\partial\Omega$, let $\mathbf{u} \in L^2(\Omega, \mathbb{R}^3)$ be a vector field and L be a line with vector of direction $\theta_L \in S^2$. The mapping

$$(\mathbf{D}\mathbf{u})(L) = \int_L \mathbf{u}(x) \cdot \theta_L \, d\ell(x)$$

is called *longitudinal ray transform for 3D vector fields* or *Doppler transform*.

The experimental setup suggested by Juhlin is even more specific since he considers only those lines L being parallel to one of the coordinate planes $\{e_j^\perp\}$. The object Ω is then scanned slice-by-slice where in each slice we have the parallel geometry as in Figure 9.3. Taking this geometry into account when parametrizing the lines L, we end up with the operator $\mathbf{D} = (\mathbf{D}_1, \mathbf{D}_2, \mathbf{D}_3)$ stating the mathematical model of Juhlin's geometry. Here,

$$(\mathbf{D}_j\mathbf{u})(\varphi, s, a) = \int_{L_j(\varphi,s,a)} \mathbf{u}(x) \cdot \theta_j(\varphi) \, d\ell(x), \quad j = 1, 2, 3,$$

where

$$L_j(\varphi, s, a) = \{x \in \mathbb{R}^3 : x \cdot \mathcal{P}_j\omega(\varphi) = s, x \cdot e_{j+2} = a\}, \quad j = 1, 2, 3, \tag{21}$$

$\theta_j(\varphi) = e_{j+2} \times \mathcal{P}_j\omega(\varphi)$ is a vector of direction of L_j, and $\mathcal{P}_j : \mathbb{R}^2 \to \{e_{j+2}^\perp\} \subset \mathbb{R}^3$ are canonical embeddings. The index $j+2$ in e_{j+2} is always to be understood modulo 3, i. e., $e_4 = e_1$, $e_5 = e_2$. The canonical embedding $\mathcal{P}_1$, e. g., is given by $\mathcal{P}_1(x_1, x_2) = (x_1, x_2, 0)^t$. The parameters are defined for $\varphi \in [0, 2\pi]$, $s, a \in \mathbb{R}$. If we again define $\Omega = B_1(0)$, then s and a can be restricted to the interval $[-1, 1]$. The parametrization (21) means that we define a line uniquely by its distance s from the e_{j+2}-coordinate axis, its distance a from the plane $\{e_{j+2}^\perp\}$, and the polar angle relative to the e_{j+2}-coordinate axis. In this way, we collect all lines being parallel to one of the three coordinate planes $\{e_j^\perp\}$. Figure 9.3 illustrates (21).

Using well-known differentiation properties of the Radon transform, we obtain an important relation to **R** that is an analog to (14).

Lemma 9.2.8. *If* $\mathbf{u} \in C_0^\infty(\Omega, \mathbb{R}^3)$, *then*

$$(\partial_s \mathbf{D}_j \mathbf{u})(\varphi, s, a) = (\mathbf{R} \otimes I)(e_{j+2} \cdot \nabla \times \mathbf{u})(\varphi, s, a), \quad j = 1, 2, 3, \tag{22}$$

where the 2D Radon transform **R** *on the right-hand side acts on the variables* (x_j, x_{j+1}), *I is the identity mapping acting on* x_{j+2}, *and the summation* $j + 2$ *is to be understood modulo* 3.

Juhlin's measurement geometry is in fact a restriction of the parallel geometry since not all lines are considered but only those which are parallel to one of the three coordinate planes. Though this is a limited data problem, it seems surprising that this does not mean a loss of information and the full field can be reconstructed if only its divergence and boundary values are available. This is also proved by Sharafutdinov [41] and Schuster [36]. The following theorem plays a key role in the reconstruction theory of 3D vector fields. Its proof is found in [14].

Theorem 9.2.9 (Reconstruction of a 3D flow field). *Let* Ω *be a simply connected bounded domain in* $\mathbb{R}^3$ *with piecewise* C^1*-boundary* $\partial\Omega$. *Provided that* $\mathbf{q} \in L^2(\Omega, \mathbb{R}^3)$, $\rho \in L^2(\Omega)$ *and* $\zeta \in H^{1/2}(\partial\Omega)$ *are given such that*

$$\int_\Omega \rho(x)\,\mathrm{d}x = \int_{\partial\Omega} \zeta(x)\,\mathrm{d}s_x \tag{23}$$

is valid, then the boundary value problem

$$\begin{aligned} \nabla \times \mathbf{u} &= \mathbf{q} \quad \text{in } \Omega, \\ \nabla \cdot \mathbf{u} &= \rho \quad \text{in } \Omega, \\ \mathbf{n}_x \cdot \mathbf{u} &= \zeta \quad \text{on } \partial\Omega \end{aligned}$$

has a unique solution in $H^1(\Omega, \mathbb{R}^3)$.

Again, by using the close connection to the Radon transform (22) it is possible to compute reconstruction kernels as solutions of

$$\mathbf{D}_j \mathbf{D}_j^* \Phi^j = \mathbf{D}(e^j \cdot e_j), \quad j = 1, 2, 3, \tag{24}$$

where $\mathbf{e}^j(x) = e(x) \cdot e^j$ and e is a mollifier having mean value 1. By $\mathbf{D}_j^*$, we denote the L^2-adjoint of $\mathbf{D}_j$. The method of approximate inverse again yields an inversion method of filtered backprojection type,

$$\begin{aligned} \mathbf{u}_\gamma^j(x) := \frac{1}{\gamma^3} \int_{-1}^{1} \int_0^{2\pi} \int_{-1}^{1} & (\mathbf{D}_j \mathbf{u})(\varphi, s, a) \\ & \times \Phi^j\left(\varphi, \frac{s - \mathcal{P}_j^* x \cdot \omega(\varphi)}{\gamma}, \frac{a - x \cdot e_{j+2}}{\gamma}\right) \mathrm{d}s\,\mathrm{d}\varphi\,\mathrm{d}a, \end{aligned} \tag{25}$$

for a regularization parameter $\gamma > 0$. Note that to reconstruct $\mathbf{u}_\gamma^j$ we only need the data set $\mathbf{D}_j\mathbf{u}$. Hence, the three components $\mathbf{u}_\gamma^j$ can be computed in parallel by an efficient inversion scheme, which is outlined in [36].

9.3 3D cone beam vector tomography

Although the parallel geometry leads to inversion methods by using obvious and close relationships to the Radon transform, it is very elaborate and not appropriate for real-world applications. Hence, in analogy to scalar X-ray tomography, the cone beam geometry has more relevance from a practical point of view. Over the last years, numerous inversion formulas have been developed for the scalar cone beam X-ray transform by many authors; see, e. g., [10, 12, 15, 16, 19, 23, 48]. A close connection to the Radon transform is given by the famous *formula of Grangeat* [11] and it is also possible to deduce an inversion formula, if only the X-ray source trajectory satisfies *Tuy's condition.* Most popular inversion formulas for the cone beam X-ray transform use Hamaker's identity [23] or the concept of π-lines [15]. An inversion formula for the vectorial 3D cone beam transform was not known for a long time. A first attempt which uses a gerneralization of Grangeat's formula to tensor fields [37] was not successful. An approach relying on the method of approximate inverse was derived in [39]. Though this approach is inexact, it shows a good performance for vector fields of a particular structure. It was a big surprise that the key ideas to finally obtain an inversion formula was a splitting of the 3D Radon transform into its normal and tangential part on the unit sphere and a formula of Kazantsev and Schuster [19] that seems to be similar to Grangeat's formula but in fact is quite different from the latter. We outline the complete history but omit the proofs that can be found in the corresponding literature which we refer to.

We consider the problem of reconstructing a smooth vector field $\mathbf{f}$, supported in the open unit ball $\mathbb{B} := B_1(0) \subset \mathbb{R}^3$, i. e., $\Omega = \mathbb{B}$, from its cone beam data

$$(\mathbf{Xf})(y(p), \eta) = \int_0^\infty \mathbf{f}(y(p) + t\eta) \cdot \eta \, \mathrm{d}t. \tag{26}$$

Here, $y(p)$, $p \in \Lambda \subset \mathbb{R}$, denotes the parametrization of a source trajectory $\Gamma \subset (\mathbb{R}^3 \backslash \overline{\mathbb{B}})$ and $\eta \in S^2 = \partial\mathbb{B}$ is the unit vector along the ray that emanates from the source $y(p)$. It is assumed that $r\eta \in C$ for all $r \geq 0$, where $C \subset \mathbb{R}^3$ is a cone, and that $\mathbb{B} \subset y(p) + C$, i. e., the unit ball is completely contained inside the union of the rays emanating from any source position $y(p)$. The situation is illustrated in Figure 9.4 for Γ being a single circle.

It is obvious that $\mathbf{X}$ has a non-trivial null space; we have $\nabla H_0^1(\mathbb{B}) \in N(\mathbf{X})$. Hence, to reconstruct the solenoidal part $\mathbf{f}^s$ of $\mathbf{f}$ from $\mathbf{Xf}$ is the best we can hope for. We refer

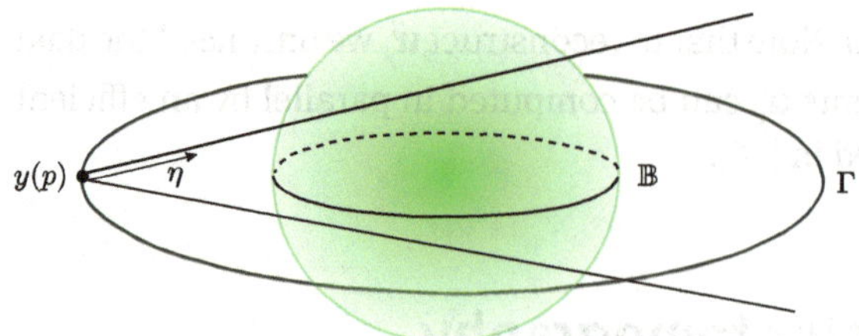

Figure 9.4: Sketch of the cone beam geometry. The source trajectory Γ consists of a single circle.

to [40] for proofs. More information about **f** can be recovered only with the help of additional data such as the divergence of **f**. We address this issue in Section 9.3.1. One key ingredient of the inversion formulas in Sections 9.3.1 and 9.3.2 is a splitting of the 3D Radon transform for vector fields on $\mathbb{B}$,

$$(\mathbf{Rf})(\eta, s) = \int_{x\cdot\eta=s} \mathbf{f}(x)\,dx = \int_{\eta^\perp} \mathbf{f}(s\eta + y)\,dy \tag{27}$$

into its tangential and normal part,

$$\mathbf{R} = \mathbf{R}^{(\tan)} + \mathbf{R}^{(\mathrm{nor})} \tag{28}$$

with

$$(\mathbf{R}^{(\mathrm{nor})}\mathbf{f})(\eta, s) := (\eta \cdot (\mathbf{Rf})(\eta, s))\eta,$$
$$(\mathbf{R}^{(\tan)}\mathbf{f})(\eta, s) := (\mathbf{Rf})(\eta, s) - (\mathbf{R}^{(\mathrm{nor})}\mathbf{f})(\eta, s),$$

where $\eta \in S^2$, $s \in [-1, 1]$. Before recapitulating the formula of Kazantsev and Schuster from [19], we recall some concepts of differential operators on the sphere S^2.

Let η be a unit vector in spherical coordinates

$$\eta = \begin{pmatrix} \cos\varphi \sin\theta \\ \sin\varphi \sin\theta \\ \cos\theta \end{pmatrix} \in S^2, \quad \varphi \in [0, 2\pi),\ \theta \in [0, \pi].$$

There are two orthogonal vectors spanning the tangent plane $\eta^\perp$,

$$\mathbf{e}_1(\eta) := \frac{1}{\sin\theta}\frac{\partial}{\partial\varphi}\eta = \begin{pmatrix} -\sin\varphi \\ \cos\varphi \\ 0 \end{pmatrix}, \quad \mathbf{e}_2(\eta) := \frac{\partial}{\partial\theta}\eta = \begin{pmatrix} \cos\varphi\cos\theta \\ \sin\varphi\cos\theta \\ -\sin\theta \end{pmatrix}. \tag{29}$$

The *surface gradient* of a scalar function $u(\eta)$ on the sphere S^2 is denoted by $\nabla_\eta u(\eta)$ and defined as

$$\nabla_\eta u := \frac{1}{\sin\theta}\frac{\partial u}{\partial\varphi}\mathbf{e}_1(\eta) + \frac{\partial u}{\partial\theta}\mathbf{e}_2(\eta).$$

The surface gradient hence neglects the derivative with respect to the normal direction η. In that sense, we have $\nabla_\eta = \nabla - (\eta \cdot \nabla)\eta$. A vector field $\mathbf{f}(\eta)$ on the plane $\eta^\perp$ has a unique representation

$$\mathbf{f}(\eta) = f_\varphi \mathbf{e}_1(\eta) + f_\theta \mathbf{e}_2(\eta).$$

The *surface divergence* div_η of $\mathbf{f}$ is then defined as

$$\operatorname{div}_\eta \mathbf{f} = \nabla_\eta \cdot \mathbf{f}(\eta) = \frac{1}{\sin\theta}\left(\frac{\partial}{\partial\theta}(\sin\theta f_\theta) + \frac{\partial}{\partial\varphi} f_\varphi\right)$$

and fulfills

$$\operatorname{div}_\eta \mathbf{f} = \nabla_\eta \cdot \mathbf{f} = \nabla \cdot \mathbf{f} - \frac{\partial}{\partial\eta}(\mathbf{f} \cdot \eta) \tag{30}$$

provided that $\mathbf{f}$ is defined in a neighborhood of S^2.

A last notation, that is necessary, is the function

$$(\mathbf{X}^{(\mathrm{even})}\mathbf{f})(y, \eta) := \frac{(\mathbf{Xf})(y, \eta) + (\mathbf{Xf})(y, -\eta)}{2}$$

which is the even part of $\mathbf{Xf}$ and can easily be computed from the data.

Theorem 9.3.1. *Let* $\mathbf{f} \in C_0^\infty(\mathbb{B}, \mathbb{R}^3)$. *Then the identity*

$$\left(\operatorname{div}_\eta\left[\frac{\partial}{\partial s}\mathbf{R}^{(\tan)}\mathbf{f}\right](s, \eta)\right)\bigg|_{s=\eta\cdot y} + y \cdot \frac{\partial^2}{\partial s^2}(\mathbf{R}^{(\tan)}\mathbf{f})(s, \eta)\bigg|_{s=\eta\cdot y} = G(y, \eta) \tag{31}$$

holds true, where

$$G(y, \eta) := -\int_{S^2} (\mathbf{X}^{(\mathrm{even})}\mathbf{f})(y, \Theta)\delta''(\eta \cdot \Theta)\, \mathrm{d}\Theta, \quad y \in \Gamma. \tag{32}$$

Here, δ denotes Dirac's delta-distribution.

The proof of Theorem 9.3.1 is found in [19]. Regarding (31), the idea to obtain an inversion method for $\mathbf{X}f$ is obvious: If the div_η-term on the left-hand side cancels out by subtracting (31) from each other with different y corresponding to the same values of s, then we can compute $\partial_s^2 \mathbf{R}^{(\tan)}\mathbf{f}$ from $G(y, \eta)$. If we were able to also compute $\partial_s^2 \mathbf{R}^{(\mathrm{nor})}\mathbf{f}$ in some way, then we would obtain

$$\frac{\partial^2}{\partial s^2}(\mathbf{Rf})(\eta, s) = \frac{\partial^2}{\partial s^2}(\mathbf{R}^{(\tan)}\mathbf{f})(\eta, s) + \frac{\partial^2}{\partial s^2}(\mathbf{R}^{(\mathrm{nor})}\mathbf{f})(\eta, s) \tag{33}$$

and use the inversion formula for $\mathbf{R}$,

$$\mathbf{f}(x) = -\frac{1}{8\pi^2}\left(\mathbf{R}^* \frac{\partial^2}{\partial s^2}\mathbf{Rf}\right)(x) \tag{34}$$

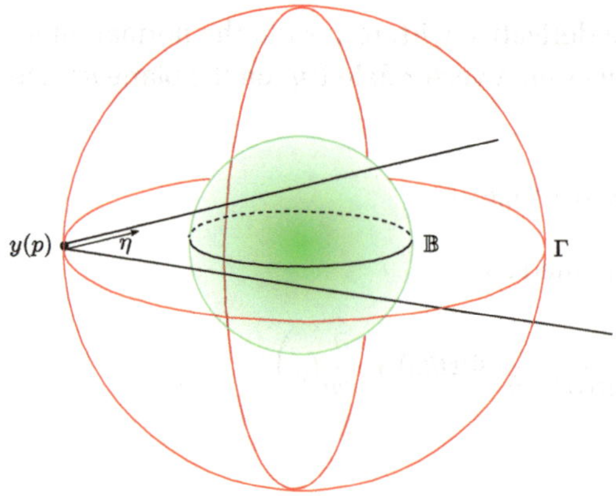

Figure 9.5: Source trajectory Γ consisting of three circles that are perpendicular to each other. This Γ is an example for a curve satisfying Tuy's condition of order 3 if the circles' radii are sufficiently large.

to finally obtain $\mathbf{f}$. As a matter of fact, step (a) of computing $\partial_s^2 \mathbf{R}^{(\tan)}\mathbf{f}$ from (31) is possible if the source trajectory satisfies *Tuy's condition of order* 3; see Definition 9.3.2, and step (b) of computing $\partial_s^2 \mathbf{R}^{(\mathrm{nor})}\mathbf{f}$ out of $\partial_s^2 \mathbf{R}^{(\tan)}\mathbf{f}$ is possible if $\mathbf{f}$ is solenoidal, i. e., $\nabla \cdot \mathbf{f} = 0$. Note that for numerical computations it is appropriate to replace δ'' in (32) by w_b'' with a smooth approximation w_b of δ, e. g., a smoothing kernel, as it was done in [17].

Definition 9.3.2 (Tuy's condition for vector tomography). We say that a source trajectory $\Gamma \subset (\mathbb{R}^3 \backslash \overline{\mathbb{B}})$ satisfies *Tuy's condition of order* 3, if any plane that passes through $\mathbb{B}$ intersects the trajectory Γ in at least 3 points that are not located on a line. That means to any $s \in [-1, 1]$ and $\eta \in S^2$ there exist at least three parameters $\lambda_i \in \Lambda$, $i = 1, 2, 3$ with

$$\mathbf{a}_1 \cdot \eta = \mathbf{a}_2 \cdot \eta = \mathbf{a}_3 \cdot \eta = s, \quad \mathbf{a}_i := \mathbf{a}(\lambda_i(s, \eta)), \quad i = 1, 2, 3 \tag{35}$$

and $\mathbf{a}_1 - \mathbf{a}_3$ and $\mathbf{a}_2 - \mathbf{a}_3$ are not collinear.

Figure 9.5 shows a trajectory Γ consisting of three circles that are perpendicular to each other and which fulfills Tuy's condition of order 3. We silently assume for the remainder of this section that the source trajectory Γ satisfies Tuy's condition for vector tomography, i. e., (35) is valid. In the following subsections, we sketch how to obtain inversion procedures that both rely on the fundamental relation (31) between Radon transform (27) and cone beam vector transform (26). We emphasize that the achievements of the articles [17–19] would not have been possible without having this very relationship.

9.3.1 An asymptotic inversion formula

Equation (31) implies that for s, η *fixed* we have three identities:

$$\mathrm{div}_\eta \left[\frac{\partial}{\partial s} \mathbf{R}^{(\tan)}\mathbf{f} \right](s, \eta) + \mathbf{a}_1 \cdot \frac{\partial^2}{\partial s^2} [\mathbf{R}^{(\tan)}\mathbf{f}](s, \eta) = G(\mathbf{a}_1, \eta), \tag{36}$$

$$\operatorname{div}_\eta\left[\frac{\partial}{\partial s}\mathbf{R}^{(\tan)}\mathbf{f}\right](s,\eta) + \mathbf{a}_2\cdot\frac{\partial^2}{\partial s^2}[\mathbf{R}^{(\tan)}\mathbf{f}](s,\eta) = G(\mathbf{a}_2,\eta), \tag{37}$$

$$\operatorname{div}_\eta\left[\frac{\partial}{\partial s}\mathbf{R}^{(\tan)}\mathbf{f}\right](s,\eta) + \mathbf{a}_3\cdot\frac{\partial^2}{\partial s^2}[\mathbf{R}^{(\tan)}\mathbf{f}](s,\eta) = G(\mathbf{a}_3,\eta). \tag{38}$$

Subtracting (38) from (36) and (37), respectively, leads to the system of linear equations

$$\begin{cases}\mathbf{d}_1\cdot\frac{\partial^2}{\partial s^2}[\mathbf{R}^{(\tan)}\mathbf{f}](s,\eta) = G(\mathbf{a}_1,\eta) - G(\mathbf{a}_3,\eta),\\ \mathbf{d}_2\cdot\frac{\partial^2}{\partial s^2}[\mathbf{R}^{(\tan)}\mathbf{f}](s,\eta) = G(\mathbf{a}_2,\eta) - G(\mathbf{a}_3,\eta)\end{cases} \tag{39}$$

with $\mathbf{d}_1 = \mathbf{a}_1 - \mathbf{a}_3$ and $\mathbf{d}_2 = \mathbf{a}_2 - \mathbf{a}_3$, that has to be solved for $\partial_s^2\mathbf{R}^{(\tan)}\mathbf{f}(s,\eta)$. Since $\partial_s^2\mathbf{R}^{(\tan)}\mathbf{f} \in \eta^\perp$, there exist coefficients $P(s,\eta), Q(s,\eta) \in \mathbb{R}$ such that

$$\frac{\partial^2}{\partial s^2}[\mathbf{R}^{(\tan)}\mathbf{f}](s,\eta) = P(s,\eta)\mathbf{e}_1(\eta) + Q(s,\eta)\mathbf{e}_2(\eta)$$

with $\mathbf{e}_1(\eta)$, $\mathbf{e}_2(\eta)$ as in (29). Putting this into system (39) gives

$$\begin{cases}P(s,\eta)\mathbf{d}_1\cdot\mathbf{e}_1(\eta) + Q(s,\eta)\mathbf{d}_1\cdot\mathbf{e}_2(\eta) = G(\mathbf{a}_1,\eta) - G(\mathbf{a}_3,\eta),\\ P(s,\eta)\mathbf{d}_2\cdot\mathbf{e}_1(\eta) + Q(s,\eta)\mathbf{d}_2\cdot\mathbf{e}_2(\eta) = G(\mathbf{a}_2,\eta) - G(\mathbf{a}_3,\eta)\end{cases} \tag{40}$$

and the determinant of the system matrix

$$\begin{vmatrix}\mathbf{d}_1(s,\eta)\cdot\mathbf{e}_1(\eta) & \mathbf{d}_1(s,\eta)\cdot\mathbf{e}_2(\eta)\\ \mathbf{d}_2(s,\eta)\cdot\mathbf{e}_1(\eta) & \mathbf{d}_2(s,\eta)\cdot\mathbf{e}_2(\eta)\end{vmatrix} \neq 0$$

since $\mathbf{d}_1$, $\mathbf{d}_2$ are not collinear, which implies that the solution of (40) is unique. Hence we can compute $\partial_s^2\mathbf{R}^{(\tan)}\mathbf{f}$ from (40) since the right-hand side is given by the data $\mathbf{X}^{(\text{even})}\mathbf{f}$.

In order to get $\partial_s^2\mathbf{Rf}(s,\eta)$ from (33), it remains to recover $\partial_s^2\mathbf{R}^{(\text{nor})}\mathbf{f}(s,\eta)$. If $\mathbf{f}$ is solenoidal, smooth, and $\mathbf{n}_x\cdot\mathbf{f} = 0$ for $x \in S^2$, then $\partial_s^2\mathbf{R}^{(\text{nor})}\mathbf{f} = 0$ and we are done. A proof of this assertion is found in [17, Section 3]. Hence we consider now a smooth vector field $\mathbf{f}$ with $\nabla\cdot\mathbf{f} = 0$ on $\mathbb{B}$, but with boundary values, that do not necessarily vanish. In fact, Theorem 9.3.1 remains valid if only $\mathbf{f} \in C^2(\mathbb{B},\mathbb{R}^3)$ by setting $\mathbf{f} = 0$ in $\mathbb{R}^3\backslash\overline{\mathbb{B}}$.

Starting with the spherical harmonics $Y_{n,l} \in L_2(S^2)$, we define *vector spherical harmonics* $\mathbf{y}_{n,l}^{(1)}(\xi)$, $\mathbf{y}_{n,l}^{(2)}(\xi)$, $\mathbf{y}_{n,l}^{(3)}(\xi)$ for $\xi \in S^2$ by

$$\mathbf{y}_{n,l}^{(1)}(\xi) = Y_{n,l}(\xi)\xi, \quad \mathbf{y}_{n,l}^{(2)}(\xi) = \nabla_\xi Y_{n,l}(\xi), \quad \mathbf{y}_{n,l}^{(3)}(\xi) = \xi\times\mathbf{y}_{n,l}^{(2)}(\xi), \tag{41}$$

where ∇_ξ denotes the surface gradient on $\xi^\perp$. The system

$$\{\mathbf{y}_{0,0}^{(1)}, \mathbf{y}_{n,l}^{(j)} : n \in \mathbb{N},\ |l| \leq n,\ j = 1,2,3\}$$

forms a complete orthogonal system in $L^2(S^2, \mathbb{R}^3)$ and we have

$$\xi \times \mathbf{y}^{(1)}_{n,l}(\xi) = \mathbf{0}, \quad \xi \cdot \mathbf{y}^{(2)}_{n,l}(\xi) = 0, \quad \xi \cdot \mathbf{y}^{(3)}_{n,l}(\xi) = 0. \tag{42}$$

We define furthermore the spaces

$$H_0(\mathrm{div}; \mathbb{B}) := \{\mathbf{f} \in H^1(\mathbb{B}, \mathbb{R}^3) : \mathrm{div}\, \mathbf{f} = 0,\ \xi \cdot \mathbf{f}(\xi) = 0,\ \xi \in S^2\}$$

of solenoidal vector fields having tangential flow at the boundary and the space of harmonic fields $\nabla\,\mathrm{Harm}(\mathbb{B})$, where

$$\mathrm{Harm}(\mathbb{B}) := \left\{h \in H^1(\mathbb{B}) : \int_{\mathbb{B}} \nabla h(x) \cdot \nabla v(x)\, \mathrm{d}x = 0, \text{ for all } v \in C_0^\infty(\mathbb{B})\right\}$$

is the space of harmonic functions. Then $L^2(\mathbb{B}, \mathbb{R}^3)$ allows for an orthogonal decomposition as

$$L^2(\mathbb{B}, \mathbb{R}^3) = \nabla H_0^1(\mathbb{B}) \oplus H_0(\mathrm{div}; \mathbb{B}) \oplus \nabla\,\mathrm{Harm}(\mathbb{B}), \tag{43}$$

which is the *Helmholtz–Hodge decomposition*. Relying on complete orthonormal systems for each of the spaces that appear in this decomposition, we find coefficients $b^{(n)}_{n+1-2k,l}, c^{(n)}_{n-2k,l} \in \mathbb{R}$ such that

$$\begin{aligned}
&\frac{\partial^2}{\partial s^2}(\mathbf{Rf})(s,\eta) \\
&\quad = \sum_{n=0}^{\infty} \frac{n+2}{2n+3} q_n(s) \sum_{|l|\le n+1} b^{(n)}_{n+1,l}\big((n+1)\mathbf{y}^{(1)}_{n+1,l}(\eta) + \mathbf{y}^{(2)}_{n+1,l}(\eta)\big) \\
&\quad + \sum_{n=2}^{\infty} q_n(s) \sum_{k=1}^{[\frac{n}{2}]} \sum_{|l|\le n+1-2k} b^{(n)}_{n+1-2k,l}\mathbf{y}^{(2)}_{n+1-2k,l}(\eta) \\
&\quad + \sum_{n=1}^{\infty} q_n(s) \sum_{k=0}^{[\frac{n-1}{2}]} \sum_{|l|\le n-2k} c^{(n)}_{n-2k,l}\mathbf{y}^{(3)}_{n-2k,l}(\eta)
\end{aligned} \tag{44}$$

with

$$q_n(s) = \frac{8\pi^2}{(n+1)(n+2)} \frac{\mathrm{d}^2}{\mathrm{d}s^2}\{(1-s^2)C_n^{(3/2)}(s)\}$$

is satisfied. Here, $C_n^{(3/2)}(s)$ denote the Gegenbauer polynomials of order 3/2. From (33), (42), and the fact that $\partial_s^2(\mathbf{R}^{(\mathrm{nor})}\mathbf{f})(s,\eta)$ only contains the normal component of $\partial_s^2(\mathbf{Rf})(s,\eta)$ we deduce that

$$\frac{\partial^2}{\partial s^2}(\mathbf{R}^{(\mathrm{nor})}\mathbf{f})(s,\eta) = \sum_{n=0}^{\infty} \frac{(n+1)(n+2)}{2n+3} q_n(s) \sum_{|l|\le n+1} b^{(n)}_{n+1,l}\mathbf{y}^{(1)}_{n,l}(\eta), \tag{45}$$

where the coefficients $b^{(n)}_{n+1,l}$ compute as

$$\begin{aligned}
&\frac{n+2}{2n+3}\, b^{(n)}_{n+1,l}\, \|\mathbf{y}^{(2)}_{n+1,l}\|^2_{L^2(S^2,\mathbb{R}^3)} \int\limits_{-1}^{1} q_n(s) C_n^{(3/2)}(s)(1-s^2)\,\mathrm{d}s \\
&= \int\limits_{S^2}\int\limits_{-1}^{1} \frac{\partial^2}{\partial s^2}(\mathbf{Rf})(s,\eta)\cdot \mathbf{y}^{(2)}_{n+1,l}(\eta) C_n^{(3/2)}(s)(1-s^2)\,\mathrm{d}s\,\mathrm{d}\eta \\
&= \int\limits_{S^2}\int\limits_{-1}^{1} \frac{\partial^2}{\partial s^2}(\mathbf{R}^{(\mathrm{tan})}\mathbf{f})(s,\eta)\cdot \mathbf{y}^{(2)}_{n+1,l}(\eta) C_n^{(3/2)}(s)(1-s^2)\,\mathrm{d}s\,\mathrm{d}\eta.
\end{aligned}$$

Since $\|\mathbf{y}^{(2)}_{n+1,l}\|^2_{L^2(S^2,\mathbb{R}^3)} = (n+1)(n+2)$ (see [8]), we get

$$b^{(n)}_{n+1,l} = -\frac{1}{8\pi^2}\frac{2n+3}{n+2}\, c(n) \int\limits_{-1}^{1} \frac{\partial^2}{\partial s^2}(\mathbf{R}^{(\mathrm{tan})}\mathbf{f})(s,\eta)\cdot \mathbf{y}^{(2)}_{n+1,l}(\eta) C_n^{(3/2)}(s)(1-s^2)\,\mathrm{d}s\,\mathrm{d}\eta, \tag{46}$$

where

$$c(n) = \left(\int\limits_{-1}^{1} \left[\frac{\mathrm{d}}{\mathrm{d}s}\{(1-s^2) C_n^{(3/2)}(s)\} \right]^2 \mathrm{d}s \right)^{-1}.$$

All these investigations result in the following reconstruction algorithm for a divergence-free **f** from cone beam data, where the source trajectory Γ is supposed to fulfill the Tuy condition of order 3.

Reconstruction algorithm from cone beam data:

Given: Measured data $\mathbf{X}^{(\mathrm{even})}\mathbf{f}(y,\eta)$ for $y \in \Gamma$, $\eta \in C$

Output: $\mathbf{f}(x)$ for solenoidal **f** and $x \in \mathbb{B}^3$

Compute:
- For any $s \in [-1,1]$, $\eta \in S^2$ compute
  - $G(\mathbf{a}_i,\eta)$ where $\mathbf{a}_i \cdot \eta = s$, $i = 1,2,3$
  - Solve system (40) for $P(s,\eta)$, $Q(s,\eta)$
  - Compute $\partial_s^2(\mathbf{R}^{(\mathrm{tan})}\mathbf{f})(s,\eta)$ from $P(s,\eta)$ and $Q(s,\eta)$
  - Compute the coefficients $b^{(n)}_{n+1,l}$ by (46)
  - Compute $\partial_s^2(\mathbf{R}^{(\mathrm{nor})}\mathbf{f})(s,\eta)$ by (45)
- Compute $\mathbf{f}(x)$ from Radon's inversion formula (34)

Remark 9.3.3. The well-known fact that **f** can only be recovered from **Xf**, if it is solenoidal, is also found in this reconstruction approach. If $\nabla\cdot\mathbf{f} \neq 0$, then we cannot compute the coefficients $b^{(n)}_{n+1,l}$, and thus $\partial_s^2\mathbf{R}^{(\mathrm{nor})}\mathbf{f}$. This is also illustrated in Figure 9.6.

$$\begin{array}{ccccccc} \mathbf{Xf} & \overset{(31)}{\Rightarrow} & \partial_s^2\mathbf{R}^{(tan)}\mathbf{f} & \overset{(*)}{\Rightarrow} & \partial_s^2\mathbf{R}^{(nor)}\mathbf{f} & \overset{+}{\Rightarrow} & \partial_s^2\mathbf{Rf} \\ & & \Downarrow (34) & & \Downarrow (34) & & \Downarrow \\ & & \mathbf{f}_1 & & \mathbf{f}_2 & \overset{+}{\Rightarrow} & \mathbf{f} \end{array}$$

Figure 9.6: Diagram of the inversion procedure for solenoidal $\mathbf{f} \in C^2(\mathbb{B}, \mathbb{R}^3)$. The vector fields $\mathbf{f}_1$, $\mathbf{f}_2$ are computed by applying Radon's inversion formula (34) to $\partial_s^2\mathbf{R}^{(\tan)}\mathbf{f}$, $\partial_s^2\mathbf{R}^{(\mathrm{nor})}\mathbf{f}$, respectively. Step (*) is only possible for solenoidal **f**.

Two drawbacks of this inversion approach become obvious. At first, we have to solve systems (40) for all pairs (s, η). Second, in applications the infinite series (45) for computing $\partial_s^2\mathbf{R}^{(\mathrm{nor})}\mathbf{f}$ has to be cut off leading to unavoidable reconstruction artifacts. This is why we call the inversion formula outlined in this section as *asymptotic*. An exact inversion formula would be preferred and is deduced in the next section. Also, that formula relies on the relationship (31).

9.3.2 An exact inversion formula

The idea to derive an exact inversion formula for **X** follows two steps: (a) using Cramer's rule for (40) as well as concepts from [16] to obtain an integral representation for $\partial_s^2\mathbf{R}^{(\tan)}\mathbf{f}$ of filtered backprojection type, and (b) deducing an analytic integral representation for the infinite series (45) to compute $\partial_s^2\mathbf{R}^{(\mathrm{nor})}\mathbf{f}$.

Let $\mathbf{f} \in C_0^\infty(\mathbb{B}, \mathbb{R}^3)$ be a smooth vector field with compact support in $\mathbb{B}$. As indicated in Figure 9.6 by $\mathbf{f}_1$, we denote the vector field, that we obtain by substituting $\partial_s^2\mathbf{R}^{(\tan)}\mathbf{f}$ in Radon's inversion formula (34) and analogously the field that we get by substituting $\partial_s^2\mathbf{R}^{(\mathrm{nor})}\mathbf{f}$ in (34) is denoted by $\mathbf{f}_2$. In this way, the solenoidal part $\mathbf{f}^s$ of **f** is obtained by

$$\mathbf{f}^s := \mathbf{P}_V\mathbf{f} = \mathbf{f}_1 + \mathbf{f}_2.$$

Here, $\mathbf{P}_V$ is the orthogonal projection onto the space of divergence-free vector fields

$$\begin{aligned} V := H(\mathrm{div}; \mathbb{B}) &:= \left\{\mathbf{f} \in L^2(\mathbb{B}, \mathbb{R}^3) : \int_{\mathbb{B}} \mathbf{f}(x) \cdot \nabla v(x)\, dx = 0 \text{ for all } v \in C_0^\infty(\mathbb{B})\right\} \\ &= H_0(\mathrm{div}; \mathbb{B}) \oplus \nabla\, \mathrm{Harm}(\mathbb{B}). \end{aligned}$$

Furthermore we assume that Tuy's condition of order 3 is fulfilled according to Definition 9.3.2.

We sketch the inversion formula for $\mathbf{f}_1$ as outlined in [18]. For $x \in \mathbb{B}$ and $\eta \in S^2$, consider the plane

$$\Pi(x, \eta) := \{z \in \mathbb{R}^3 : z \cdot \eta = x \cdot \eta\}$$

which passes through x and is perpendicular to η. Let $s_j = s_j(x, \eta)$ be points of intersection of the source trajectory Γ and $\Pi(x, \eta)$. Since there can be more than three intersection points, in general we can form multiple triples from these points. Denote those triples by $\mathcal{S}_m = \mathcal{S}_m(x,\eta)$, where the subscript m denotes the index of a triple. From the results described in Section 9.3.1, we derive

$$\frac{\partial^2}{\partial s^2}(\mathbf{R}^{(\tan)}\mathbf{f})(s,\eta)\bigg|_{s=\eta\cdot x} = \sum_{s_j\in\mathcal{S}_m} \phi(s_j,\mathcal{S}_m)G(y(s_j),\eta) \tag{47}$$

with $G(y,\eta)$ from (32). The coefficients $\phi(s_j,\mathcal{S}_m)$ can be computed analytically by applying Cramer's rule to the systems (39). Applying Radon's inversion formula yields

$$\begin{aligned}\mathbf{f}_1(x) &= -\frac{1}{8\pi^2}\int_{S^2}\frac{\partial^2}{\partial s^2}(\mathbf{R}^{(\tan)}\mathbf{f})(s,\eta)\bigg|_{s=\eta\cdot x}\,\mathrm{d}\eta\\ &= -\frac{1}{8\pi^2}\int_{S^2}\sum_{\text{all triples } m}\bigg[\sum_{s_j\in\mathcal{S}_m}\phi(s_j,\mathcal{S}_m)G(s_j,\eta)\bigg]n_m(x,\eta)\,\mathrm{d}\eta,\end{aligned} \tag{48}$$

where $n_m(x,\eta)$ is a weight assigned to each triple. Using a partition of unity on S^2, we can choose a smooth weight function so that the coefficients $\phi(s_j,\mathcal{S}_m)$ are well-defined on the support of $n_m(x,\eta)$. Clearly, in order to achieve exact reconstruction the weight needs to be normalized:

$$\sum_m n_m(x,\eta) = 1. \tag{49}$$

Similar to [16], we replace one of the summations in (48) by the integral with respect to s:

$$\mathbf{f}_1(x) = -\frac{1}{8\pi^2}\int_{S^2}\int_I|\eta\cdot\dot{y}(s)|\delta(\eta\cdot(x-y(s)))\bigg[\sum_{m:s\in\mathcal{S}_m}\phi(s,\mathcal{S}_m)n_m(x,\eta)\bigg]G(s,\eta)\,\mathrm{d}s\,\mathrm{d}\eta. \tag{50}$$

In (50), the sum with respect to m is over all triples that involve $y(s)$ and use intersection points in $\Gamma\cap\Pi(y(s),\eta)$. Using the homogeneity of the delta-distribution and changing the order of integration, we rewrite (50) as follows:

$$\begin{aligned}\mathbf{f}_1(x) &= -\frac{1}{8\pi^2}\int_{S^2}\int_I\frac{|\eta\cdot\dot{y}(s)|}{|x-y(s)|}\delta(\eta\cdot\beta(s,x))\Phi(s,\eta)G(y(s),\eta)\,\mathrm{d}s\,\mathrm{d}\eta\\ &= -\frac{1}{8\pi^2}\int_I\frac{1}{|x-y(s)|}\bigg[\int_{S^2}\Phi(s,\eta)G(y(s),\eta)\delta(\eta\cdot\beta(s,x))\,\mathrm{d}\eta\bigg]\mathrm{d}s,\end{aligned} \tag{51}$$

where

$$\beta(s,x) := \frac{x-y(s)}{|x-y(s)|},\quad \Phi(s,\eta) := |\eta\cdot\dot{y}(s)|\sum_{m:s\in\mathcal{S}_m}\phi(s,\mathcal{S}_m)n_m(x,\eta). \tag{52}$$

The integral in brackets in (51) can be simplified by using appropriate coordinate transforms and some calculus so that we finally arrive at the filtered backprojection formula for $\mathbf{f}_1$,

$$\mathbf{f}_1(x) = \frac{1}{8\pi^2} \int_I \frac{1}{|x - y(s)|} \int_0^{2\pi} [\partial_\theta^2 \Phi(s, \eta(\theta)) + \Phi(s, \eta(\theta))] \times \int_0^{2\pi} \frac{(\mathbf{X}^{(\mathrm{even})}\mathbf{f})(y(s), \cos\gamma\, \eta^\perp(\theta) + \sin\gamma\, \beta)}{\cos^2\gamma}\, \mathrm{d}\gamma\, \mathrm{d}\theta\, \mathrm{d}s, \tag{53}$$

where $\theta \in [0, 2\pi]$ denotes the polar angle of the unit vector $\eta(\theta) \in (S^2 \cap \beta^\perp)$, $\beta^\perp$ is the plane perpendicular to $\beta = \beta(s, x)$ from (52) and $\eta^\perp(\theta) = \beta \times \eta(\theta)$.

The deduction of $\mathbf{f}_2$ is more intricate. The statement of the following lemma follows by straightforward calculations.

Lemma 9.3.4. *For $v \in H_0^1(\mathbb{B})$, we have $\mathbf{R}^{(\mathrm{tan})}\nabla v = 0$.*

We multiply (44) by $\chi(s)\overline{\mathbf{y}}^{(2)}_{n+1,l}(\eta)$ for a reasonable function $\chi(s)$ defined on $[-1, 1]$ to be specified later, and integrate over S^2. Here, the overbar denotes complex conjugation. Since vector spherical harmonics are orthogonal, we get

$$\int_{S^2} \int_{-1}^{1} (\partial_s^2 \mathbf{R}\mathbf{f}^{\mathrm{s}})(s, \eta) \cdot [\chi(s)\overline{\mathbf{y}}^{(2)}_{n+1,l}(\eta)]\, \mathrm{d}s\, \mathrm{d}\eta = -\frac{8\pi^2 \check{\chi}_n (n+2) b^{(n)}_{n+1,l}}{2n+3} \|\mathbf{y}^{(2)}_{n+1,l}\|^2,$$

$$\check{\chi}_n := \int_{-1}^{1} C_n^{3/2}(s)\chi(s)\, \mathrm{d}s, \tag{54}$$

where the norms of vector spherical harmonics are always to be understood with respect to $L^2(S^2, \mathbb{R}^3)$. Since $\mathbf{y}^{(2)}_{n+1,l}(\eta)$ is orthogonal to the normal component of $\mathbf{Rf}$, we can replace $\partial_s^2 \mathbf{R}\mathbf{f}^{\mathrm{s}}$ with $\partial_s^2 \mathbf{R}^{(\mathrm{tan})}\mathbf{f}^{\mathrm{s}}$ in (54). The latter, of course, is equal to $\partial_s^2 \mathbf{R}^{(\mathrm{tan})}\mathbf{f}$ because of Lemma 9.3.4. Multiply (54) by

$$(n+1)\frac{C_n^{3/2}(p)}{\check{\chi}_n} \frac{\mathbf{y}^{(1)}_{n+1,l}(\eta)}{\|\mathbf{y}^{(2)}_{n+1,l}\|^2}$$

and sum over n, l to get the derivative $\partial_s^2 \mathbf{R}^{(\mathrm{nor})}\mathbf{f}^{\mathrm{s}}$ as

$$(\partial_p^2 \mathbf{R}^{(\mathrm{nor})}\mathbf{f}^{\mathrm{s}})(q, \alpha) = \sum_{n \geq 0} \sum_{|l| \leq n+1} \int_{S^2} \int_{-1}^{1} (\partial_s^2 \mathbf{R}^{(\mathrm{tan})}\mathbf{f})(s, \eta) \cdot [\chi(s)\overline{\mathbf{y}}^{(2)}_{n+1,l}(\eta)]\, \mathrm{d}s\, \mathrm{d}\eta \times (n+1)\frac{C_n^{3/2}(q)}{\check{\chi}_n} \frac{\mathbf{y}^{(1)}_{n+1,l}(\alpha)}{\|\mathbf{y}^{(2)}_{n+1,l}\|^2}, \tag{55}$$

for any $q \in [-1,1]$, $\alpha \in S^2$. Using again that $\|\mathbf{y}^{(2)}_{n+1,l}\|^2 = (n+1)(n+2)$, we write the sum with respect to l in the form of a rank-one matrix

$$\alpha \otimes \nabla_\eta \sum_{|l|\le n+1} Y_{n+1,l}(\alpha)\overline{Y}_{n+1,l}(\eta) = \frac{2n+3}{4\pi}\eta \otimes \nabla_\eta P_{n+1}(\alpha\cdot\eta). \tag{56}$$

Here, P_n are the Legendre polynomials, and we used the addition theorem for spherical harmonics. The operator in (56) acts on vectors by computing the dot product of an input vector with $\nabla_\eta P_{n+1}(\alpha\cdot\eta)$ and then multiplying the result by the vector $(2n+3)\eta/4\pi$. Using (56) in (55), substituting the result into the Radon transform inversion formula, and changing the order of integration and summation we get

$$\mathbf{f}^{(2)}(x) = -\frac{1}{4(2\pi)^3}\int_{S^2} K(x,\eta)\int_{-1}^{1}\chi(s)\partial_s^2(\mathbf{R}^{(\tan)}\mathbf{f})(s,\eta)\,\mathrm{d}s\,\mathrm{d}\eta, \tag{57}$$

where the kernel $K(x,\eta)$ is given by

$$K(x,\eta) := \int_{S^2}\alpha\otimes\nabla_\eta K_1(x,\alpha;\eta)\,\mathrm{d}\alpha,$$

$$K_1(x,\alpha;\eta) := \sum_{n\ge 0}\frac{2n+3}{\check{\phi}_n(n+2)}P_{n+1}(\alpha\cdot\eta)C_n^{3/2}(\alpha\cdot x),$$

and we define

$$\chi(s) := (1-s^2)\sum_{n\ge 0}\frac{\check{\chi}_n}{\|C_n^{3/2}\|^2}C_n^{3/2}(s). \tag{58}$$

In [18], the authors prove that the change of integration order in the sense of distributions actually is justified. From the orthogonality of the Gegenbauer polynomials (equation (22.2.3) in [1]), $\chi(s)$ satisfies (54). In view of (54), any function given by (58) can be used in (57). The goal is then to choose the coefficients $\check{\chi}_n$ so we could use the following identity (see (5.10.2.2) in [33]), which we write here in a symmetric form:

$$\begin{aligned}\sum_{n\ge 1}\frac{2n+1}{n(n+1)}P_n(s)P_n(t) &= 2\ln 2 - 1 - \ln(1+|t-s|-st)\\ &= 2\ln 2 - 1 - \ln((1+\max(s,t))(1-\min(s,t))),\end{aligned} \tag{59}$$

$-1 < s,t < 1$. A deeper analysis as outlined in [18] shows that an appropriate selection is $\check{\chi}_n = n+1$ resulting in

$$\chi(s) = (1-s^2)\left[(2-2s)^{-3/2} - \frac{1}{2}((2-2s)^{-1/2}(1-s+\sqrt{2-2s})^{-1})\right],$$

which obviously is a function in $L^2(-1,1)$. A technical derivation then gives us an explicit representation of the kernel K from (57) by

$$K(x,\eta) = 2\int_{S^2}\alpha\otimes[\alpha-(\alpha\cdot\eta)\eta]\frac{\delta(\alpha\cdot(\eta-x))}{1-(\alpha\cdot x)^2}\,\mathrm{d}\alpha.$$

Finally, we derive at

$$\mathbf{f}_2(x) = -\frac{1}{8\pi^2}\int_{S^2}\frac{(\Psi(\eta)\cdot e_1)u + (\Psi(\eta)\cdot e_2)e_2}{|\eta - x| + \eta\cdot(\eta - x)}\,d\eta, \tag{60}$$

$$\Psi(\eta) := \int_{-1}^{1}\chi(s)(\partial_s^2\mathbf{R}^{(\tan)}\mathbf{f})(s,\eta)\,ds. \tag{61}$$

A detailed derivation of all these representations is found in [17]. Note that an implementation of representation (61) of $\mathbf{f}_2$ is of the same order as the filtered backprojection formula (53) for $\mathbf{f}_1$.

In [17], the authors sketch a generalization of the inversion formula for $\mathbf{f}_2$ (60), (61) for general convex domains $\Omega \subset \mathbb{R}^3$ by computing a solution h of a corresponding Dirichlet problem for the Laplace equation and then setting $\mathbf{f}_2 := \nabla h$. Here, we summarize the proposed procedure. Since $\mathbf{f}_2 = \mathbf{f}^s - \mathbf{f}_1$, we have $\nabla\cdot\mathbf{f}_2 = \nabla\times\mathbf{f}_2 = 0$. Hence $\mathbf{f}_2 = \nabla h$ for a harmonic function h. The function h then is computed by solving a corresponding Dirichlet problem as follows:

1. Compute $\mathbf{f}^{(1)}$ by inversion formula (53)
2. Compute the cone beam transform $(\mathbf{X}\mathbf{f}_1)(y(p),\eta)$ by forward projecting the computed $\mathbf{f}_1$
3. Subtract $\mathbf{X}\mathbf{f} - \mathbf{X}\mathbf{f}_1 = \mathbf{X}\nabla h$ and obtain the cone beam transform of ∇h. Observe that

$$(\mathbf{X}\nabla h)(y(p),\eta) = h(x_{\text{out}}(y(p),\eta)) - h(x_{\text{in}}(y(p),\eta)), \tag{62}$$

   where x_{out}, x_{in} denote the points where the line parametrized by $(y(p),\eta)$ exits the domain Ω and enters the domain Ω, respectively.
4. Get the Dirichlet data $h_{|\partial\Omega}$ up to a constant. This is possible if the collection of lines is sufficiently rich, which is the case if the trajectory Γ satisfies the Tuy condition of order 3. It can be done by solving a linear system of equations based on (62).
5. Solve the Dirichlet problem for the Laplacian $\Delta h(x) = 0$ for $x \in \Omega$, $h(x) = h_{|\partial\Omega}(x)$ for $x \in \partial\Omega$.
6. Set $\mathbf{f}_2 = \nabla h$.

Note that the inversion formula of filtered backprojection type for $\mathbf{f}_1$ (53) is valid in convex domains anyway. A first reconstruction for $\Omega = [0,1]^3$ is seen in Figure 9.7 together with the exact solution. The result is very promising to pursue this approach further.

9.4 Conclusion

This article proves that although more than 100 years old, the Radon transform still has exciting applications in current tomographic problems. Of course, the mathemat-

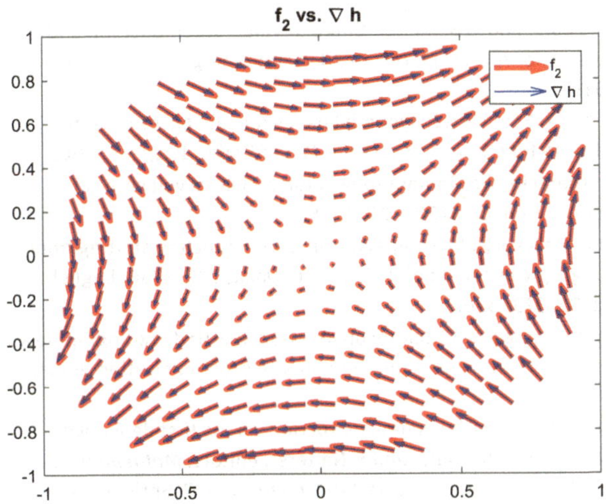

Figure 9.7: Comparison of the exact $\mathbf{f}_2$ and the reconstruction ∇h at $x_3 = 0$ using a solution h of a Dirichlet problem.

ical models of vector field tomography are closely related to the Radon transform. But nevertheless it is amazing that the 2D Radon transform is useful for reconstruction techniques in 3D Doppler tomography and that a splitting of the 3D Radon transform into a tangential and a normal part is a key ingredient to obtain inversion formulas for the vectorial cone beam transform. It is surprising that the formula of Kazantsev and Schuster, which relies on that splitting, is totally different by its nature from Grangeat's famous formula which opened the door for exact inversion formulas for the scalar cone beam transform. A current research field consists of finding analytical and numerical inversion schemes for the Radon as well as related transforms for vector fields on Riemannian manifolds. So, there is good hope that the story of the importance of Radon's inversion formula for research in vector field tomography is not at its end.

Bibliography

[1] M. Abramowitz and I. Stegun. *Handbook of Mathematical Functions*. Dover, New York, 1970.

[2] A. L. Balandin and Y. Ono. The method of series expansion for 3-D vector tomography reconstruction. *J. Comput. Phys.*, 202(1):52–64, 2005.

[3] M. Barth and A. Raabe. Acoustic tomographic imaging of temperature and flow fields in air. *Meas. Sci. Technol.*, 22:035102, 2011.

[4] M. Barth, A. Raabe, K. Arnold, C. Resagk, and R. du Puits. Flow field detection using acoustic travel time tomography. *Meteorol. Z.*, 16(4):443–450, 2007.

[5] H. Braun and A. Hauck. Tomographic reconstruction of vector fields. *IEEE Trans. Signal Process.*, 39(2):464–471, 1991.

[6] M. Defrise and G. T. Gullberg. 3D reconstruction of tensors and vectors. Technical report, Lawrence Berkely National Laboratory, 2005.

[7] E. Y. Derevtsov, A. V. Efimov, A. K. Louis, and T. Schuster. Singular value decomposition and its application to numerical inversion for ray transforms in 2D vector tomography. *J. Inverse Ill-Posed Probl.*, 19:689–715, 2011.

[8] E. Y. Derevtsov, S. G. Kazantsev, and T. Schuster. Polynomial bases for subspaces of vector fields in the unit ball: method of ridge functions. *J. Inverse Ill-Posed Probl.*, 15:19–55, 2007.

[9] N. Efremov, N. Pamentov, and V. Kharachenko. Tomography of ion and atom velocities in plasmas. *J. Quant. Spectrosc. Radiat. Transf.*, 6:723–728, 1995.

[10] I. M. Gel'fand and A. B. Goncharov. Recovery of a compactly supported function starting from its integrals over lines intersecting a given set of points in space. *Dokl.*, 290, 1986. English translation in Soviet. Math. Dokl. 34, 373–376.

[11] P. Grangeat. Analyse d'un système d'imagerie 3D par reconstruction àpartir de radiographie X en geometrie conique. PhD Thesis, École Normale Superieure des Télécommunication, Paris, France, 1987.

[12] P. Grangeat. Mathematical framework of cone-beam reconstruction via the first derivative of the Radon transform. In G. T. Herman, A. K. Louis, and F. Natterer, editors, *Mathematical Methods in Tomography*. Lecture Notes in Math., volume 1497, pages 66–97. Springer, New York, 1991.

[13] I. Jovanović, L. Sbaiz, and M. Vetterli. Acoustic tomography for scalar and vector fields: theory and application to temperature and wind estimation. *J. Atmos. Ocean. Technol.*, 26(8):1475–1492, 2009.

[14] P. Juhlin. Principles of Doppler tomography. Technical report, Center for Mathematical Sciences, Lund Institute of Technology, SE-221 00 Lund, Schweden, 1992.

[15] A. Katsevich. Theoretically exact filtered back projection-type inversion algorithm for spiral CT. *SIAM J. Appl. Math.*, 62(6):2002–2026, 2002.

[16] A. Katsevich. A general scheme for constructing inversion algorithms for cone beam CT. *Int. J. Math. Sci.*, 21:1305–1321, 2003.

[17] A. Katsevich, D. Rothermel, and T. Schuster. An improved exact inversion formula for solenoidal fields in cone beam vector tomography. *Inverse Probl.*, 33:064001, 19 pp., 2017.

[18] A. Katsevich and T. Schuster. An exact inversion formula for cone beam vector tomography. *Inverse Probl.*, 29(6):065013, 13 pp., 2013.

[19] S. G. Kazantsev and T. Schuster. Asymptotic inversion formulas in 3D vector field tomography for different geometries. *J. Inverse Ill-Posed Probl.*, 19:769–799, 2011.

[20] W. R. B. Lionheart and P. J. Withers. Diffraction tomography of strain. *Inverse Probl.*, 31:045005, 17 pp., 2015.

[21] A. K. Louis. Approximate inverse for linear and some nonlinear problems. *Inverse Probl.*, 12:175–190, 1996.

[22] A. K. Louis. A unified approach to regularization methods for linear ill-posed problems. *Inverse Probl.*, 15:489–498, 1999.

[23] A. K. Louis. Development of algorithms in computerized tomography. In G. Ólafsson and E. T. Quinto, editors, *The Radon Transform, Inverse Problems, and Tomography*. Proceedings of Symposia in Applied Mathematics, volume 63, pages 25–42. AMS, 2006.

[24] A. K. Louis and P. Maass. A mollifier method for linear operator equations of the first kind. *Inverse Probl.*, 6:427–440, 1990.

[25] F. Monard. Numerical implementation of geodesic X-ray transforms and their inversion. *SIAM J. Imaging Sci.*, 7(2):1335–1357, 2014.

[26] F. Natterer. *The Mathematics of Computerized Tomography*. Wiley, Chichester, 1986.

[27] F. Natterer and F. Wübbeling. *Mathematical Methods in Image Reconstruction*. SIAM, Philadelphia, 2001.

[28] S. J. Norton. Tomographic reconstruction of 2-D vector fields: application to flow imaging. *Geophys. J.*, 97:161–168, 1988.

[29] N. Osman and J. Prince. 3D vector tomography on bounded domains. *Inverse Probl.*, 14:185–196, 1998.

[30] G. P. Paternain, M. Salo, and G. Uhlmann. The attenuated ray transform for connections and Higgs fields. *Geom. Funct. Anal.*, 22(5):1460–1489, 2012.

[31] T. Pfitzenreiter and T. Schuster. Tomographic reconstruction of the curl and divergence of 2D vector fields taking refractions into account. *SIAM J. Imaging Sci.*, 4:40–56, 2011.

[32] C. Phatak and D. Gürsoy. Iterative reconstruction of magnetic induction using Lorentz transmission electron tomography. *Ultramicroscopy*, 150:54–64, 2015.

[33] A. P. Prudnikov, Y. A. Brychkov, and O. I. Marichev. *Integrals and Series. Volume 2. Special Functions*. Gordon and Breach, New York, 1988.

[34] D. Rouseff, K. B. Winters, and T. E. Ewart. Reconstruction of oceanic microstructure by tomography: a numerical feasibility study. *J. Geophys. Res.*, 96(C5):8823–8833, 1991.

[35] U. Schröder and T. Schuster. An iterative method to reconstruct the refractive index of a medium from time-of-flight measurements. *Inverse Probl.*, 32:085009, 35 pp., 2016.

[36] T. Schuster. An efficient mollifier method for three-dimensional vector tomography: convergence analysis and implementation. *Inverse Probl.*, 17:739–766, 2001.

[37] T. Schuster. The formula of Grangeat for tensor fields of arbitrary order in n dimensions. *Int. J. Biomed. Imaging*, 2007:12839, 4 pp., 2007. DOI: 10.1155/2007/12839.

[38] T. Schuster. 20 years of imaging in vector field tomography: a review. In *Mathematical Methods in Biomedical Imaging and Intensity-Modulated Dariation Therapy (IMRT)*. Publications of the Scuola Normale Superiore, CRM Series, volume 7. Birkhäuser, 2008.

[39] T. Schuster, D. Theis, and A. K. Louis. A reconstruction approach for imaging in 3D cone beam vector field tomography. *Int. J. Biomed. Imaging*, 2008:174283, 2008. DOI: 10.1155/2008/174283.

[40] V. A. Sharafutdinov. *Integral Geometry of Tensor Fields*. VSP, Utrecht, 1994.

[41] V. A. Sharafutdinov. Slice-by-slice reconstruction algorithm for vector tomography with incomplete data. *Inverse Probl.*, 23:2603–2627, 2007.

[42] H. Sielschott. Measurement of horizontal flow in a large scale furnace using acoustic vector tomography. *Flow Meas. Instrum.*, 8:191–197, 1997.

[43] G. Sparr and K. Stråhlén. Vector field tomography: an overview. Technical report, Mathematical Imaging Group, Centre for Mathematical Sciences, Lund Institute of Technology, Lund, Sweden, 1998.

[44] G. Sparr, K. Stråhlén, K. Lindström, and H. W. Persson. Doppler tomography for vector fields. *Inverse Probl.*, 11:1051–1061, 1995.

[45] P. Stefanov and G. Uhlmann. Stability estimates for the X-ray transform of tensor fields and boundary rigidity. *Duke Math. J.*, 123(3):445–467, 2004.

[46] P. Stefanov and G. Uhlmann. Local lens rigidity with incomplete data for a class of non-simple Riemannian manifolds. *J. Differ. Geom.*, 82(2):383–409, 2009.

[47] I. E. Svetov, E. Y. Derevtsov, Y. S. Volkov, and T. Schuster. A numerical solver based on B-splines for 2D vector field tomography in a refracting medium. *Math. Comput. Simul.*, 97:207–223, 2014.

[48] H. K. Tuy. An inversion formula for cone-beam reconstruction. *SIAM J. Appl. Math.*, 43:546–552, 1983.

[49] G. Uhlmann, P. Stefanov, and A. Vasy. Inverting the local geodesic X-ray transform on tensors, 2014. arXiv:1410.5145.

[50] S. N. Vecherin, V. E. Ostashev, and K. D. Wilson. Three-dimensional acoustic travel-time tomography of the atmosphere. *Acta Acust. Acust.*, 94(3):349–358, 2008.

G. T. Herman

10 Iterative reconstruction techniques and their superiorization for the inversion of the Radon transform

Abstract: The problem of image reconstruction is to produce an image of a distribution that is compatible with constraints provided by measurements of its line integrals (i. e., values of its Radon transform) along a finite number of lines of known locations and with other possible constraints, such as nonnegativity. Algebraic reconstruction techniques (ART) form a family of iterative algorithms used for image reconstruction. Their distinguishing features are: (1) they assume that the image is represented as a linear combination of some known basis functions and (2) the unknown coefficients in this linear combination are estimated by an iterative process in which just one of the measured line integrals is used in any one iterative step. The first section of the chapter gives a tutorial overview of variants of ART, concentrating on the mathematical results regarding them and mentioning their computational efficiency in various applications. All variants of ART aim at producing an image that is compatible with the constraints, such constraints-compatibility is mathematically defined by what is referred to as the primary criterion. The second section of the chapter is devoted to the superiorization methodology, which is a recently-developed heuristic approach to optimization, and to a discussion of its applicability to improving the results of iterative approaches to inverting the Radon transform. The underlying idea is that many iterative algorithms for finding such an inverse are perturbation resilient in the sense that, even if certain kinds of changes are made at the end of each iterative step, the algorithm still produces a constraints-compatible solution. This property is exploited by using permitted changes to steer the algorithm to a solution that is not only constraints-compatible, but is also desirable according to a specified secondary criterion. The approach is very general, it is applicable to many iterative procedures and secondary criteria used in the inversion of the Radon transform. Superiorization produces automatically from any given iterative algorithm its superiorized version. If the original iterative algorithm satisfies certain mathematical conditions, then the output of its superiorized version is guaranteed to be as good as the output of the original algorithm from the point of view of the primary criterion, but it is superior to the latter according to the secondary criterion. This intuitive description is made more precise and is illustrated in the second section of the chapter.

Keywords: Image reconstruction, iterative algorithms, superiorization, inverse problems, Radon transform

G. T. Herman, Department of Computer Science, City University of New York, 365 Fifth Avenue, New York, NY 10016, USA, e-mail: gabortherman@yahoo.com

https://doi.org/10.1515/9783110560855-010

MSC 2010: 44A12, 68U10, 65J22, 65K10, 92C55

10.1 Algebraic reconstruction techniques (ART)

Let $f(r,\phi)$ denote the value of a two-dimensional function f at the point with polar coordinates (r,ϕ). We use the real-number pair (ℓ,θ) to define a line (ℓ is the distance from the origin and θ is the angle of tilt) and $[\mathscr{R}f](\ell,\theta)$ to denote the line integral of f along that line; in other words, $\mathscr{R}f$ is the *Radon transform* of f.

The input to a *reconstruction algorithm* is a set of (in practice, physically-obtained) estimates of the values of $[\mathscr{R}f](\ell,\theta)$ for a finite number of pairs $(\ell_1,\theta_1),\dots,(\ell_I,\theta_I)$. Let, for $1 \le i \le I$,

$$\mathscr{R}_i f = [\mathscr{R}f](\ell_i,\theta_i). \tag{1.1}$$

Let y_i denote the estimate of $\mathscr{R}_i f$ and y the I-dimensional *measurement vector* whose ith component is y_i. The *image reconstruction* problem is: *given* the measurement vector y, *estimate* f.

In this chapter, we restrict our attention to one of the major categories of reconstruction algorithms: the *series expansion methods*, in which we select a finite set $\{b_1,\dots,b_J\}$ of two-dimensional *basis functions* and we aim at a reconstruction $\hat{f}$ that is an *approximation* of f (notation: $\hat{f} \simeq f$) and can be expressed as

$$\hat{f}(r,\phi) = \sum_{j=1}^{J} x_j b_j(r,\phi), \tag{1.2}$$

where the x_j are real numbers (in shorthand, $\hat{f} = \sum_{j=1}^{J} x_j b_j$). We use x to denote the J-dimensional vector whose jth component is x_j and refer to x as the *image vector*. This approach results in a finite-dimensional setting; generalization to infinite-dimensional setting is possible, but will not be dealt with in this chapter.

It is easy to see that, under some mild mathematical assumptions,

$$\mathscr{R}_i f \simeq \mathscr{R}_i \hat{f} = \sum_{j=1}^{J} x_j \mathscr{R}_i b_j, \tag{1.3}$$

for $1 \le i \le I$. Since the b_j are user-defined, usually the $\mathscr{R}_i b_j$ can be calculated easily by analytical means; we use $r_{i,j}$ to denote the calculated value of $\mathscr{R}_i b_j$. Since y_i is an estimate of $\mathscr{R}_i f$, we get that, for $1 \le i \le I$,

$$y_i \simeq \sum_{j=1}^{J} r_{i,j} x_j. \tag{1.4}$$

The right-hand-side of this equation can also be written as the inner product $\langle r_i, x\rangle$, where r_i is the J-dimensional vector whose jth component is $r_{i,j}$. From now on, we will use such inner product notation

Let R denote the matrix whose (i,j)th element is $r_{i,j}$. We refer to this matrix as the *projection matrix*. Let e be the I-dimensional column vector whose ith component, e_i, is the difference between the left- and right-hand sides of (1.4). We refer to this as the *error vector*. Then (1.4) can be rewritten as

$$y = Rx + e. \tag{1.5}$$

The series expansion approach leads us to the *discrete reconstruction problem*: based on (1.4), *given* the measurement vector y, *estimate* the image vector x. If the solution to this problem is x^*, then the estimate f^* of f is given by $f^* = \langle x^*, b\rangle$, which is interpreted as the function $\sum_{j=1}^{J} x_j^* b_j$.

Clearly, in order for an I-dimensional vector x^* to be considered a "solution" to the discrete reconstruction problem (1.5), it has to be the case that in some sense the difference between the vectors y and Rx is "small." Several precise mathematical definitions may be considered to be the appropriate for "small" in this context; the chosen definition is referred to as the *primary criterion*. A frequently-used choice is the squared-norm of the error; that is, $\|y - Rx^*\|^2$. Whatever is the precise definition of the primary criterion, it should be one whose size reflects how badly the constraints on the solution provided by the measurements are violated by a specific proposed solution. A proposed solution is considered *constraints-compatible* if the value of the associated primary criterion is acceptably small. In fact, if the primary criterion is specified, its value for any proposed solution can be called the *constraints-compatibility* of that proposed solution.

The *algebraic reconstruction techniques* (ART) are series expansion methods. All ART methods are iterative procedures: they produce a sequence of vectors $x^{(0)}$, $x^{(1)}, \ldots$ that is supposed to *converge* to x^*. The process of producing $x^{(k+1)}$ from $x^{(k)}$ is referred to as an *iterative step*. In ART, $x^{(k+1)}$ is obtained from $x^{(k)}$ by considering a single one of the I approximate equations in (1.4). We use i_k to denote the index i of that equation. There are many ways of choosing the sequence of equations, one commonly-used method is the *cyclic ordering* in which $i_k = (k \bmod I) + 1$; the resulting sequence is $1, 2, \ldots, I, 1, 2, \ldots, I, 1, 2, \ldots$.

The rest of this section presents variants of ART; its organization is highly influenced by the chronological development. The first publication that proposed the use of ART as a reconstruction method is [10]. At the time of this writing, it is reported by Google Scholar that [10] has been cited 2622 times. The simplified mathematical model used in [10] for the approximation of the Radon transform results in each of the $r_{i,j}$ in (1.4) having the value 0 or 1; however, this assumption was abandoned in the later literature (and here) in favor of the more general one of real-valued $r_{i,j}$. Also, in

[10] $x^{(0)}$ was defined to be a vector whose components are positive and uniform; this assumption was also relaxed in the later literature.

A particularly simple variant (also called *Kaczmarz's method*) of ART is provided by

$$x^{(k+1)} = x^{(k)} + c^{(k)} r_{i_k}, \tag{1.6}$$

where, using a sequence of real-valued *relaxation parameters* $\lambda^{(k)}$,

$$c^{(k)} = \lambda^{(k)} \frac{y_{i_k} - \langle r_{i_k}, x^{(k)} \rangle}{\|r_{i_k}\|^2}. \tag{1.7}$$

In [10], the only allowed choice was $\lambda^{(k)} = 1$, in which case it is easy to check that, for $k \geq 0$,

$$y_{i_k} = \langle r_{i_k}, x^{(k+1)} \rangle. \tag{1.8}$$

This variant is referred to as *additive*, because $x^{(k+1)}$ in (1.6) is obtained by adding a correction term to $x^{(k)}$. Specifically, the algorithm in [10] that is called the *direct additive method* is the one we have just described except for the introduction of non-negativity into the algorithm: all negative components of $x^{(k+1)}$ are set to zero before proceeding to the next iterative step.

Following [18], we now discuss some interesting mathematical properties associated with this method. Let $L = \{x \mid Rx = y\}$ and $S = \{x \mid x = \sum_{i=1}^{I} \alpha_i r_i$, for some real numbers $\alpha_i\}$. The *Minimum Norm Theorem* of [18] states the following: "If L is not empty, then there exists one and only one element x^* in $L \cap S$. Furthermore, for all x in L other than x^*, $\|x^*\| < \|x\|$." We use $\|x\|$ to denote the *norm* (more precisely, the ℓ_2*-norm*) of x; $\|x\| = \sqrt{\sum_{j=1}^{J} x_j^2}$; thus x^* is the *minimum norm solution* of $Rx = y$. It is proved in [18] that a sequence $x^{(0)}, x^{(1)}, x^{(2)}, \ldots$ generated by (1.6) and (1.7) using cyclic ordering converges to a vector in L, provided that L is not empty and that, for some ε_1 and ε_2 and for all k,

$$0 < \varepsilon_1 \leq \lambda^{(k)} \leq \varepsilon_2 < 2. \tag{1.9}$$

Furthermore, if $x^{(0)}$ is a vector in S, then the sequence converges to the minimum norm solution x^*.

An alternative variant of ART introduced in [10], for which (1.8) also holds, is the *direct multiplicative method* (usually abbreviated as MART) in which (1.6) is replaced by: For $1 \leq j \leq J$,

$$x_j^{(k+1)} = \frac{y_{i_k}}{\langle r_{i_k}, x^{(k)} \rangle} x_j^{(k)}. \tag{1.10}$$

The statement of the discrete reconstruction problem is not sufficient to determine a unique solution. This was recognized in the earliest ART paper [10], where it was suggested that one should look for an estimate x (with nonnegative components) that is not only consistent with the measurement vector y, as in (1.4), but is also desirable according to some *prior* (also called *secondary*) criterion. One such criterion that was proposed in [10] is that of having a large *entropy*, which is defined as

$$-\sum_{j=1}^{J}\left(\frac{x_j}{J\bar{x}}\right)\ln\left(\frac{x_j}{J\bar{x}}\right), \tag{1.11}$$

where $\bar{x}$ denotes the *average* value of the components of x. Similar to the convergence of additive ART to the minimum norm solution, MART can be shown to converge to the *maximum entropy solution* of $Rx = y$; a set of conditions under which this claim is true is given in Theorem 4 of [16].

It was claimed in [10] that the presented algorithms are applicable to the reconstruction of three-dimensional objects from electron micrographs and are also applicable to "X-ray photography." The latter is interesting, since [10] was published two years prior to the introduction of the first commercial CT (*computerized tomography*) scanner, which in fact used ART to produce reconstructions from X-ray projections. The applicability of ART is quite wide-ranging; for example, in the early paper [11] ART was used to reconstruct the three-dimensional distribution of acoustic impedance within tissue (in particular, in an intact canine heart) from their two-dimensional projections of ultrasonic energy.

Development of the mathematics of ART, way beyond that of the simple Kaczmarz's method of (1.6) and (1.7), was published in the 1970s [12, 15–19]. We now present some aspects this development.

One aspect has to do with the set of permissible solutions. In the *unconstrained* case, there are no prior constraints on the components of x. In the *partially constrained* case, the components of the permissible solutions are nonnegative; that case is the one that motivated the introduction of nonnegativity into the direct additive method discussed above. In the *fully constrained* case, the components of x are expected to be bounded both from below and from above (say by 0 and 1, resp.). Matching these cases, there are three algorithms: *unconstrained ART* (in which the iterative step is given by (1.6)), *partially constrained ART* (which is essentially the direct additive method) and *fully constrained ART* (in which all components of $x^{(k+1)}$ that are smaller than the lower bound, respectively larger than the upper bound, are set to the lower bound, respectively to the upper bound, before proceeding to the next iterative step). It has been shown in [19] that, in all three cases, if there is a solution at all (i. e., an x that satisfies both (1.4) with equality and the relevant constraints), then the corresponding version of ART with $\lambda^{(k)} = 1$ (this restriction may be considerably weakened, see (1.9)) converges to a solution.

As already discussed, if $x^{(0)} \in S$, then unconstrained ART converges to the minimum norm solution. Unless there is some alternative prior information, such a choice is reasonable, especially since (under some conditions discussed in [19]) the minimal norm solution also minimizes the *variance*, which is defined as

$$\frac{1}{J}\sum_{j=1}^{J}(x_j - \bar{x})^2, \tag{1.12}$$

where, as before, $\bar{x}$ denotes the average value of the components of x. This secondary criterion is a measure of the "roughness" of the image.

The Bayesian approach to image reconstruction [15] combines the desires that both the error vector e in (1.5) and the variance defined by (1.12) should be small. This leads to the formulation of the discrete image reconstruction problem in which we aim at finding the *Bayesian estimate*, which is the vector x that minimizes

$$t^2\|y - Rx\|^2 + \|x - \mu\|^2, \tag{1.13}$$

where t is a constant (can be chosen to be the so-called *signal-to-noise ratio)* and μ is a J-dimensional vector (all of whose components may be equal to $\bar{x}$). The results presented in [15] show that the following version of ART using cyclic control converges to the minimizer of (1.13), provided only that (1.9) holds:

$$\begin{aligned} &u^{(0)} \text{ is the } I\text{-dimensional zero vector,}\\ &x^{(0)} = \mu,\\ &u^{(k+1)} = u^{(k)} + c^{(k)} o_{i_k},\\ &x^{(k+1)} = x^{(k)} + t c^{(k)} r_{i_k}, \end{aligned} \tag{1.14}$$

where o_{i_k} is the I-dimensional vector with i_kth component 1 and all other components zero and

$$c^{(k)} = \lambda^{(k)} \frac{t(y_{i_k} - \langle r_{i_k}, x^{(k)}\rangle) - u_{i_k}^{(k)}}{1 + t^2\|r_{i_k}\|^2}. \tag{1.15}$$

Note that in both (1.6) and (1.14) the updating of $x^{(k)}$ is simple: we add to it a multiple of the vector r_{i_k}.

One significant way that the Bayesian approach is more general than the ART approaches discussed earlier is that its convergence result does not need in its statement a requirement such as "if L is not empty." This is because, for every J-dimensional vector x, (1.13) has a well-defined value; its size indicates how bad (from a Bayesian point of view) x is as a solution to the discrete reconstruction problem. Removal of such a requirement is important, because in practice L may very well be empty since the error vector e in (1.5) is typically not the I-dimensional zero vector. This is due to two kinds of approximations made during the derivation of (1.5): the discretization in (1.2) and the ignoring of the *noise* in the measurement vector that is due to the difference between the mathematical ideal $\mathscr{R}_i f$ and its physically-obtained estimate y_i.

An alternative way to get rid of the requirement that L should be not empty is to replace it by a less restrictive one of the form that there should be an x such that

$$y_i \le \langle r_i, x \rangle \le \delta_i, \tag{1.16}$$

for $1 \le i \le I$, for some suitably selected y_i and δ_i. Noting the use of inequalities in (1.16), we observe that constraining can be similarly introduced by

$$w_j \le x_j \le v_j, \tag{1.17}$$

for $1 \le j \le J$. We can specify that the solution of the discrete reconstruction problem is the minimum norm x that satisfies all these inequalities. It is proved in [17, Theorem 2] that if there exists a J-dimensional vector x that satisfies (1.16) for $1 \le i \le I$ and (1.17) for $1 \le j \le J$, then the following algorithm using cyclic control produces a sequence a $x^{(0)}, x^{(1)}, \ldots$ that converges to such an x with minimal norm:

$$\begin{aligned} &u^{(0)} \text{ is the } I\text{-dimensional zero vector,}\\ &\hat{x}^{(0)} \text{ is the } J\text{-dimensional zero vector,}\\ &u^{(k+1)} = u^{(k)} - c^{(k)} o_{i_k},\\ &\hat{x}^{(k+1)} = \hat{x}^{(k)} + c^{(k)} r_{i_k}, \end{aligned} \tag{1.18}$$

where o_{i_k} is the I-dimensional vector with i_kth component 1 and all other components zero,

$$c^{(k)} = \text{mid}\left\{ u_{i_k}^{(k)}, \frac{\delta_{i_k} - \langle r_{i_k}, x^{(k)} \rangle}{\|r_{i_k}\|^2}, \frac{y_{i_k} - \langle r_{i_k}, x^{(k)} \rangle}{\|r_{i_k}\|^2} \right\}, \tag{1.19}$$

where mid when applied to three real numbers returns the one with middle value and, for $1 \le j \le J$,

$$x_j^{(k)} = \begin{cases} w_j, & \text{if } \hat{x}_j^{(k)} < w_j,\\ \hat{x}_j^{(k)}, & \text{if } w_j \le \hat{x}_j^{(k)} \le v_j,\\ v_j, & \text{if } v_j < \hat{x}_j^{(k)}. \end{cases} \tag{1.20}$$

Note, especially if you wish to implement this algorithm, that there is something clever going on here. The sequence $\hat{x}^{(0)}, \hat{x}^{(1)}, \ldots$ can be calculated without calculating the sequence $x^{(0)}, x^{(1)}, \ldots$. If ever during the iterative process we would like to look at an $x^{(k)}$, we need to apply (1.20) only once to the corresponding $\hat{x}^{(k)}$. The algorithm for generating the sequence $\hat{x}^{(0)}, \hat{x}^{(1)}, \ldots$ was introduced in the literature in [16] under the name ART4. That the corresponding sequence $x^{(0)}, x^{(1)}, \ldots$ converges to a minimal norm solution subject to the constraints (1.16) for $1 \le i \le I$ and (1.17) for $1 \le j \le J$ was first presented in [17].

An alternative approach, named ART3, to finding a J-dimensional vector x that satisfies the constraints (1.16) for $1 \le i \le I$ was proposed in [12]. The important aspect of ART3 is that, under some reasonable assumptions, it finds a desired x in a *finite number of steps*. Following the presentation in [5], the details are as follows. For $1 \le i \le I$, we define the ith *hyperslab* to be the set

$$K_i = \{x \mid y_i \le \langle r_i, x\rangle \le \delta_i\} \tag{1.21}$$

and the *signed distance* of a point x from the ith *median hyperplane* to be

$$d_i(x) = \frac{\langle r_i, x\rangle - \frac{1}{2}(y_i + \delta_i)}{\|r_i\|}, \tag{1.22}$$

with the corresponding *sign*

$$t_i(x) = \begin{cases} 1, & \text{if } d_i(x) \ge 0, \\ -1, & \text{if } d_i(x) < 0. \end{cases} \tag{1.23}$$

With these definitions, ART3 is specified by

$$\begin{aligned} &x^{(0)} \text{ is an arbitrary } J\text{-dimensional vector,} \\ &x^{(k+1)} = x^{(k)} + c^{(k)} r_{i_k}, \end{aligned} \tag{1.24}$$

where

$$c^{(k)} = -\frac{s_{i_k}(x)}{\|r_{i_k}\|}, \tag{1.25}$$

with $s_i(x)$ defined, for $1 \le i \le I$, as follows. Let

$$w_i = \frac{\delta_i - y_i}{2\|r_i\|} \tag{1.26}$$

(this is the half-width of the ith hyperslab), then

$$s_i(x) = \begin{cases} 0, & \text{if } |d_i(x)| \le w_i, \\ 2(d_i(x) - t_i(x)w_i), & \text{if } w_i < |d_i(x)| < 2w_i, \\ d_i(x), & \text{if } 2w_i \le |d_i(x)|. \end{cases} \tag{1.27}$$

In words, $x^{(k+1)}$ is the same as $x^{(k)}$ if $x^{(k)}$ is in the i_kth hyperslab, it is the projection of $x^{(k)}$ onto the i_kth median hyperplane if the distance of $x^{(k)}$ from the i_kth median hyperplane is at least the width of the i_kth hyperslab and, otherwise, $x^{(k+1)}$ is obtained from $x^{(k)}$ by reflecting it in the nearer bounding hyperplane of the i_kth hyperslab. The basic mathematical theorem regarding ART3 with cyclic ordering is the following (proved in [12]): Provided that the intersection $K = \bigcap_{i=1}^{I} K_i$ is not a subset of a hyperplane in the J-dimensional vector space (such a K is referred to as *full dimensional*), there is a nonnegative integer k such that $x^{(k)} \in K$.

We now turn to a discussion to the ordering of the sequence of equations (also referred to as *data access ordering*) as determined by i_k. Until now, we have concentrated on cyclic ordering, but that is often not necessary or even desirable. For example, the statement and proof in [17] of the convergence of ART4 to the minimum norm solution subject to the constraints (1.16) for $1 \le i \le I$ and (1.17) for $1 \le j \le J$ are in fact for the more general case of *almost cyclic ordering*, which is defined by the existence of an integer C such that, for all $k \ge 0$, $\{1, \dots, I\} \subseteq \{i_{k+1}, \dots, i_{k+C}\}$. Also, as demonstrated in [20] and elsewhere in the literature, data access ordering can have a significant effect on the practical performance of an ART algorithm. The underlying principle of the *efficient ordering* advocated in [20] is that the action taken by the iterative step of the type (1.6) should be as independent as possible of recent preceding actions or, to phrase this slightly more mathematically, we would like the vector r_{i_k} to be as orthogonal as possible to the space generated by the recently used vectors. That the specific data access ordering advocated in [20] is indeed efficient has been demonstrated experimentally both in [20] and in the later literature. Adjustment of data access ordering has been successful also in conjunction with finitely convergent versions of ART. For example, it is reported in [2, Section 2.1] that ART3+ (a version of ART3 with a data access ordering designed to avoid unnecessary checks on constraints that are likely to be satisfied) solved the problem named ISRAEL from the *linear programming* problem set Netlib/CUTEr over 500 times faster than what was reported for the same problem in the previous literature.

In more recent applications, it was found advantageous to use a three-dimensional mathematical model instead of the two-dimensional one introduced in the first sentence of this section (and used in the acoustic application [11]). Also, a particular choice of three-dimensional basis functions (the so-called *blobs*, which are spherically symmetric functions that decay smoothly to zero at the boundaries of their bounded supports) were found to be particularly efficacious. This was reported to be the case, based on careful comparisons of the task-oriented performance several reconstruction algorithms, in *positron emission tomography* and in *electron microscopy*; see [22, 23]. The somewhat recent publication [2] reports on the relative efficacy of ART as compared to alternatives in a number of fields: linear programming (as discussed in the last sentence of the previous paragraph), *image representation* (calculating the image vector for a blob basis from that for a more traditional basis, such as pixels or voxels), *radiation therapy planning*, computerized tomography (CT), and the reconstruction of the *orientation distribution function* (used to characterize crystallographic texture) from diffraction data. For the last of these applications, the reported discrete reconstruction problem involves $I = 1{,}000{,}000{,}000$ and $J = 1{,}372{,}000{,}000$; ART provides a solution in spite of the enormous size of the problem.

We complete this section with a discussion of a significant application, namely *multicriteria intensity-modulated proton therapy planning*, for which ART3+ has been demonstrated to provide a solution orders of magnitude faster than the previously industry-standard general purpose algorithms [6]. Intensity-modulated radiation therapy (IMRT) is a tool for treating cancer. The goal of IMRT is to deliver sufficient doses to tumors to kill them, but without causing irreparable damage to critical organs. IMRT differs from older approaches to radiation therapy in as much that it uses a multi-leaf collimator that allows the treatment planner to control the intensity of radiation within each of J relatively thin beamlets. The therapy requirements can be formulated as a linear feasibility problem (1.16), in which x_j is the to-be-determined intensity of the jth beamlet and the y_i and δ_i are determined by the constraints on the doses $\langle r_i, x\rangle$ to be delivered to the ith of I locations. Using the fact that, in the full dimensional case, ART3+ finds in a finite number of steps an x that satisfies (1.16), for $1 \le i \le I$, an algorithm has been proposed and demonstrated in [6] for multicriteria intensity-modulated proton therapy planning that is orders of magnitude faster than the previously-used industry-standard algorithms.

10.2 Superiorization of iterative reconstruction techniques

As seen in the last section, optimization (i. e., minimization or maximization of a function of the image vector x such as its norm, entropy (1.11), variance (1.12) or Bayesian estimate (1.13)) arises naturally in the mathematical discussion of image reconstruction. In fact, we are likely to use constrained optimization, with the constraints arising either from the measurements, as in (1.4) and in (1.16), or from prior knowledge regarding the nature of permissible solutions, as in (1.17). Intuitively speaking, constraints-compatibility is a primary requirement for an image vector x to be considered a solution to an image reconstruction problem. This is translated into something mathematically precise by introducing a function on x, called the *primary criterion*, that measures the incompatibility of x with the given constraints. For example, if the constraints are those of (1.4), then the primary criterion may be defined as $\|y - Rx\|^2$. In addition, we may select a real number ε such that we consider an x a *constraints-compatible solution* if the value of the primary criterion for x is not larger than ε. It is frequently the case in practical image reconstruction problems that there is a large number of constraints-compatible solutions. In such a case, a *secondary criterion* may be made use of to help us distinguish *superior* constraints-compatible solutions.

The superiorization methodology is a relatively recently-developed heuristic approach to constrained optimization. The underlying idea is that in many applications there exist computationally-efficient iterative algorithms that produce solutions that are constraints-compatible. An extreme example of this is ART3 that, as discussed in the last section, produces in a finite number of steps an image vector x that satisfies the constraints (1.16), for $1 \leq i \leq I$. More typically, compatibility involves an approximate satisfaction of the constraints, as in (1.4). Often the algorithm is *perturbation resilient* in the sense that, even if certain kinds of changes are made at the end of each iterative step, the algorithm still produces a constraints-compatible solution. This property is exploited in superiorization by using such perturbations to steer the algorithm to a solution that is not only constraints-compatible, but is also desirable according to a specified secondary criterion (such as entropy (1.11)). The approach is very general, it is applicable to many iterative procedures and secondary criteria. Most importantly, superiorization is an automatic procedure that turns an iterative algorithm into its superiorized version. This, and its practical consequences in various application areas, have been investigated for a variety of constrained optimization tasks.

The number of combinations of a set of constraints and an optimization criterion that my arise is limitless. In spite of the great deal of knowledge that exists regarding constrained optimization, we keep coming across new combinations for which the existing knowledge is insufficient and a new algorithm has to be developed. It is often the case that the mathematical discovery and implementation of an algorithm for such a new constrained optimization problem is far from trivial, it requires a major investment of time of the researcher and even then success is not guaranteed. On the other hand, it is generally a much easier task to find a computationally-efficient algorithm just for constraints-compatibility (without optimization). Typically, such algorithms are iterative; they produce a potentially infinite sequence of images from which we can select the first one that is a constraints-compatible solution. Furthermore, often the algorithm is perturbation resilient. This property is exploited in the superiorization approach by using such perturbations to steer the algorithm to an output that is as constraints-compatible as the output of the original algorithm, but it is superior to it according to the secondary criterion. The aim is to present a totally automatic procedure that turns the iterative algorithm into such a superiorized version. This has been done for a very large class of iterative algorithms and for secondary criteria in general; typical restrictions (such as convexity) on the secondary optimization criterion are not needed. Thus superiorization can be significantly helpful, because it has the potential of saving a lot of time and effort for the researcher when the application of interest gives rise to a new constrained optimization problem. This essential contribution deserves repeating in alternate words: If we have an iterative algorithm for constraints-compatibility (which, in general, is much easier to produce than one for constrained optimization) and if we are given a secondary criterion, then the production of the superiorized version of the algorithm is automatic.

As an example, consider *likelihood maximization* using the iterative algorithm of *expectation maximization* (ML-EM) for positron emission tomography (PET). Details for our summary discussion that follows can be found in [8, 21]. Here, *likelihood* (the probability of observing the measurement vector y if the image vector is x) is used as a measure of constraints-compatibility. In a commonly-used model for the physics of

PET, given the measurement vector y, the x that maximizes the likelihood is the one that minimizes

$$\sum_{i=1}^{I}(\langle r_i, x\rangle - y_i \ln\langle r_i, x\rangle), \tag{2.1}$$

which is the primary criterion in case of PET. It has been proved, under some reasonable assumptions, that the following ML-EM algorithm produces a sequence a $x^{(0)}, x^{(1)}, \dots$ that converges to a minimizer of (2.1):

$$\begin{aligned} &x^{(0)} \text{ is a } J\text{-dimensional vector with arbitrary positive components,}\\ &x_j^{(k+1)} = \frac{x_j^{(k)}}{\sum_{i=1}^{I} r_{i,j}} \sum_{i=1}^{I} \frac{r_{i,j} y_i}{\langle r_i, x^{(k)}\rangle}, \quad \text{for } 1 \le j \le J. \end{aligned} \tag{2.2}$$

However, it turns out that the mathematical formalization does not reflect the physical reality exactly. It was observed that the images $x^{(k)}$ deteriorate with a large number of iterations, in the sense that they present some undesirable high amplitude patterns; see Figure 10.1(Top left). There have been various approaches to deal with the appearance of these artifacts. One approach is by reformulating the PET reconstruction problem into one of *maximum a posteriori* (MAP) estimation that aims at minimizing

$$t^2 \sum_{i=1}^{I}(\langle r_i, x\rangle - y_i \ln\langle r_i, x\rangle) + \psi(x), \tag{2.3}$$

where t is a constant (compare (2.3) with (1.13)) and

$$\psi(x) = \sum_{r\in M}\left(x_r - \frac{1}{8}\sum_{l\in M_r} x_l\right)^2, \tag{2.4}$$

with M being the set of indices r such that the rth pixel is not on the border of the image region and M_r, for $r \in M$, is the set of indices associated with the eight pixels neighboring the rth pixel. This was first done in [21], which contains a MAP-EM algorithm that is very similar in its computational details to the ML-EM algorithm and yet converges to the minimizer of (2.3). However, to guarantee this desirable behavior it was necessary to develop new mathematics; the effort invested in doing that for [21] was significant. In contrast, as discussed in [8] and illustrated in Figure 10.1, the undesirable behavior of ML-EM can be automatically eliminated by superiorizing it for the secondary criterion provided in (2.4). This last point is worth restating: We are not claiming here that the superiorized version of ML-EM is the best known way to perform PET reconstruction; rather, we are illustrating that the general approach of superiorization can be in particular applied automatically to improve the performance of ML-EM. This is

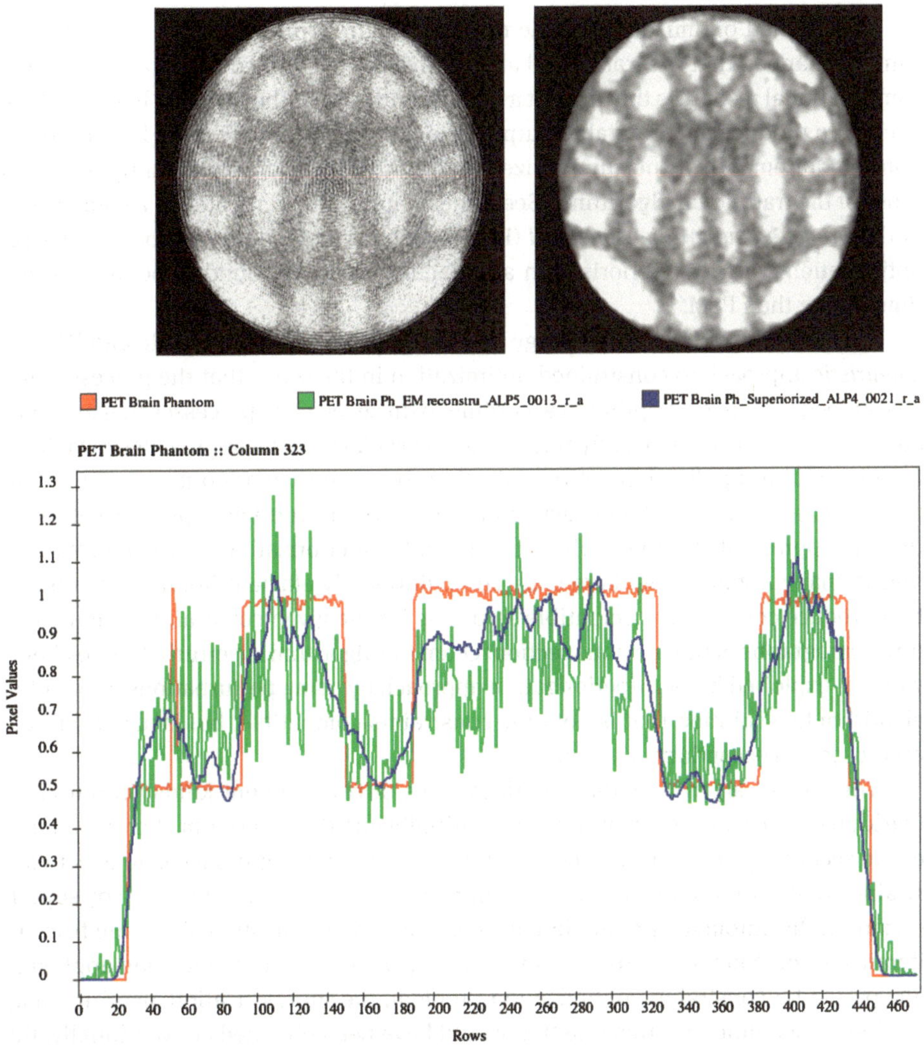

Figure 10.1: PET reconstructions from simulated data. (Reproduced with permission from G. T. Herman. Superiorization for image analysis. In: *Combinatorial image analysis*, Lecture notes in computer science, volume 8466, pages 1–7. Springer, 2014.) Top left: Reconstruction using ML-EM. Top right: Reconstruction using superiorization. Bottom: Activity along the 323rd column in the phantom (red), in the ML-EM reconstruction (green) and in the superiorization reconstruction (blue).

just one of many possible applications. Below we provide the general scheme for the *Superiorized Version of Algorithm* **P**. The important thing to appreciate about superiorization is that the same scheme is efficacious in many applications, such as **P** being ML-EM for PET, as here, or SART for industrial nondestructive testing, as illustrated at the end of this chapter.

Constrained optimization problems that arise in applications are often huge. It can then happen that the traditional algorithms for constrained optimization require computational resources that are not available and, even if they are, the length of time needed to produce an acceptable output is too long to be practicable. The computational requirements of the superiorized algorithms are often significantly less than those of the traditional algorithms. See, for example, [3] that reports on a comparison of the *projected subgradient method* (PSM), which is a standard method of classical optimization, with a superiorization approach, which was found to be over twenty times faster than PSM.

The superiorization methodology was first proposed over a decade ago [1]. It is a *heuristic* approach to constrained optimization in the sense that the process is not guaranteed to lead to an optimum according to the given task; processes that are guaranteed in that sense are usually referred to as *exact*. Heuristic approaches have been found useful in applications of optimization, because they are often computationally less expensive than their exact counterparts, but nevertheless provide solutions that are appropriate for the application at hand. Superiorization is based on the fact that in many applications there exist computationally-efficient iterative algorithms that produce constraints-compatible solutions for the given constraints; see the last three paragraphs of the previous section. Often the algorithm is perturbation resilient, which is exploited in the superiorization approach by using perturbations to steer the algorithm to a solution that is not only constraints-compatible, but is also desirable according to a secondary criterion.

The idea of algorithms that interlace steps of two different kinds (in our case, one kind of steps aim at constraints-compatibility and the others aim at improvement of the secondary criterion) has been made use of in many approaches to exact constrained optimization. However, those approaches cannot do what is done by superiorization: the automatic production of a constrained optimization algorithm from an iterative algorithm for constraints-compatibility. While it is true that when applied to any particular constrained optimization problem, the superiorization algorithm may not look much different from one that would have been obtained conventionally, the essential difference is the following: The conventional approach designs an algorithm that depends simultaneously on both the constraints and the optimization criterion (and this may be difficult to do), while the superiorization algorithm takes an iterative algorithm for the constraints (this is often easy to design) and produces from it, in an automatic fashion, the superiorized version using the given optimization criterion as the secondary criterion of the superiorization methodology.

The world-view of superiorization is different from that of classical constrained optimization. Both in superiorization and in classical constrained optimization we assume the existence of a domain Ω and an optimization/secondary criterion, specified by a function ϕ that has a real-number value for every x in Ω. In the following discussion, we will use as the domain Ω the set of all J-dimensional vectors with nonnegative components. This is just one possible choice, the superiorization methodology applies to many other choices as well. An example for the secondary criterion ϕ is the ψ of (2.4).

In classical optimization, it is assumed that there is a constraints set C and the optimization task is to find an $x \in C$ for which $\phi(x)$ is minimal. One problem with this is that the constraints may not be consistent and so C could be empty and the optimization task as stated would not have a solution. Another is that iterative methods of classical constrained optimization typically converge to a solution only in the limit and so some stopping rule has to be applied to terminate the process. The actual output at that time may not be in C and, even if it is in C, it is most unlikely to be a minimizer of ϕ over C.

Both of these problems are handled in the superiorization approach by replacing the C by a nonnegative real-valued function $\mathcal{P}r$ that indicates of how incompatible a given $x \in \Omega$ is with the constraints. ($\mathcal{P}r$ is our general notation for the function to which we have been referring as the primary criterion up to now.) The merit of an output x of an algorithm is given by the smallness of the two numbers $\mathcal{P}r(x)$ and $\phi(x)$. Roughly summarizing (details follow soon), it was shown in [14] that if an iterative algorithm produces an output x, then its superiorized version will produce an output x' for which $\mathcal{P}r(x')$ is not larger then $\mathcal{P}r(x)$, but (in general) $\phi(x')$ is much smaller than $\phi(x)$. It is also the case that the superiorization approach is often more efficacious than classical optimization, in the sense that if the latter produces an actual output x, then the superiorization approach will produce an output x' for which $\mathcal{P}r(x')$ is not larger then $\mathcal{P}r(x)$, $\phi(x')$ is much smaller than $\phi(x)$ and the computer time required to obtain x' is less than that required by the classical approach to obtain x (see the above-quoted result from [3] comparing PSM with superiorization).

Now we give a more detailed and mathematically more rigorous treatment of the claim in the previous paragraph on the comparative behavior (as measured by $\mathcal{P}r$ and ϕ) of the outputs of an iterative algorithm and its superiorized version. This treatment is based on [14].

In order to explore superiorization in the spirit of [14], we need to reconsider the notion of an iterative step. We first consider the special case of ART, for which $x^{(k+1)}$ depends not only on $x^{(k)}$ but also on $\lambda^{(k)}$ and on i_k; see (1.6) and (1.7). In common practice, the first of these is not really a dependence on k, since typically the relaxation parameter $\lambda^{(k)}$ is chosen to have a constant value λ. However, the dependence of i_k on k is at the unavoidable essence of the method. For a straightforward explanation of how an iterative method is superiorized, it is advantageous to present the method in a way in which the image at the end of one "step" depends only on the image at the beginning of that step. Such a presentation for ART can be given, for example, as follows. Assume cyclic ordering and a constant relaxation parameter λ. Consider now the mapping $\mathbf{P} : \Omega \to \Omega$ that is defined by $s = \mathbf{P}x$, with $x^{(0)} = x$ and, for $1 \le i \le I$,

$$x^{(i)} = x^{(i-1)} + \lambda \frac{y_i - \langle r_i, x^{(i-1)} \rangle}{\|r_i\|^2} r_i \qquad (2.5)$$

and, for $1 \le j \le J$,

$$s_j = \max\{x_j^I, 0\}. \tag{2.6}$$

Note that (2.5) and (2.6) is equivalent to cycling through (1.6) and (1.7) once for each of the approximate equalities in (1.4) and then setting negative components of the vector to zero. From this, it follows that the iterative sequence $(\mathbf{P}^k x)_{k=0}^{\infty} = (x^{(0)}, \mathbf{P}x^{(0)}, \mathbf{PP}x^{(0)}, \dots)$ is very similar to the subsequence $(x^{(0)}, x^{(I)}, x^{(2I)}, \dots)$ of the sequence of iterates produced by produced by the direct additive method of [10]. Since the mapping $\mathbf{P}$ is uniquely defined (given the measurement vector y, the projection matrix R and the relaxation parameter λ), it is reasonable to refer to it as the *algorithm* $\mathbf{P}$. From now on, we consider unless otherwise stated, the general case in which the algorithm $\mathbf{P}$ is an arbitrary mapping from Ω into Ω.

The specification of the superiorized version of $\boldsymbol{P}$ assumes that we have available a summable sequence $(\gamma_\ell)_{\ell=0}^{\infty}$ of positive real numbers (e. g., $\gamma_\ell = a^\ell$, where $0 < a < 1$) and it generates, simultaneously with the sequence $(x^{(k)})_{k=0}^{\infty}$, sequences $(v^{(k)})_{k=0}^{\infty}$ and $(\beta_k)_{k=0}^{\infty}$. The latter is generated as a subsequence of $(\gamma_\ell)_{\ell=0}^{\infty}$, resulting in a summable sequence $(\beta_k)_{k=0}^{\infty}$. The algorithm further depends on a specified initial image $\bar{x} \in \Omega$ and on a positive integer N. It makes use of a logical variable called *loop*.

Superiorized Version of Algorithm $\mathbf{P}$

```
set k = 0
set x^(k) = x̄
set ℓ = −1
repeat
    set n = 0
    set x^(k,n) = x^(k)
    while n < N
        set v^(k,n) to be a nonascending vector for φ at x^(k,n)
        set loop = true
        while loop
            set ℓ = ℓ + 1
            set β_{k,n} = γ_ℓ
            set z = x^(k,n) + β_{k,n} v^(k,n)
            if z ∈ Ω and φ(z) ≤ φ(x^k) then
                set n = n + 1
                set x^(k,n) = z
                set loop = false
    set x^(k+1) = Px^(k,N)
    set k = k + 1
```

In the next paragraphs, we discuss some details of this specification of the superiorized version of an algorithm **P**. Prior to getting into its step-by-step behavior, we note that Line 8 makes use of the essential notion of a *nonascending vector*. A J-dimensional vector d is nonascending for ϕ at x if $\|d\| \le 1$ and

$$\begin{aligned} &\text{there is a } \delta > 0 \text{ such that for all } \lambda \in [0, \delta], \\ &(x + \lambda d) \in \Omega \text{ and } \phi(x + \lambda d) \le \phi(x). \end{aligned} \tag{2.7}$$

Ways of finding nonascending vectors are provided, for example, by [8, Theorem 1].

In the superiorized version of an algorithm **P**, the iteration number k is set to 0 by line 1 and $x^{(k)} = x^{(0)}$ is set to its initial value $\bar{x}$ by line 2. The integer index ℓ for picking the next element from the sequence $(\gamma_\ell)_{\ell=0}^{\infty}$ is initialized to -1 by line 3 (it is repeatedly increased by line 11). The lines 5–19 that follow the **repeat** in line 4 perform a complete iterative step from $x^{(k)}$ to $x^{(k+1)}$. During one iterative step, there is one application of the operator **P**, in line 18, but there are N steering steps aimed at reducing the value of ϕ; the latter are done by lines 5–17. These lines produce images $x^{(k,n)}$, for $0 \le n \le N$, with $x^{(k,0)} = x^{(k)}$, $x^{(k,n)} \in \Omega$, and $\phi(x^{(k,n)}) \le \phi(x^{(k)})$. For the proof of the validity of the last sentence, see [14].

We now return to the earlier roughly-made claim that if an iterative algorithm **P** produces an output x, then its superiorized version will produce an output x' for which $\mathcal{P}r(x')$ is not larger then $\mathcal{P}r(x)$, but $\phi(x')$ is much smaller than $\phi(x)$. The second part of this claim is essentially not mathematical and so it cannot be mathematically proved; it is believed on the basis of the designed behavior of the superiorized algorithm, which aims at producing images for which the secondary criterion has a small value. In particular, it follows from what is stated in the previous paragraph that. for all $k \ge 0$, $\phi(x^{(k,N)}) \le \phi(x^{(k)})$. Unfortunately, this does not strictly imply that $\phi(x^{(k+1)}) = \phi(\mathbf{P}x^{(k,N)}) \le \phi(x^{(k)})$; nevertheless, the validity of the second part of the claim has been demonstrated in many specific instances in the published literature. The first part of the claim can be made strictly mathematical, and thus be proved. We now sketch how this is done.

First, we need a mathematical definition of output in our context. Slightly modifying a definition in [14], we propose the following. If ε is a real number and $\boldsymbol{R} = (x^{(k)})_{k=0}^{\infty}$ is a sequence of images in Ω, then $O(\varepsilon, \boldsymbol{R})$ denotes the $x \in \Omega$ that has the following properties: $\mathcal{P}r(x) \le \varepsilon$ and there is a nonnegative integer K such that $x^{(K)} = x$ and, for all nonnegative integers $k < K$, $\mathcal{P}r(x^{(k)}) > \varepsilon$. Clearly, if there is such an x, then it is unique. If there is no such x, then we say that $O(\varepsilon, \boldsymbol{R})$ is *undefined*, otherwise we say that it is *defined*. The intuition behind this definition is the following: If we think of $\boldsymbol{R}$ as the (infinite) sequence of images that is produced by an iterative algorithm without a termination criterion, then $O(\varepsilon, \boldsymbol{R})$ is the *output* produced by that algorithm when we add to it instructions that make it terminate as soon as it reaches an image for which the value of the primary criterion is not greater than ε.

Second, we have not been able to prove the desired claim for all algorithms $\mathbf{P}$: $\Omega \to \Omega$, but only for ones that are *strongly perturbation resilient* in the following sense:

1. there exists a nonnegative real number ε such that $O(\varepsilon, (\mathbf{P}^k x)_{k=0}^{\infty})$ is defined for every $x \in \Omega$; and
2. for all nonnegative real numbers ε such that $O(\varepsilon, (\mathbf{P}^k x)_{k=0}^{\infty})$ is defined for every $x \in \Omega$, we also have that $O(\varepsilon', R)$ is defined for every $\varepsilon' > \varepsilon$ and all sequences $R = (x^k)_{k=0}^{\infty}$ of images in Ω generated by

$$x^{k+1} = \mathbf{P}(x^{(k)} + \beta_k v^{(k)}), \quad \text{for all } k \geq 0, \tag{2.8}$$

where $\beta_k v^{(k)}$ are *bounded perturbations*, meaning that the sequence $(\beta_k)_{k=0}^{\infty}$ of nonnegative real numbers is summable, the sequence $(v^k)_{k=0}^{\infty}$ of J-dimensional vectors is bounded and, for $k \geq 0$, $x^{(k)} + \beta_k v^{(k)} \in \Omega$.

While this definition looks complicated, sufficient conditions for strong perturbation resilience of an algorithm $\mathbf{P}$ have been published that involve only the checking of such mathematically simple and standard notions as the uniform continuity of $\mathcal{P}r$ and the bounded convergence and nonexpansiveness of $\mathbf{P}$; see, for example, [14, Theorem 1]. Such checks are often easy to do.

We are now ready to state the mathematical theorem (for proof see [14]) that corresponds to the first part of the roughly-stated claim above: If $\mathbf{P}$ is strongly perturbation resilient, and the positive real number ε is such that $O(\varepsilon, (\mathbf{P}^k x)_{k=0}^{\infty})$ is defined for every $x \in \Omega$, then $O(\varepsilon', \boldsymbol{R})$ is also defined for every $\varepsilon' > \varepsilon$ and every sequence $\boldsymbol{R} = (x^{(0)}, x^{(1)}, x^{(2)}, \ldots)$ produced by the superiorized version of $\mathbf{P}$.

We conclude this chapter with a brief overview of some additional publications relevant to the superiorization of iterative reconstruction methods.

As mentioned earlier, the superiorization methodology was first proposed in [1]. There perturbation resilience was proved for the general class of *string-averaging methods*, in one “step” of which the end-points of several finite strings of sequential projections onto the constraints (e. g., each projection may be specified by (1.6) and (1.7)) are averaged [4]. ART is a special case of string averaging in which there is only one string and so there is no need for averaging. In the superiorization demonstrations of [1], $\|x\|$ is discussed as a possible secondary criterion; however, it is pointed out that it may not be a particularly good measure of the undesirability of the image x as a solution of the reconstruction problem and that many research workers in image processing have been advocating the use of *total variation* (*TV*) instead of the norm. For an image x using pixels for the basis functions, we define (see, e. g., [8])

$$TV(x) = \sum_{c \in C} \sqrt{(x_c - x_{\rho(c)})^2 + (x_c - x_{\xi(c)})^2}, \tag{2.9}$$

where C is the set of all indices of pixels that are not in the rightmost column or the bottom row of the pixel array. For any pixel with index c in C, $\rho(c)$ and $\xi(c)$ are the indices of the pixels to its right and below it, respectively. A way of obtaining nonascending vectors for TV is presented in [1] and it is shown, for both unconstrained and partially constrained ART, that the superiorized versions of those algorithms are more efficacious than not only the original versions, but also than a previously published classical constrained optimization algorithm. The use of TV as a optimization/secondary criterion has been claimed to be particularly useful in image reconstruction if the number I of measurements is small, meaning that it is (possibly much) smaller than the number J of basis functions. This situation is discussed (using superiorization) in [13]. An alternative secondary criterion that has been claimed to be efficacious is the *ℓ_1-norm* of the *Haar transform*; however, in superiorization experiments on simulated CT data of the head, reconstructions whose Haar transform has a small ℓ_1-norm were not found to be not more efficacious than reconstructions that have a small TV value [9].

A large class of iterative reconstruction algorithms for which perturbation resilience and superiorization has been carefully studied in [7, 24] is the class of so-called *block-iterative projection* (BIP) methods. Such methods differ from ART, in which $x^{(k+1)}$ is obtained from $x^{(k)}$ by considering a single one of the I approximate equations in (1.4), by making use of a "block" of equations in each iterative step. Clearly, ART is a special case of BIP with each block having size one.

X-ray CT is an important and widespread inspection technique also in *industrial nondestructive testing*. However, large-sized and heavily absorbing objects cause artifacts due to either the lack of penetration of the specimen in specific directions or by having data from only a limited angular range of views. In such cases, valuable information about the specimen is not revealed by the CT measurements alone; see Figure 10.2(Top). Further imaging modalities, such as optical scanning and ultrasonic testing, are able to provide data (such as an edge map) that are complementary to the CT acquisition. In [25], a superiorization approach is used in which complementary data is incorporated into a secondary criterion; this allows precise localization of edges that are not resolvable from the CT data by itself. Superiorization, as presented in [25], exploits the fact that the so-called simultaneous *algebraic reconstruction technique* (SART), which is frequently used for CT reconstruction, is resilient to perturbations, and so it can be modified to produce an output that is as compatible with the constraints provided by the CT measurements as the output of unmodified SART, but is more consistent with the complementary data. We demonstrate in Figure 10.2 that using the superiorized version of the SART algorithm on physically measured data of a turbine blade results in a clear improvement over the quality of the image reconstructed by SART without superiorization.

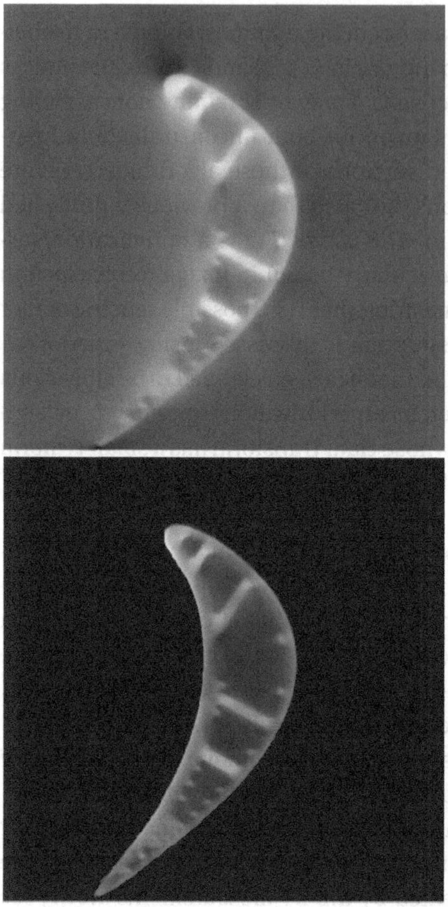

Figure 10.2: CT reconstructions of a turbine blade. (Reproduced with permission from M. J. Schrapp and G. T. Herman. Data fusion in x-ray computed tomography using a superiorization approach. *Review of scientific instruments*, 85:053701, 2014.) Top: Reconstruction by the SART algorithm with clearly visible artifacts due to lack of penetration; especially, concave edges are smeared out reflecting a low contrast-to-noise ratio in these image regions. Bottom: Reconstruction using a superiorized SART algorithm in which the secondary criterion has been provided based on optical and ultrasonic measurements of the turbine blade.

Bibliography

[1] D. Butnariu, R. Davidi, G. T. Herman, and I. G. Kazantsev. Stable convergence behavior under summable perturbations of a class of projection methods for convex feasibility and optimization problems. *IEEE J. Sel. Top. Signal Process.*, 1:540–547, 2007.

[2] Y. Censor, W. Chen, P. L. Combettes, R. Davidi and G. T. Herman. On the effectiveness of projection methods for convex feasibility problems with linear inequality constraints. *Comput. Optim. Appl.*, 51:1065–1088, 2012.

[3] Y. Censor, R. Davidi, G. T. Herman, R. W. Schulte and L. Tetruashvili. Projected subgradient minimization versus superiorization. *J. Optim. Theory Appl.*, 160:730–747, 2014.

[4] Y. Censor, T. Elfving, and G. T. Herman. Averaging strings of sequential iterations for convex feasibility problems. In: D. Butnariu, Y. Censor, and S. Reich, editors, *Inherently parallel algorithms in feasibility and optimization and their applications*, Elsevier Science Publishers, 101–113, 2000.

[5] Y. Censor and G. T. Herman. On some optimization techniques in image reconstruction from projections. *Appl. Numer. Math.*, 3:365–391, 1987.

[6] W. Chen, D. Craft, T. M. Madden, K. Zhang, H. M. Kooy and G. T. Herman. A fast optimization algorithm for multicriteria intensity modulated proton therapy planning. *Med. Phys.*, 37:4938–4945, 2010.

[7] R. Davidi, G. T. Herman and Y Censor. Perturbation-resilient block-iterative projection methods with application to image reconstruction from projections. *Int. J. Oper. Res.*, 16:505–524, 2009.

[8] E. Garduño and G. T. Herman. Superiorization of the ML-EM algorithm. *IEEE Trans. Nucl. Sci.*, 61:162–172, 2014.

[9] E. Garduño, G. T. Herman and R. Davidi. Reconstruction from a few projections by ℓ_1-minimization of the Haar transform. *Inverse Probl.*, 27:055006, 2011.

[10] R. Gordon, R. Bender and G. T. Herman. Algebraic Reconstruction Techniques (ART) for three-dimensional electron microscopy and X-ray photography. *J. Theor. Biol.*, 29:471–481, 1970.

[11] J. F. Greenleaf, S. A. Johnson, S. L. Lee, G. T. Herman and E. H. Wood. Algebraic reconstruction of spatial distributions of acoustic absorption within tissue from their two-dimensional acoustic projections. In: P. Green, editor, *Acoustical holography*, volume 5, pages 591–603, 1974.

[12] G. T. Herman. A relaxation method for reconstructing objects from noisy x-rays. *Math. Program.*, 8:1–19, 1975.

[13] G. T. Herman and R. Davidi. Image reconstruction from a small number of projections. *Inverse Probl.*, 24:045011, 2008.

[14] G. T. Herman, E. Garduño, R. Davidi and Y Censor. Superiorization: An optimization heuristic for medical physics. *Med. Phys.*, 39:5532–5546, 2012.

[15] G. T. Herman, H. Hurwitz, A. Lent and H-P. Lung. On the Bayesian approach to image reconstruction. *Inf. Control*, 42:60–71, 1979.

[16] G. T. Herman and A. Lent. Iterative reconstruction algorithms. *Comput. Biol. Med.*, 6:319–332, 1976.

[17] G. T. Herman and A. Lent. A family of iterative quadratic optimization algorithms for pairs of inequalities, with application in diagnostic radiology. *Math. Program. Stud.*, 9:15–29, 1978.

[18] G. T. Herman, A. Lent and P. H. Lutz. Relaxation methods for image reconstruction. *Commun. ACM*, 21:152–158, 1978.

[19] G. T. Herman, A. Lent and S. W. Rowland. ART: Mathematics and applications (A report on the mathematical foundations and on the applicability to real data of the algebraic reconstruction techniques). *J. Theor. Biol.*, 42:1–32, 1973.

[20] G. T. Herman and L. B. Meyer. Algebraic reconstruction techniques can be made computationally efficient. *IEEE Trans. Med. Imaging*, 12:600–609, 1993.

[21] E. Levitan and G. T. Herman. A maximum a posteriori probability expectation maximization algorithm for image reconstruction in emission tomography. *IEEE Trans. Med. Imaging*, 6:185–192, 1987.

[22] R. Marabini, G. T. Herman and J.-M. Carazo. 3D reconstruction electron microscopy using ART with smooth spherically-symmetric volume elements (blobs). *Ultramicroscopy*, 72:53–65, 1998.

[23] S. Matej, G. T. Herman, T. K. Narayan, S. S. Furuie, R. M. Lewitt and P.Kinahan. Evaluation of the task-oriented performance of several fully 3-D PET reconstruction algorithms. *Phys. Med. Biol.*, 39:355–367, 1994.

[24] T. Nikazad, R. Davidi and G. T. Herman. Accelerated perturbation-resilient block-iterative projection methods with application to image reconstruction. *Inverse Probl.*, 28:035005, 2012.

[25] M. J. Schrapp and G. T. Herman. Data fusion in X-ray computed tomography using a superiorization approach. *Rev. Sci. Instrum.*, 85:053701, 2014.

Tanja Tarvainen

11 Quantitative photoacoustic tomography in Bayesian framework

Abstract: Quantitative photoacoustic tomography is an imaging modality based on the photoacoustic effect generated by the absorption of an externally introduced light pulse. The technique aims at estimation of the distribution of light absorbing molecules in three-dimensional tissue volume. This is an ill-posed problem which needs to be approached in the framework of inverse problems. In this work, a Bayesian approach to inverse problems is taken. The forward and inverse problems of quantitative photoacoustic tomography are reviewed, and two case studies related to the acoustic and optical inverse problems of quantitative photoacoustic tomography are considered.

Keywords: Photoacoustics, tomography, Bayesian inverse problems, quantitative imaging, uncertainty quantification

MSC 2010: 92C55, 68U10, 62F15

11.1 Introduction

Photoacoustic tomography (PAT) is an imaging modality based on the photoacoustic effect generated through the absorption of an externally introduced light pulse. The method combines optical contrast with high spatial resolution of ultrasound. The optical contrast is provided by distinctive absorption spectra by different chromophores, i. e., light absorbing molecules. The chromophores of interest are, e. g., hemoglobin, melanin, and optionally various contrast agents. In soft biological tissue, the ultrasonic waves carry this optical information to the surface of tissue with minimal scattering, thus retaining accurate spatial information. PAT can be used to provide images of soft biological tissues with high spatial resolution. It has successfully been applied to the visualization of different structures in biological tissues, such as human blood vessels, microvasculature of tumors, and the cerebral cortex in small animals. For more information about PAT see, e. g., the reviews [12, 15, 52, 58, 68, 101, 103–105, 107, 109] and the references therein.

Acknowledgement: Collaboration with Aki Pulkkinen, Jenni Tick, Jari Kaipio, Ben Cox, and Simon Arridge is gratefully acknowledged. The work has been supported by Jane and Aatos Erkko Foundation and Academy of Finland (projects 314411 and 312342 Finnish Centre of Excellence of Inverse Modelling and Imaging).

Tanja Tarvainen, Department of Applied Physics, University of Eastern Finland, P.O. Box 1627, 70211 Kuopio, Finland, e-mail: tanja.tarvainen@uef.fi

https://doi.org/10.1515/9783110560855-011

In PAT measurement situation, a short (nanosecond scale) pulse of visible or near-infrared light is used to illuminate the tissue region of interest. As light propagates within the tissue, it is absorbed leading to temperature rise which results in localized increases in pressure. The generated pressure propagates through the tissue as an acoustic wave and can be measured on the surface of the tissue using ultrasound sensors. Due to the large difference in time scales between the optical and acoustic parts of the problem, they can be decoupled and treated separately. In the acoustic inverse problem of PAT, the initial pressure distribution caused by the light absorption is estimated from the pressure waves measured on the surface of the tissue. Further, in the optical inverse problem, which is sometimes referred to as quantitative photoacoustic tomography, one aims at estimating the absolute concentrations of the chromophores from the reconstructed initial pressure distribution, i. e., from photoacoustic images [18].

In this chapter, we concentrate on PAT with an aim at quantitative imaging. This problem is approached in the framework of Bayesian inverse problems. The chapter is organized as follows. Modeling of the photoacoustic effect, i. e., the forward problem, is considered in Section 11.2. Then the inverse problem including a review of the Bayesian approach is considered in Section 11.3. Computational examples of two case studies are shown in Section 11.4, and the approaches are discussed in Section 11.5.

11.2 Forward model

The forward problem in quantitative photoacoustic tomography (QPAT) is to solve the photoacoustic time series at the acoustic sensor locations when the parameters of input light and the optical and acoustic properties of the target are given.

11.2.1 Light propagation and absorption

In optical imaging, light transport through a medium containing scattering particles is described by transport theory [39]. In transport theory, light conservation within a small volume element of phase space is investigated. The wave phenomenon is ignored. The transport theory can be modeled through stochastic and deterministic methods. In the stochastic approach, such as the Monte Carlo method, individual photon interactions are modeled as the they are scattered and absorbed within the medium. In the deterministic approach, light transport is described with integro-differential equations which can be solved either analytically or numerically. In QPAT, the radiative transfer equation (RTE) and its approximations such as the diffusion approximation (DA) have been utilized.

11.2.1.1 Radiative transfer equation

The radiative transfer equation is a one-speed approximation of the transport equation, and thus it basically assumes that the energy (or speed) of photons does not change in collisions and that the refractive index is constant within the medium. For discussion of photon transport in medium with spatially varying refractive index and for studies considering piecewise constant refractive index see, e. g., [8, 57] and the references therein.

Let $\Omega \subset \mathbb{R}^d$, $d = 2$ or 3 denote the domain with boundary $\partial\Omega$ where d is the (spatial) dimension of the domain and $\hat{s} \in S^{d-1}$ is a unit vector in the direction of interest. Light transport in medium with scattering particles can be described using the RTE as

$$\begin{aligned}\frac{1}{c}\frac{\partial\phi(r,\hat{s},t)}{\partial t} &+ \hat{s}\cdot\nabla\phi(r,\hat{s},t) + (\mu_s(r)+\mu_a(r))\phi(r,\hat{s},t)\\ &= \mu_s(r)\int_{S^{d-1}} \Theta(\hat{s}\cdot\hat{s}')\phi(r,\hat{s}',t)\mathrm{d}\hat{s}', \quad r\in\Omega \end{aligned} \tag{1}$$

$$\phi(r,\hat{s},t) = \begin{cases}\phi_0(r,\hat{s},t), & r\in\xi_j, \quad \hat{s}\cdot\hat{n}<0\\ 0, & r\in\partial\Omega\backslash\xi_j, \quad \hat{s}\cdot\hat{n}<0\end{cases} \tag{2}$$

where $\phi(r,\hat{s},t)$ is the radiance at point r, direction $\hat{s}$ and time instance t, c is the speed of light in the medium, $\mu_s(r)$ is the scattering coefficient, $\mu_a(r)$ is the absorption coefficient, $\Theta(\hat{s}\cdot\hat{s}')$ is the scattering phase function, $\phi_0(r,\hat{s},t)$ is a boundary light source at a source position $\xi_j \subset \partial\Omega$, and $\hat{n}$ is an outward unit normal. The radiance can be defined such that the amount of power transfer in the infinitesimal angle $\mathrm{d}\hat{s}$ in direction $\hat{s}$ at time t through an infinitesimal area $\mathrm{d}S$ is given by

$$\phi(r,\hat{s},t)\hat{s}\cdot\hat{n}\mathrm{d}S\mathrm{d}\hat{s}$$

where $\hat{n}$ is the normal to the surface $\mathrm{d}S$ [39]. The scattering phase function $\Theta(\hat{s}\cdot\hat{s}')$ describes the probability that a photon with an initial direction $\hat{s}'$ will have a direction $\hat{s}$ after a scattering event. In optical imaging, the most commonly applied phase function is the Henyey–Greenstein scattering function [35] which is of the form

$$\Theta(\hat{s}\cdot\hat{s}') = \begin{cases}\frac{1}{2\pi}\frac{1-g^2}{(1+g^2-2g(\hat{s}\cdot\hat{s}'))}, & d=2\\ \frac{1}{4\pi}\frac{1-g^2}{(1+g^2-2g(\hat{s}\cdot\hat{s}'))^{3/2}}, & d=3\end{cases} \tag{3}$$

where g is the scattering anisotropy parameter that defines the shape of the probability density. It gets values between $-1 < g < 1$, such that, if $g = 0$, the scattering probability density is a uniform distribution, $g > 0$ for forward dominated scattering, and $g < 0$ for backward dominated scattering.

As light propagates within the tissue, it is absorbed leading to localized increase in pressure and generation of a pressure wave. In practical applications, the propagation of the acoustic wave occurs on a microsecond time scale, about five orders of magnitude slower than the optical propagation, so only the total absorbed optical energy density is of interest and not the rate of the absorption. Thus, in QPAT, light propagation can be modeled using a time-independent model of light transport. The time-independent form of the RTE is

$$\begin{aligned}\hat{s}\cdot\nabla\phi(r,\hat{s}) + (\mu_s(r)+\mu_a(r))\phi(r,\hat{s}) \\ = \mu_s(r)\int_{S^{d-1}}\Theta(\hat{s}\cdot\hat{s}')\phi(r,\hat{s}')\mathrm{d}\hat{s}', \quad r\in\Omega\end{aligned} \tag{4}$$

$$\phi(r,\hat{s}) = \begin{cases}\phi_0(r,\hat{s}), & r\in\xi_j,\quad \hat{s}\cdot\hat{n}<0\\ 0, & r\in\partial\Omega\backslash\xi_j,\quad \hat{s}\cdot\hat{n}<0\end{cases} \tag{5}$$

where $\phi(r,\hat{s}) = \int_{-\infty}^{\infty}\phi(r,\hat{s},t)\mathrm{d}t$ is the time-independent radiance and $\phi_0(r,\hat{s})$ is the time-independent boundary light source.

11.2.1.2 Diffusion approximation

In optical imaging, light propagation in tissue is often modeled with the diffusion approximation (DA) to the RTE. The most typical approach to derive the DA from the RTE is to expand the radiance, source term, and phase function into series using the spherical harmonics and truncate the series [5]. The first-order spherical harmonics approximation is referred to as the P_1 approximation and the DA can be regarded as a special case for that.

In the DA framework, the radiance is approximated by

$$\phi(r,\hat{s},t) \approx \frac{1}{|S^{n-1}|}\Phi(r,t) + \frac{n}{|S^{n-1}|}\hat{s}\cdot J(r,t) \tag{6}$$

where $\Phi(r,t)$ and $J(r,t)$ are the photon fluence and photon current

$$\Phi(r,t) = \int_{S^{d-1}}\phi(r,\hat{s},t)\mathrm{d}\hat{s} \tag{7}$$

$$J(r,t) = \int_{S^{d-1}}\hat{s}\phi(r,\hat{s},t)\mathrm{d}\hat{s}. \tag{8}$$

By inserting the approximation (6) and similar approximations written for the source term and phase function into equation (1)–(2) and following the derivation in [5, 39], the DA can be derived

$$\frac{1}{c}\frac{\partial\Phi(r,t)}{\partial t} - \nabla\cdot\frac{1}{d(\mu_a(r)+\mu_s'(r))}\nabla\Phi(r,t) + \mu_a(r)\Phi(r,t) = 0, \quad r\in\Omega \tag{9}$$

$$\Phi(r,t) + \frac{1}{2\gamma_d}\frac{1}{d(\mu_a(r)+\mu_s'(r))}\frac{\partial\Phi(r,t)}{\partial\hat{n}} = \begin{cases}\frac{I_s(r,t)}{\gamma_d}, & r\in\xi_j\\ 0, & r\in\partial\Omega\setminus\xi_j\end{cases} \tag{10}$$

where $\mu_s'(r) = (1-g)\mu_s(r)$ is the (reduced) scattering coefficient, $I_s(r,t)$ is a diffuse boundary current at the source position $\xi_j \subset \partial\Omega$, and γ_d is a dimension-dependent constant which takes values $\gamma_2 = 1/\pi$ and $\gamma_3 = 1/4$.

Similarly, as in the case of the RTE, the time-independent form of the DA can be formed. It takes the form

$$-\nabla \cdot \frac{1}{d(\mu_a(r)+\mu_s'(r))}\nabla\Phi(r) + \mu_a(r)\Phi(r) = 0, \quad r \in \Omega \tag{11}$$

$$\Phi(r) + \frac{1}{2\gamma_d}\frac{1}{d(\mu_a(r)+\mu_s'(r))}\frac{\partial\Phi(r)}{\partial\hat{n}} = \begin{cases} \frac{I_s(r)}{\gamma_d}, & r \in \xi_j \\ 0, & r \in \partial\Omega \setminus \xi_j \end{cases} \tag{12}$$

where $\Phi(r) = \int_{-\infty}^{\infty}\Phi(r,t)dt$ is the time-independent fluence and $I_s(r)$ is a time-independent boundary light source.

The DA is a valid approximation in situations in which the radiance is almost a uniform distribution, i. e., in a scattering dominated medium further than a few scattering lengths from the light source [39]. In QPAT, however, imaging depth can be small compared to the average scattering length, and thus the DA is not always a valid approximation.

11.2.1.3 Light absorption and photoacoustic efficiency

As light propagates in tissue, it is absorbed by chromophores. The absorbed optical energy density $H(r)$ can be computed from the fluence as

$$H(r) = \mu_a(r)\Phi(r) \tag{13}$$

where $\Phi(r)$ is the fluence as in equations (4)–(5), (7), and (11)–(12). The energy turns into heat which results into pressure increase. The absorbed energy density is connected to the initial acoustic pressure distribution $p_0(r)$ through the photoacoustic efficiency which can be identified with the Grüneisen parameter $G(r)$ for an absorbing fluid

$$p_0(r) = G(r)H(r) = \frac{\beta v^2}{C_p}H(r) \tag{14}$$

where β is the volume thermal expansivity of the fluid, v is the speed of sound and C_p is the specific heat capacity at constant pressure [12, 18].

11.2.1.4 Numerical implementations

In this work, the solution of the RTE (4)–(5) and the DA (11)–(12) are numerically approximated using the finite element method (FEM); for details of the implementation see, e. g., [89, 92, 93]. In the FE-approximation of the RTE, both spatial and angular discretizations are implemented in piecewise linear bases. Furthermore, a streamline diffusion modification is utilized to stabilize the numerical solution of the RTE [45, 92].

11.2.2 Acoustic initial value problem

In photoacoustic tomography, propagation of the photoacoustic wave can be modeled using equations of linear acoustics. For soft biological tissue, it is generally assumed that the medium is isotropic and quiescent and that the shear waves can be neglected. In a lossless medium, it can be written for the equation of motion, equation of continuity, and equation of state

$$\frac{\partial u(r,t)}{\partial t} = -\frac{1}{\rho_0(r)}\nabla p(r,t) \tag{15}$$

$$\frac{\partial \rho(r,t)}{\partial t} = -\rho_0(r)\nabla \cdot u(r,t) \tag{16}$$

$$p(r,t) = v(r)^2\rho(r,t) \tag{17}$$

with initial conditions

$$p(r,t=0) = p_0(r) \tag{18}$$

$$u(r,t=0) = 0 \tag{19}$$

where $p(r,t)$ is the acoustic pressure with an initial pressure $p_0(r)$ as in equation (13), $u(r,t)$ is the acoustic particle velocity, $v(r)$ is the speed of sound, and $\rho(r,t)$ and $\rho_0(r)$ are the acoustic and ambient densities, respectively, [48, 63, 99]. These equations can be combined to give wave equation

$$\frac{1}{v(r)^2}\frac{\partial^2 p(r,t)}{\partial t^2} + \frac{1}{\rho_0(r)}\nabla\rho_0(r)\cdot\nabla p(r,t) - \nabla^2 p(r,t) = 0 \tag{20}$$

which, in the case of constant ambient density and constant speed of sound v_0, leads to

$$\frac{1}{v_0^2}\frac{\partial^2 p(r,t)}{\partial t^2} - \nabla^2 p(r,t) = 0 \tag{21}$$

with initial conditions (18) and

$$\frac{\partial p(r,t=0)}{\partial t} = 0. \tag{22}$$

In this work, propagation of photoacoustic wave is modeled using (18), (21), and (22). That is, the speed of sound is assumed constant and acoustic attenuation is neglected. Modeling ultrasound propagation in PAT in the presence of acoustic attenuation have been considered, e. g., in [38, 51, 98, 99]. Furthermore, modeling wave propagation in elastic media such as bones has been considered, e. g., in [47, 62, 74, 106].

11.2.2.1 Numerical implementations

In this work, the k-space time domain method implemented in the k-Wave MATLAB (The MathWorks Inc., Natick, MA) Toolbox is used to solve the initial value problem (18), (21)–(22) [97]. A perfectly matched layer (PML) is applied at the boundaries to simulate free space propagation. In the k-space method, the spectral calculation of spatial derivatives is combined with a temporal propagator expressed in the spatial frequency domain, the k-space. This allows field gradients to be calculated efficiently using the fast Fourier transform. Therefore, the k-space method enables a computationally efficient way to simulate propagation of photoacoustic wave [21, 22].

11.3 Inverse problem

In the inverse problem of quantitative photoacoustic tomography, the aim is to estimate the distribution of light absorbing molecules in the medium when the measured photoacoustic time series on the detectors and the parameters of input light are given. As mentioned earlier, the difference in time scales of light absorption and ultrasound propagation allows the optical and acoustic parts of the inverse problem to be decoupled and treated separately. The two inverse problems in QPAT then are: (1) acoustic inverse problem, i. e., reconstruct the initial acoustic pressure distribution from measured acoustic time series, and (2) optical inverse problem, i. e., reconstruct the distributions of the optical parameters from the absorbed optical energy density (or the initial pressure).

11.3.1 Acoustic inverse problem

The first inverse problem in QPAT is the acoustic inverse initial value problem in which the initial acoustic pressure distribution is estimated when the measured photoacoustic time series on the detectors is given. The acoustic inverse problem is not ill-posed if the object is fully surrounded by detectors on a closed surface. In such case, good quality reconstructions of the whole target volume can be provided. However, in practice, such an experimental setting cannot typically be constructed and one is restricted to perform the measurements on a limited surface. It has been shown that, in limited view measurement geometries, the target regions that are enclosed by the detection surface can be reconstructed accurately [111]. Those inclusions within the object, that are not enclosed by the detection surface, suffer from distortions apart from those inclusion boundaries whose normals intersect the detection surface [111].

Various methods for the solution of the acoustic inverse problem have been developed. These include, e. g., back-projection algorithms, time reversal, regularized least squares, and Bayesian approach. The back-projection algorithms and eigenfunction expansion approach are based on algebraic inversion formulae. The back-projection algorithms, such as filtered back-projection, relate to the inverse Radon transform. For more information on their usage in PAT see, e. g., [27, 53, 108]. In the eigenfunction expansion approach, the initial pressure is obtained as series solution, and the series coefficients are calculated from measured pressure signals [1, 54]. The back-projection and eigenfunction expansion based methods are, however, limited to specific geometries such as spherical, cylindrical and planar acoustic detection surfaces.

Time reversal, regularized least squares, and Bayesian approaches allow performing image reconstruction in more general imaging scenarios. They utilize numerical solution of the problem, and thus are computationally more intensive as the photoacoustic wave within the entire domain needs to be computed. The time reversal algorithms are based on reciprocity principles in acoustics. In the approach, the image reconstruction is performed by simulating the propagation of the time-reversed measured signals back into the volume [16, 23, 37, 99, 110]. Approaches based on regularized least squares, minimize error between the measured acoustic signals and those computed using the photoacoustic forward acoustic model regularized with a penalty functional such as Tikhonov regularization or total variation [4, 6, 13, 14, 24, 30, 40, 62, 69, 102] Recently, using model based learning for teaching the regularization term in PAT has been considered [33].

In Bayesian approach, all parameters are modeled as random variables and the formal solution of the inverse problem consists of a probability density for the initial pressure in each pixel of the reconstruction domain [94, 95]. It combines the information obtained through the measurements, forward model, and prior model for unknown parameters. In addition, Bayesian approach facilitates representing and taking into account the uncertainties of parameters, models and geometries. Bayesian approach to inverse problems is reviewed later in Section 11.3.3 and utilizing it in PAT is studied in Section 11.4.

The most approaches for the acoustic inverse problem of PAT have been using models (18), (21)–(22) as the forward model. The inverse problem becomes significantly more challenging in cases in which the speed of sound, density or acoustic attenuation within the tissue are varying. These have been considered, e. g., in [3, 13, 23, 25, 30, 37, 38, 42, 51, 83, 84, 99]. Furthermore, estimation of the initial pressure in elastic medium has been considered in [62].

11.3.2 Optical inverse problem

The optical inverse problem of QPAT is to estimate the distribution of the light absorbing molecules from the reconstructed initial pressure distribution with given input illumination. In practice, in many studies, the Grüneisen parameter has been assumed to be known and one estimates the optical parameters when the absorbed optical energy density and input illumination are given. However, the recovery of the Grüneisen parameter simultaneously with the optical parameters in the case of spectral QPAT has been addressed, e. g., in [71]. Further, estimation of absorption and scattering together with mapping of the temperature distribution has been studied [116].

The concentrations of the chromophores can be estimated either directly from absorbed optical energy density obtained at various wavelengths [10, 19, 55, 61, 71, 76, 77] or by first recovering the absorption coefficients at different wavelengths and then calculating the concentrations from the absorption spectra [10, 18, 19, 71].

Most of the studies on the inverse problem of QPAT have considered estimation of the distribution of absorption and scattering (or diffusion). The problem has been approached using various methods see, e. g., [9, 10, 20, 29, 32, 61, 67, 72, 73, 82, 85, 89, 91, 114, 115]. A different approach has been taken in [2, 80] where the absorption and the photon fluence were extracted using a sparse signal representation. As an alternative approach, the inverse problem was formulated as a problem of finding boundaries between piecewise constant optical parameters in [66]. Furthermore, a combined reconstruction-classification approach was considered in [60] by alternating reconstruction and segmentation steps in an iterative process.

In the optical inverse problem of QPAT, estimation of more than one optical parameter is nonunique if only one light illumination is used. To overcome this problem, one approach has been to assume the scattering as known and to estimate only the absorption [11, 20, 41, 79, 112, 113]. This, however, is unrealistic since in practical applications scattering is usually not known. The approach has been improved by modeling the errors caused by the fixed scattering assumption by using a Bayesian approximation error modeling [73]. In [9], it was shown that the nonuniqueness can be overcome by using multiple optical illuminations. Generally, this has been achieved by illuminating the target from different directions [9, 29, 61, 82, 85, 89, 114]. However, in [72] it was numerically demonstrated that also spatially modulated light patterns from one direction can be utilized. Also combining QPAT and diffuse optical tomography (DOT) data types can be used to overcome the nonuniqueness [59, 78]. In addition, the approach can be used to increase the resolution and imaging depth of QPAT [67].

The most of the approaches for the optical inverse problem of QPAT have been based on using the diffusion approximation (11)–(12) as light transport model. However, utilizing the RTE has been studied, e. g., in [31, 61, 72, 82, 89]. In addition to DA, utilizing other approximation such as δ-Eddington approximation have been proposed [81]. Furthermore, recently, approaches based on Monte Carlo method have been proposed [36, 46].

11.3.2.1 One step approaches

As an alternative to the conventional two stage approach, estimation of the optical parameters directly from the photoacoustic time series has also been considered recently. The single stage approach was formulated and studied utilizing Born approximation in [86, 87]. In [28], the ℓ_1 sparsity regularization was utilized and the minimization problem was solved with BFGS algorithm. The approach was studied also in a limited view measurement geometry. Furthermore, the RTE was used as the light transport model in [31] where, in the inverse problem, Tikhonov regularization solved with a proximal gradient algorithm was utilized. In [26], the singe stage approach was used to recover the speed of sound and the optical parameters simultaneously. The Bayesian approach was taken in [70], and the approach was studied also in a limited view measurement geometry and in three dimensions. In addition, the reliability of the estimated optical parameters was addressed. In [100], stochastic search algorithms were proposed for single stage QPAT.

11.3.3 Bayesian formulation for the inverse problem

Let us denote the distribution of the unknown parameters of interest at a point r by $x(r)$. Further, denote the measurements by a finite dimensional vector $y = (y_1, \ldots, y_m)^{\mathrm{T}} \in \mathbb{R}^m$ where m is the number of measurements. The observation model with an additive noise model is of the form

$$y = A(x(r)) + e \tag{23}$$

where A is the forward model which maps the parameters to the measurable data, and e denotes the noise. Typically, in practical implementations, the parameters and the forward mapping are represented in discrete vector spaces $x(r) \mapsto x \in \mathbb{R}^N$, $A \mapsto A_h : \mathbb{R}^N \to \mathbb{R}^m$ and the continuous model (23) is replaced by an approximate equation

$$y = A_h(x) + e \tag{24}$$

where x is a discretized parameter distribution, A_h is a discretized forward model and $h > 0$ is a mesh parameter controlling the level of the discretization.

In the Bayesian approach, the inverse problem is treated as a problem of statistical inference [17, 43, 88]. The approach combines information obtained through measurements, forward model and prior model for unknown parameters as well as their uncertainties.

Assume that x and y are random variables in finite-dimensional spaces $\mathbb{R}^N$ and $\mathbb{R}^m$ called parameter and data space, respectively. The joint probability density of x and y can be written in terms of conditional probability densities as

$$\pi(x, y) = \pi(x)\pi(y|x) = \pi(y)\pi(x|y). \tag{25}$$

The solution of the inverse problem is the posterior probability density $\pi(x|y)$ which according to equation (25) is of the form

$$\pi(x|y) = \frac{\pi(x)\pi(y|x)}{\pi(y)} \tag{26}$$

where $\pi(x)$ is the prior probability density and $\pi(y|x)$ is the likelihood density [43]. Equation (26) is the Bayes' formula, and it is often written in the nonnormalized form

$$\pi(x|y) \propto \pi(y|x)\pi(x) \tag{27}$$

since $\pi(y)$ is constant for fixed measurements y. If we assume that the noise e and the unknown x are mutually independent, formula (24) leads to likelihood density

$$\pi(y|x) = \pi_e(y - A_h(x)) \tag{28}$$

where π_e is the probability distribution of the noise e.

If the unknown x and the noise can be modeled as Gaussian random variables, we have

$$x \sim \mathcal{N}(\eta_x, \Gamma_x), \quad e \sim \mathcal{N}(\eta_e, \Gamma_e)$$

where $\eta_x \in \mathbb{R}^N$ and $\eta_e \in \mathbb{R}^m$ are the means and $\Gamma_x \in \mathbb{R}^{N\times N}$ and $\Gamma_e \in \mathbb{R}^{m\times m}$ are the covariance matrices. In this case, the posterior density (27) becomes

$$\begin{aligned}\pi(x|y) &\propto \exp\left\{-\frac{1}{2}(y - A_h(x) - \eta_e)^{\mathrm{T}}\Gamma_e^{-1}(y - A_h(x) - \eta_e)\right.\\ &\quad \left. - \frac{1}{2}(x - \eta_x)^{\mathrm{T}}\Gamma_x^{-1}(x - \eta_x)\right\}\\ &\propto \exp\left\{-\frac{1}{2}\|L_e(y - A_h(x) - \eta_e)\|^2 - \frac{1}{2}\|L_x(x - \eta_x)\|^2\right\}\end{aligned} \tag{29}$$

where L_e and L_x are matrix square roots such as Cholesky decompositions of the inverse covariance matrices of the noise and the prior, respectively, i. e., $\Gamma_e^{-1} = L_e^{\mathrm{T}}L_e$ and $\Gamma_x^{-1} = L_x^{\mathrm{T}}L_x$.

In principle, the posterior distributions of the unknown parameters can be estimated using Markov chain Monte Carlo (MCMC) methods [56]. However, these methods can be computationally prohibitively too expensive in large dimensional tomographic inverse problems. Therefore, point estimates such as the maximum a posteriori (MAP) estimate are computed. The MAP estimate is defined as

$$\begin{aligned}x_{\mathrm{MAP}} &= \arg\max_x\{\pi(x|y)\}\\ &= \arg\min_x\left\{\frac{1}{2}\|L_e(y - A_h(x) - \eta_e)\|^2 + \frac{1}{2}\|L_x(x - \eta_x)\|^2\right\}.\end{aligned} \tag{30}$$

11.3.3.1 Bayesian approximation error modeling

In Bayesian approximation error approach [43, 44], one aims at modeling the errors and uncertainties present in imaging situations and including that information in the solution of the inverse problem. Let us assume that the continuous model $x(r) \mapsto A(x(r))$, equation (23), can be approximated by a densely discretized finite-dimensional model

$$A_\delta : \mathbb{R}^N \to \mathbb{R}^m, \quad x \mapsto A_\delta(x), \quad \delta > 0 \quad \text{small}$$

where the mesh parameter δ is so small that the forward model can be considered to be exact within measurement accuracy. Thus, the discretized observation model, that is exact within the measurement accuracy, is of the form

$$y = A_\delta(x) + e. \tag{31}$$

In the approximation error approach [43, 44], the observation model is written in the form

$$\begin{aligned} y &= A_h(x) + (A_\delta(x) - A_h(x)) + e \\ &= A_h(x) + \varepsilon(x) + e \end{aligned} \tag{32}$$

where $A_h(x)$ is the reduced model and $\varepsilon(x)$ is the modeling error. The modeling error describes the discrepancy between the accurate forward model and the reduced model, e. g., a model that is an approximation to the accurate physical model and/or a model with a coarser discretization or uncertainties in geometry or parameters [43, 44].

In the approximation error approach, a Gaussian approximation is constructed for ε, and the total error $n = \varepsilon + e$ is approximated by a Gaussian distribution, i. e.,

$$\varepsilon \sim \mathcal{N}(\eta_\varepsilon, \Gamma_\varepsilon), \quad n \sim \mathcal{N}(\eta_n, \Gamma_n).$$

Furthermore, if we ignore the mutual dependence of x and ε, we get an approximation that is referred to as the enhanced error model [43] and the posterior density becomes

$$\pi(x|y) \propto \exp\left\{-\frac{1}{2}\|L_n(y - A_h(x) - \eta_n)\|^2 - \frac{1}{2}\|L_x(x - \eta_x)\|^2\right\} \tag{33}$$

where $\eta_n = \eta_\varepsilon + \eta_e$, $\Gamma_n = \Gamma_\varepsilon + \Gamma_e$ and $\Gamma_n^{-1} = L_n^{\mathrm{T}} L_n$. The MAP estimate with the enhanced error model is obtained as

$$x_{\mathrm{MAP}} = \arg\min_x \left\{\frac{1}{2}\|L_n(y - A_h(x) - \eta_n)\|^2 + \frac{1}{2}\|L_x(x - \eta_x)\|^2\right\}. \tag{34}$$

The approximation error ε can be approximated, e. g., using simulations [43, 44]. The Bayesian approximation error modeling has been utilized in various studies, for those including optical or ultrasonic methodologies see, e. g., [7, 34, 49, 50, 64, 65, 73, 90].

11.4 Case examples and numerical studies

11.4.1 Image reconstruction and uncertainty quantification in the acoustic inverse problem

Let us consider the acoustic inverse problem of QPAT in Bayesian framework. Now, in the case of the linear model for wave propagation (18), (21)–(22), the observation model (23) is linear, and can be written in the form

$$p_t = K_h p_0 + e \tag{35}$$

where $p_t \in \mathbb{R}^m$ is a vector of measured acoustic pressure waves at the detector locations and temporal samples, $p_0 \in \mathbb{R}^N$ is a discrete initial pressure distribution within the object and $K_h \in \mathbb{R}^{m\times N}$ is the discrete forward model for the pressure propagation which maps the initial pressure distribution to the measurable data, and $e \in \mathbb{R}^m$ denotes the noise which is modeled as Gaussian distributed $e \sim \mathcal{N}(\eta_e, \Gamma_e)$. The posterior density is

$$\pi(p_0|p_t) \propto \exp\left\{-\frac{1}{2}\|L_e(p_t - K_h p_0 - \eta_e)\|^2 - \frac{1}{2}\|L_{p_0}(p_0 - \eta_{p_0})\|^2\right\} \tag{36}$$

where $\eta_{p_0} \in \mathbb{R}^N$ is the mean and $\Gamma_{p_0} \in \mathbb{R}^{N\times N}$ is the covariance of the prior initial pressure distribution, and $\Gamma_{p_0}^{-1} = L_{p_0}^{\mathrm{T}} L_{p_0}$. Now, in the case of a linear observation model and Gaussian distributed noise and prior, the posterior density is also a Gaussian distribution

$$p_0|p_t \sim \mathcal{N}(\eta_{p_0|p_t}, \Gamma_{p_0|p_t})$$

where

$$\eta_{p_0|p_t} = K^{-1}b, \quad \Gamma_{p_0|p_t} = K^{-1}$$

where

$$K = K_h^{\mathrm{T}}\Gamma_e^{-1}K_h + \Gamma_{p_0}^{-1}$$
$$b = K_h^{\mathrm{T}}\Gamma_e^{-1}(p_t - \eta_e) + \Gamma_{p_0}^{-1}\eta_{p_0}.$$

The credible interval of the estimated parameter is determined by considering the marginal density of the posterior distribution. The marginal density of the kth element of $p_0|p_t$ is defined as

$$p_{0,k}|p_t \sim \mathcal{N}(\eta_{p_0|p_t,k}, \Gamma_{p_0|p_t,kk}), \tag{37}$$

where $\eta_{p_0|p_t,k}$ is the kth element of $\eta_{p_0|p_t}$ and $\Gamma_{p_0|p_t,kk}$ is the kth diagonal element of $\Gamma_{p_0|p_t}$. Now, e. g., a 99.7 %, i. e., 3 standard deviation (STD) of the mean, credible interval of the $p_{0,k}|p_t$ is

$$[\eta_{p_0|p_t,k} - 3\sqrt{\Gamma_{p_0|p_t,kk}}, \eta_{p_0|p_t,k} + 3\sqrt{\Gamma_{p_0|p_t,kk}}]. \tag{38}$$

In this work, the forward model is computed using the k-Wave toolbox [97] using MATLAB. In practice, the matrix K_h is assembled going through each pixel of the discretization for the initial pressure p_0, setting the pixel value to one while keeping the other pixels at zero, computing the acoustic output using k-Wave, and proceeding to the next pixel. The outputs of the k-Wave then form the columns of the matrix K_h. This corresponds to computing the impulse response of the discrete system approximating (18), (21)–(22).

11.4.1.1 Prior models

Utilizing three Gaussian priors: a white noise prior, a squared exponential prior and Ornstein–Uhlenbeck prior [75] in PAT were studied. The white noise prior assumes that the estimated parameters are independent of each other or have no spatial correlation. Thus, the covariance matrix of a white noise prior is a diagonal matrix with the values of variance $\sigma_{p_0}^2$ on the diagonal

$$\Gamma_{p_0} = \sigma_{p_0}^2 I$$

where $I \in \mathbb{R}^{N\times N}$ is an identity matrix. In the case of the squared exponential prior, the elements of the covariance matrix are of the form

$$\Gamma_{p_0}(i,j) = \sigma_{p_0}^2 \exp\left(-\frac{\|r_j - r_j\|^2}{2\zeta^2}\right)$$

where i and j are the indices of the discretization points (i. e., pixels) and r_i and r_j are their spatial positions. Further, ζ is the characteristic length scale which controls the range of spatial correlation. This covariance function is infinitely differentiable, and thus the process is smooth. The Ornstein–Uhlenbeck prior belongs to the class of Matérn covariance functions. The elements of the covariance matrix are of the form

$$\Gamma_{p_0}(i,j) = \sigma_{p_0}^2 \exp\left(-\frac{\|r_j - r_j\|}{\zeta}\right).$$

Thus, the process is continuous but not continuously differentiable. Sample draws from the priors are shown in Figure 11.1.

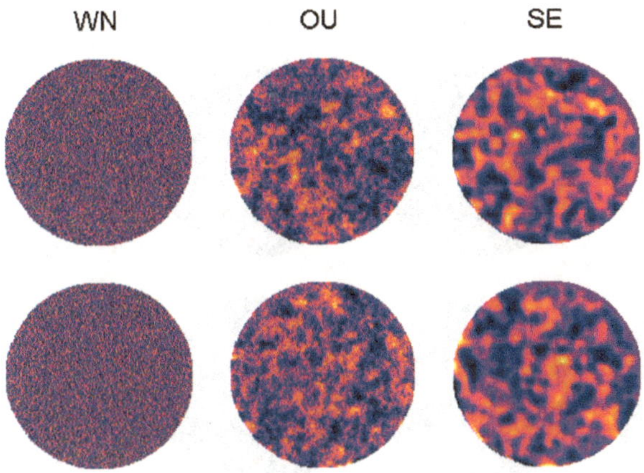

Figure 11.1: Two sample draws from the white noise (first column), Ornstein–Uhlenbeck (second column) and squared exponential (third column) priors.

11.4.1.2 Simulations

The acoustic inverse problem was investigated with numerical simulations. The simulation geometry was a circle with a radius of 2 mm. The medium was assumed to be homogeneous and nonattenuating with a constant speed of sound $v_0 = 1500$ m/s. Two different initial pressure distributions were considered: five Gaussian shaped bumps with a peak amplitude of 4 and full-width at half-maximum of 0.4, and five circles with constant initial pressure of 10 and background initial pressure of zero. The acoustic sensors were distributed at a distance of 2.25 mm from the centre of the simulation domain. Different sensor geometries were considered: 180 sensors distributed around the target (a full view scenario), 15 sensors distributed around the target (sparse view scenario), 90 sensors located on the upper half of the target (limited view scenario 1), and 30 sensors covering angular interval of 60 degrees on the upper side of the target (limited view scenario 2). The simulated targets and the acoustic sensor locations can be seen in the images on the left columns of Figures 11.2 and 11.3.

The data was simulated using the k-space time domain method implemented with the k-Wave MATLAB toolbox. In the discretization, a pixel width of $\Delta h = 14.29\,\mu$m was used. The pressure signals were recorded for 7 μs at 100 MHz (701 temporal samples). Uncorrelated Gaussian distributed noise with zero mean and a standard deviation proportional to 1 % of the peak amplitude of the simulated pressure signal was added to the simulated data.

In the inverse problem, the posterior distribution was solved. The system matrix K_h was formed as described earlier in this section. For more details, see [95, 96]. In discretization, a pixel width of $\Delta h = 25\,\mu$m was used. Three prior models: the white noise, Ornstein–Uhlenbeck and squared exponential priors were studied. The means of the priors were set to the value of the background, i. e., zero, and the standard deviations were set to 3.5. For the Ornstein–Uhlenbeck and squared exponential priors, the characteristic length scale $\zeta = 0.125$ mm was used. The measurement noise was considered

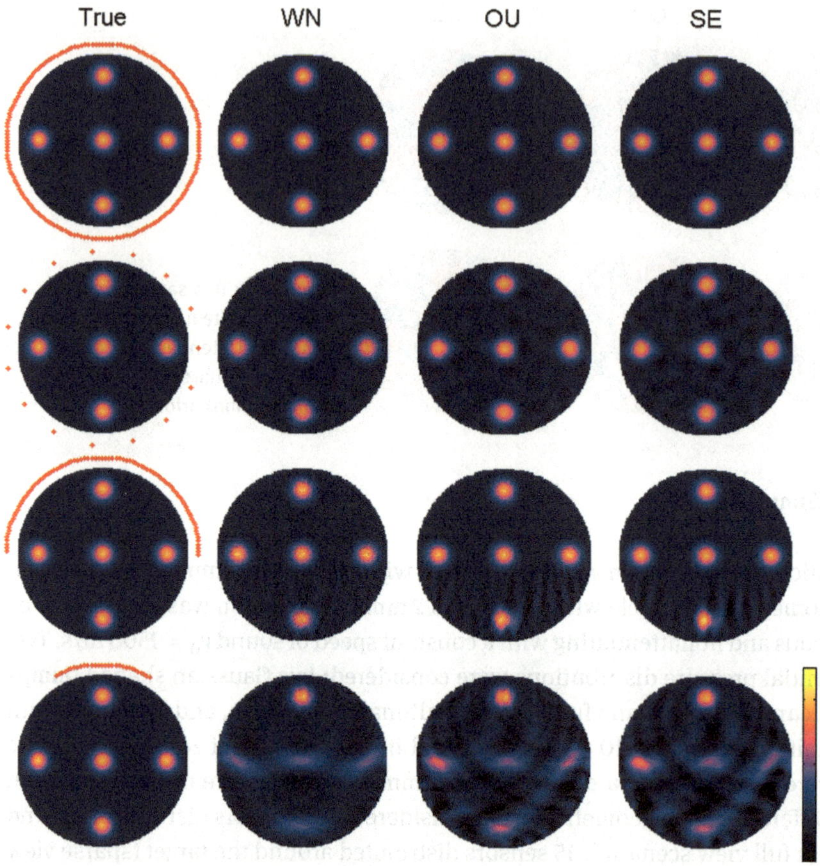

Figure 11.2: The simulated (true) initial pressure p_0 (first column) and the estimated mean of the posterior distribution obtained using white noise prior (second column), Ornstein–Uhlenbeck prior (third column) and squared exponential prior (fourth column). The results obtained using different sensor geometries are on the rows from top to bottom: full view (first row), sparse view (second row), limited view 1 (third row), and limited view 2 (fourth row). The red dots on the first column indicate the locations of the acoustic sensors.

to be uncorrelated Gaussian distributed noise with zero mean and the standard deviation set to 1 % of the peak positive amplitude of the noisy simulated data.

The means of the posterior distribution obtained with the different prior models are shown in Figures 11.2 and 11.3. Further, vertical cross sections from the upper edge to the bottom edge through the target are visualized in Figures 11.4 and 11.5 which show the simulated (true) initial pressure distribution and the mean of the posterior as well as the credible intervals of 99.7 %. The calculated mean relative errors for the posterior mean

$$\bar{E}_{p_0|p_t} = 100\,\% \cdot \frac{\|p_0 - \eta_{p_0|p_t}\|}{\|p_0\|} \tag{39}$$

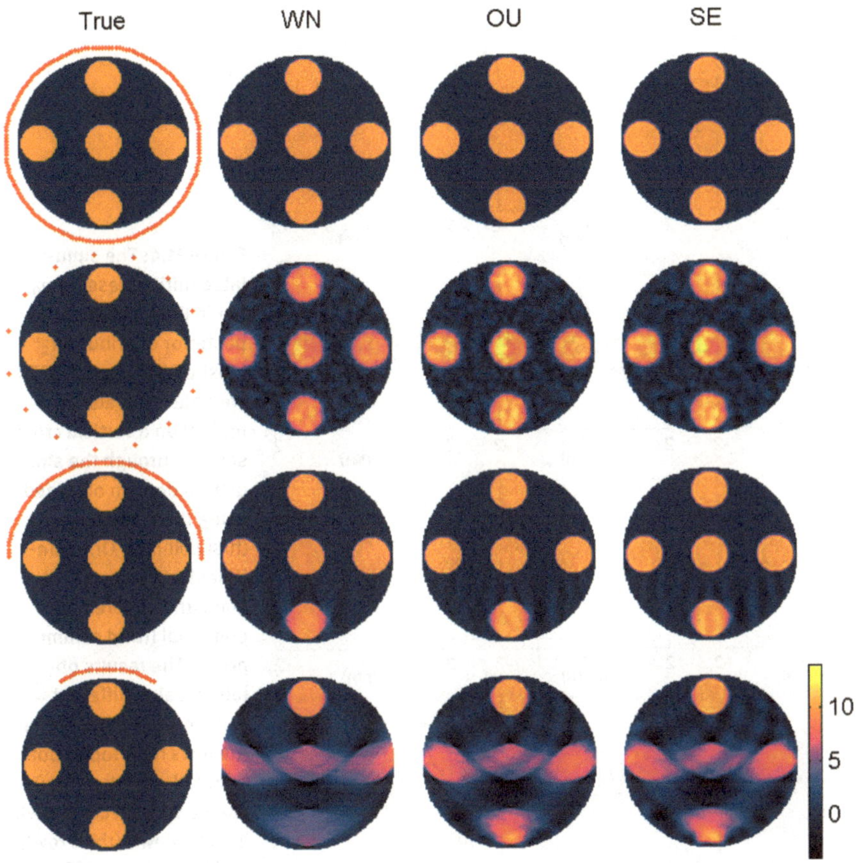

Figure 11.3: The simulated (true) initial pressure p_0 (first column) and the estimated mean of the posterior distribution obtained using white noise prior (second column), Ornstein–Uhlenbeck prior (third column) and squared exponential prior (fourth column). The results obtained using different sensor geometries are on the rows from top to bottom: full view (first row), sparse view (second row), limited view 1 (third row), and limited view 2 (fourth row). The red dots on the first column indicate the locations of the acoustic sensors.

where p_0 is the simulated initial pressure distribution and $\eta_{p_0|p_t}$ is the mean of the posterior distribution interpolated to the simulation space are given in Table 11.1.

As it can be seen from Figures 11.2–11.5, the mean of the posterior distribution resembles the original initial pressure distribution in the case of the full view imaging scenario for both targets and with different prior models. If sparse or limited view scenarios are considered, the means suffer from artefacts which are different depending on the measurement scenario and the prior model. The mean relative errors given in Table 11.1 support these findings. The errors are larger in sparse and limited view geometries than in a full view geometry and prior information can be used to alleviate these errors. It can also be noted that in the case of the Gaussian bump initial pres-

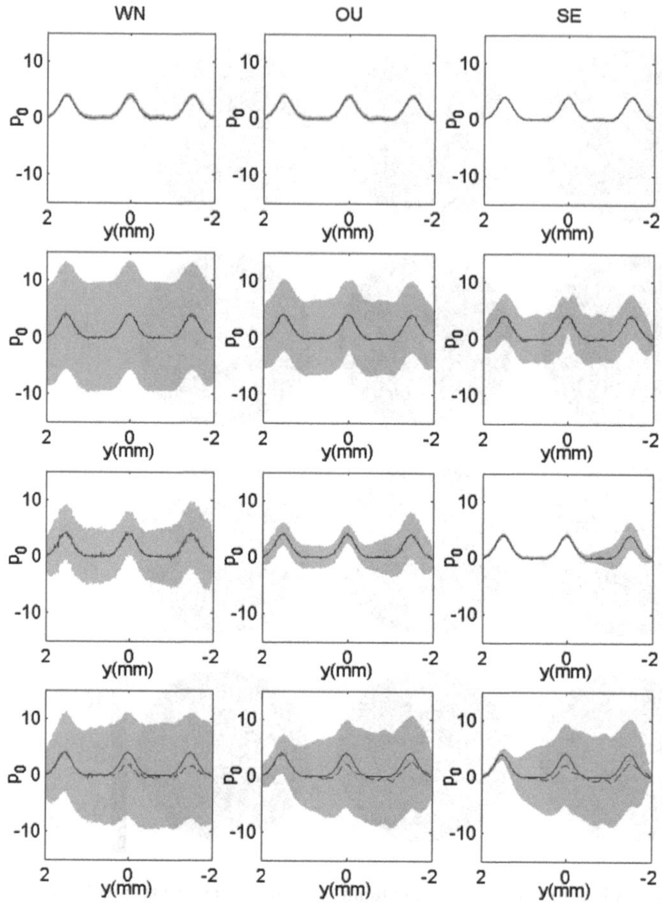

Figure 11.4: The simulated initial pressure p_0 (solid line), mean of the posterior distribution (dashed line), and 3 STD credible intervals (grey region) on a vertical cross section through the simulation domain obtained using the white noise (first column), Ornstein–Uhlenbeck (second column) and squared exponential (third column) priors. The results obtained using different sensor geometries are on the rows from top to bottom: full view (first row), sparse view (second row), limited view 1 (third row), and limited view 2 (fourth row).

sure distribution, the squared exponential prior and Ornstein–Uhlenbeck prior produced the best quality estimates, and in the case of the uniform circles, the Ornstein–Uhlenbeck prior produced the best estimates. This is due to the fact that the prior information on the target is most appropriate in those situations. However, the differences were not very large between any prior models.

In the case of evaluating the reliability of the estimates by looking at the standard deviations, the impact of the prior can be seen to be more significant. One way to inspect reliability is to consider, if the mean of the posterior is within 3 STD limits from the original value. As it can be seen, in ill-posed imaging situations, i. e., limited or partial view, this is mostly valid and the standard deviation of the posterior is large. Thus, basically in these situations, the estimated images are not very accurate and the uncertainty of these estimates is very high. On the other hand, in the case of a full-view scenario and other situations in which the inverse problem is not ill-posed or it is only mildly ill-posed, the standard deviation of the posterior is small. This is especially ev-

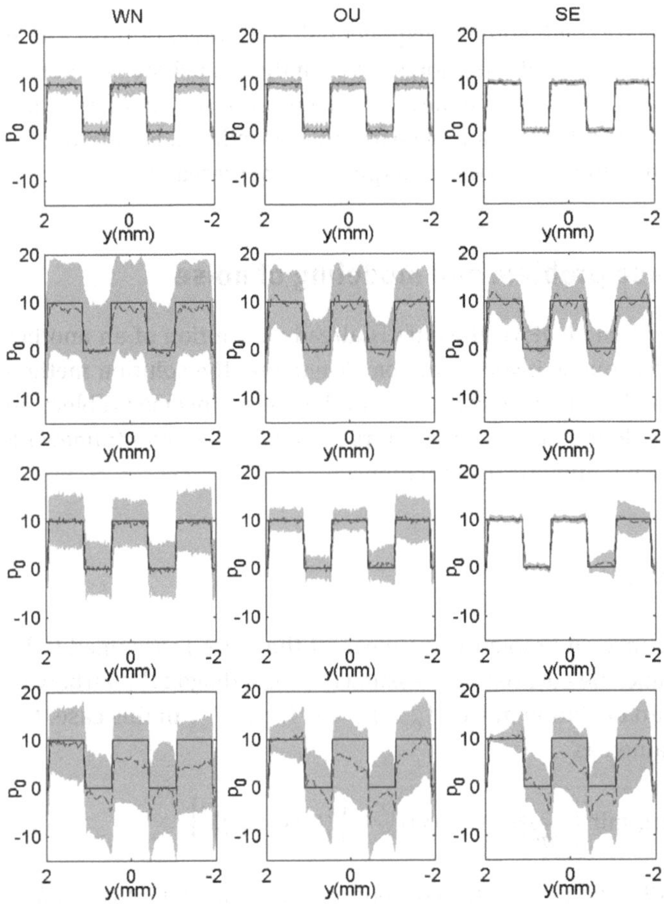

Figure 11.5: The simulated initial pressure p_0 (solid line), mean of the posterior distribution (dashed line) and 3 STD credible intervals (grey region) on a vertical cross section through the simulation domain obtained using the white noise (first column), Ornstein–Uhlenbeck (second column), and squared exponential (third column) priors. The results obtained using different sensor geometries are on the rows from top to bottom: full view (first row), sparse view (second row), limited view 1 (third row), and limited view 2 (fourth row).

Table 11.1: Mean relative errors of the posterior mean $\bar{E}_{p_0|p_t}$, equation (39), obtained from targets with circles with a Gaussian distributed initial pressure distribution and circles with an uniform initial pressure. The results obtained using white noise (WN), Ornstein–Uhlenbeck (OU) and squared exponential (SE) priors are shown on different columns and different sensor geometries are shown on different rows.

	$\bar{E}_{p_0\|p_t}$					
	Gaussian distribution			Uniform circles		
	WN	OU	SE	WN	OU	SE
Full view	9.4	7.9	4.9	13.4	12.9	14.9
Sparse view	19.1	17.5	19.2	30.6	28.5	29.8
Limited view 1	29.6	22.8	20.7	18.4	16.7	18.4
Limited view 2	62.6	62.2	67.7	58.6	52.2	53.0

ident in the full view scenario with squared exponential prior. Now, in this case, if the target is not within the support of the prior, such as in the case of the circular inclusions with uniform initial pressure, the mean of the posterior may not lay within the 3 STD limits, and thus, the estimate cannot be regarded as reliable. Evaluating the credibility of the reliability estimates, however, requires further research.

11.4.2 Optical inverse problem and modeling of noise

In the optical inverse problem of QPAT, the data is already a solution of an another inverse problem, i. e., the acoustic inverse problem. Therefore, the solution method of the acoustic inverse problem may affect the data of the optical inverse problem. In this work, we consider modeling of noise and errors in data in Bayesian framework and study the approach using numerical simulations.

11.4.2.1 Modeling of noise

Typically in tomographic inverse problems, the mean of the noise is assumed to be zero, i. e., $\eta_e = 0 \in \mathbb{R}^m$, and the covariance is assumed to be a diagonal matrix with known (constant) standard deviation σ, i. e., $\Gamma_e = \Gamma_\sigma = \sigma^2 I \in \mathbb{R}^{m\times m}$. In this case, the MAP estimate (30) is obtained as

$$x_{\text{MAP}} = \arg\min_x \left\{ \frac{1}{2\sigma^2} \|(y - A_h(x))\|^2 + \frac{1}{2} \|L_x(x - \eta_x)\|^2 \right\} \tag{40}$$

which we refer as the MAP estimate with the conventional noise model (MAP-CNM).

Now, in the optical inverse problem of QPAT, the noise is not necessarily uncorrelated and it may have a nonzero mean. A more accurate noise model can be, e. g., approximated as follows for the case of an additive acoustic noise model [91]. First, a set of noise samples of pressure measurements are simulated. Then the inverse initial value problem is solved using these noise samples as data. As a result, noise samples of the optical inverse problem $e^{(l)}$ are obtained. The mean and the covariance of the noise model can then be approximated using these noise samples as

$$\eta_e = \frac{1}{L} \sum_{l=1}^{L} e^{(l)} \tag{41}$$

$$\Gamma_e = \frac{1}{L-1} \sum_{l=1}^{L} e^{(l)} e^{(l)\,\mathrm{T}} - \eta_e \eta_e^{\mathrm{T}} \tag{42}$$

where L is the number of samples. These are then applied in the solution of the minimization problem (30). In this work, we refer to the solution of (30) together with (41)–(42) as the MAP estimate with an approximate noise model (MAP-ANM).

In practice, the numerical implementation of the acoustic inverse method has also an effect on the data. Thus, the absorbed optical energy density distribution obtained as the solution of the acoustic inverse problem contains modeling error ϵ which can be due to, e. g., discretization of the geometry and time, implementation of the boundary conditions and smoothing of the data by the acoustic solver. Therefore, let us approach modeling of the discrepancy between data, which now contains errors due to the acoustic solver, and forward model which is assumed to be accurate within measurement accuracy, using Bayesian approximation error modeling. Let us denote this modeling error as

$$\epsilon = y_h - y$$

where y is "ideal" absorbed energy density data and y_h is absorbed optical energy data containing errors due to the acoustic solver. Now, we can write the observation model (24) in the form

$$\begin{aligned} y &= A_h(x) + e \\ y_h + (y - y_h) &= A_h(x) + e \\ y_h &= A_h(x) + \epsilon + e. \end{aligned} \tag{43}$$

Then, similarly as in the framework of Bayesian approximation error modeling, a Gaussian approximation is constructed for ϵ, and the total error $n = \epsilon + e$ is approximated by a Gaussian distribution, thus

$$\epsilon \sim \mathcal{N}(\eta_\epsilon, \Gamma_\epsilon), \quad n \sim \mathcal{N}(\eta_n, \Gamma_n)$$

where $\eta_n = \eta_\epsilon + \eta_e$ and $\Gamma_n = \Gamma_\epsilon + \Gamma_e$. Then, similarly as in (34), the MAP estimate with the noise and error modeling (MAP-AEM) can be written

$$x_{\mathrm{MAP}} = \arg\min_x \left\{ \frac{1}{2} \| L_n(y - A_h(x) - \eta_n) \|^2 + \frac{1}{2} \| L_x(x - \eta_x) \|^2 \right\} \tag{44}$$

where $\Gamma_n^{-1} = L_n^{\mathrm{T}} L_n$. Note that the mutual dependence of x and ϵ was ignored above. The impact of this was, however, not studied and it may be that under some circumstances it needs to be taken into account. The modeling error can be simulated, e. g., as follows. First, a set of samples $\{x^{(\ell)}, \ell = 1, \dots, M\}$ are drawn from the prior distribution $\pi(x)$ of the optical parameters. Then the optical forward problem is solved using these samples. This results in a set samples of "ideal" data $y^{(\ell)}$ that has not been corrupted by the numerics of the acoustic inverse problem solver. Then, in order to obtain samples of the realistic data, the acoustic forward and inverse problems are solved. The obtained reconstructed initial pressure samples are used to calculate $y_h^{(\ell)}$. Then samples of the approximation error are computed as

$$\epsilon^{(\ell)} = y_h^{(\ell)} - y^{(\ell)} \tag{45}$$

and the mean and covariance of the approximation error are estimated as

$$\eta_\epsilon = \frac{1}{M}\sum_{\ell=1}^{M}\epsilon^{(\ell)} \tag{46}$$

$$\Gamma_\epsilon = \frac{1}{M-1}\sum_{\ell=1}^{M}\epsilon^{(\ell)}\epsilon^{(\ell)\,\mathrm{T}} - \eta_\epsilon\eta_\epsilon^{\mathrm{T}}. \tag{47}$$

These are then applied in the solution of the minimization problem (44) together with the noise model e.

11.4.2.2 Simulations

A rectangular domain of size 20 mm × 20 mm was considered. The Grüneisen coefficient was assumed to be a known constant with value $G = 1$. Four illuminations, one edge working as a light source in turn, were considered. The acoustic sensors were modeled as point-like detectors evenly positioned around the target domain. 316 sensors (80 detectors on each side of the domain) were used. For the forward and inverse acoustic computations, a larger computation domain of size 25 mm × 25 mm was created in order to minimize boundary effects. In all of the acoustic computations, the PML layer was located outside the computation domain.

Data simulation

Absorption and scattering distributions shown in the left columns of Figure 11.8 were studied. The anisotropy parameter was $g = 0.8$ throughout the domain. The absorption and scattering values were represented in piecewise constant bases using 16796 triangular elements. The light propagation was modeled using the RTE (4)–(5) which was solved using the FEM. The spatial FE-discretization consisted of 16796 triangular elements and 8559 nodes, and the angular discretization consisted of 64 angular directions. Then the absorbed optical energy density distribution and the initial pressure were computed using (13) and (14), respectively. The k-space method implemented with the k-Wave toolbox was used to simulate propagation of the pressure wave (18), (21)–(22) through the domain [97]. The discretization of the computation domain consisted of 1024 × 1024 pixels. The time-varying acoustic pressure was recorded at the sensors and noise with a standard deviation of 1 % of the peak amplitude of the simulated pressure signal was added to the data. Then, in order to obtain the data for the optical inverse problem, the acoustic inverse problem was solved using the time reversal method with the k-Wave toolbox. The acoustic inverse problem was solved using discretization of 128 × 128 pixels. The absorbed optical energy density in the reconstruction grid, which is the data of the optical inverse problem, was computed from the initial pressure using (14).

Noise models

For the MAP estimate with the conventional noise model (40), the noise was considered to be uncorrelated Gaussian distributed noise with zero mean. The standard deviation σ of the noise was set as 1 % of the peak amplitude of the simulated data. Sample of the noise covariance structure is shown on the left image of Figure 11.7.

For the MAP estimate with the approximate noise model (30) a simulated noise statistics (41)–(42) was used. It was simulated as follows. First, a set of time-varying pressure signals was simulated by solving the acoustic forward problem where the initial pressure was obtained by computing the FE-solution of the DA (11)–(12) with optical parameters drawn from the informative smoothness prior [7, 49] and using equations (13)–(14). The characteristic length scale of the teaching prior was 4 mm for both absorption and scattering, and the mean and standard deviations are given in Table 11.2. Then Gaussian distributed noise samples were generated with standard deviation of 1 % of the peak amplitude of the simulated time varying pressure signal. Then the inverse initial value problem was solved using the time reversal using the generated noise samples as data. As a result, noise samples of the optical inverse problem were obtained. The number of the simulated pressure signals used to define the peak amplitude was 80 and the number of the noise samples generated for each of the pressure signals was 100. Note that the optical and acoustic forward solutions were utilized only to give a realistic amplitude for the noise and no other information of the forward solutions were included into the noise model. The mean and the covariance of the noise model e were approximated from the reconstructed noise samples using (41)–(42). The noise model computations were performed in the same discretizations that were used in the solution of the inverse problem, i. e., 976 nodes and 1842 triangular elements for the optical part and 128×128 pixels for the acoustic part. The mean of the noise η_e when the target has been illuminated from the upper side is shown in the left image of Figure 11.6. A sample of the simulated noise covariance matrix Γ_e is shown in the middle image of Figure 11.7. As it can be seen, the mean of the noise does not show any structure and is close to zero. However, the covariance of the noise of the optical data is not uncorrelated. Thus, it can be expected that including information of the noise model, will improve the reconstructions.

Table 11.2: Prior mean η_x and standard deviations for the background σ_x^{bg} and the inclusions σ_x^{incl} used in the teaching and reconstruction prior distributions.

	Teaching		Reconstruction	
	μ_a	μ_s	μ_a	μ_s
η_x (mm^{-1})	0.01	1	0.01	1
σ_x^{bg} (mm^{-1})	0.001	0.1	0.002	0.2
σ_x^{incl} (mm^{-1})	0.002	0.2	0.01	1

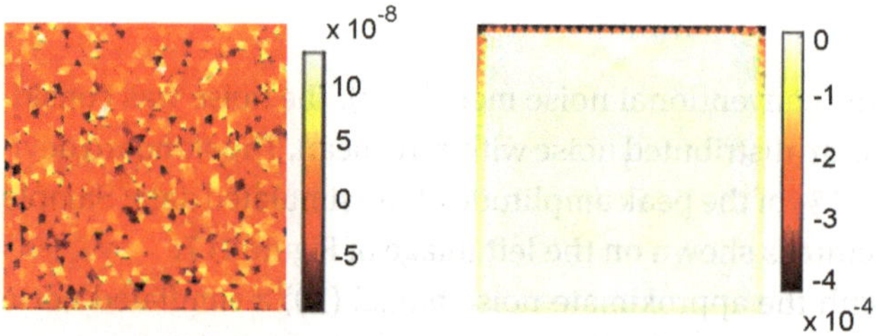

Figure 11.6: Means of the approximate noise model η_e (left image) and the approximation error model η_ε (right image) when light is illuminated from the upper side of the domain.

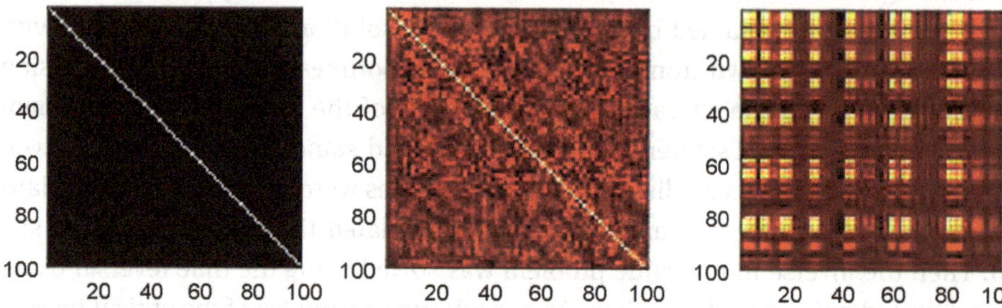

Figure 11.7: Covariance structures (100×100 sample of a covariance matrix) of the conventional noise model Γ_σ (left image), the approximate noise model Γ_e (middle image), and the approximation error model Γ_ε (right image).

The modeling error mean and covariance for the MAP estimate with noise and error modeling (44) were simulated as follows. First, 1200 samples $x^{(\ell)}$ were drawn from the teaching prior distribution given in Table 11.2 as in the case of the approximate noise model. The DA (11)–(12) was solved with the FEM in the same discretization of 976 nodes and 1842 triangular elements that is used in the solution of the inverse problem. Then samples of the absorbed optical energy density of the accurate model, i. e., ideal data, $y^{(\ell)}$ were calculated using (13). The initial pressure was computed from these samples using (14) and the acoustic forward problem was solved using the k-space method in a 512×512 pixel discretization grid. Then, the acoustic inverse problem was solved using the time reversal in the discretization of 128×128 pixels, and samples of realistic data $y_h^{(\ell)}$ were computed using (14). Then samples of the modeling error were computed from the samples of the ideal and realistic data using (45). Further, the mean and covariance of the approximation error were estimated using (46)–(47).

The mean of the approximation error η_ε when light is illuminated from the upper side of the domain is shown in the right image of Figure 11.6. Further, a sample of the simulated noise covariance matrix Γ_ε is shown in the right image of Figure 11.7. As it can be seen, the mean of the approximation error is nonzero close to the domain boundaries with the largest values close to the light source. Also, the covariance has a structure that differs from the noise model covariance. These results show that the acoustic inversion method has an effect on the data. Furthermore, including this modeling error information in the optical inverse model is likely to improve the reconstructions.

Reconstructions

In the solution of the optical inverse problem, the domain was discretized using 976 nodes and 1842 triangular elements for the FE-approximation of the DA. For the representation of the absorption and scattering, the domain was discretized into 1842 disjoint elements, and the absorption and scattering coefficients were represented in a piecewise constant basis. Furthermore, the data for the inverse problem, the absorbed optical energy density, was represented using the same piecewise constant basis.

The MAP estimates using the three approaches were computed by minimizing (40) (MAP-CNM), (30) (MAP-ANM), and (44) (MAP-AEM). A Gauss–Newton method equipped with a line search algorithm and positivity constraint was used to solve the minimization problems. The prior distribution utilized was the informative smoothness prior [7, 49] with characteristic length scale of 4 mm for both absorption and scattering. The mean and standard deviations of the prior distributions used in reconstructions are given in Table 11.2.

The reconstructed absorption and scattering distributions are shown in Figure 11.8. The calculated mean relative errors

$$\bar{E}_{\mu_\mathrm{a}} = 100\,\% \cdot \frac{\|\mu_\mathrm{a} - \hat{\mu}_\mathrm{a}\|}{\|\mu_\mathrm{a}\|}, \quad \bar{E}_{\mu_\mathrm{s}} = 100\,\% \cdot \frac{\|\mu_\mathrm{s} - \hat{\mu}_\mathrm{s}\|}{\|\mu_\mathrm{s}\|} \tag{48}$$

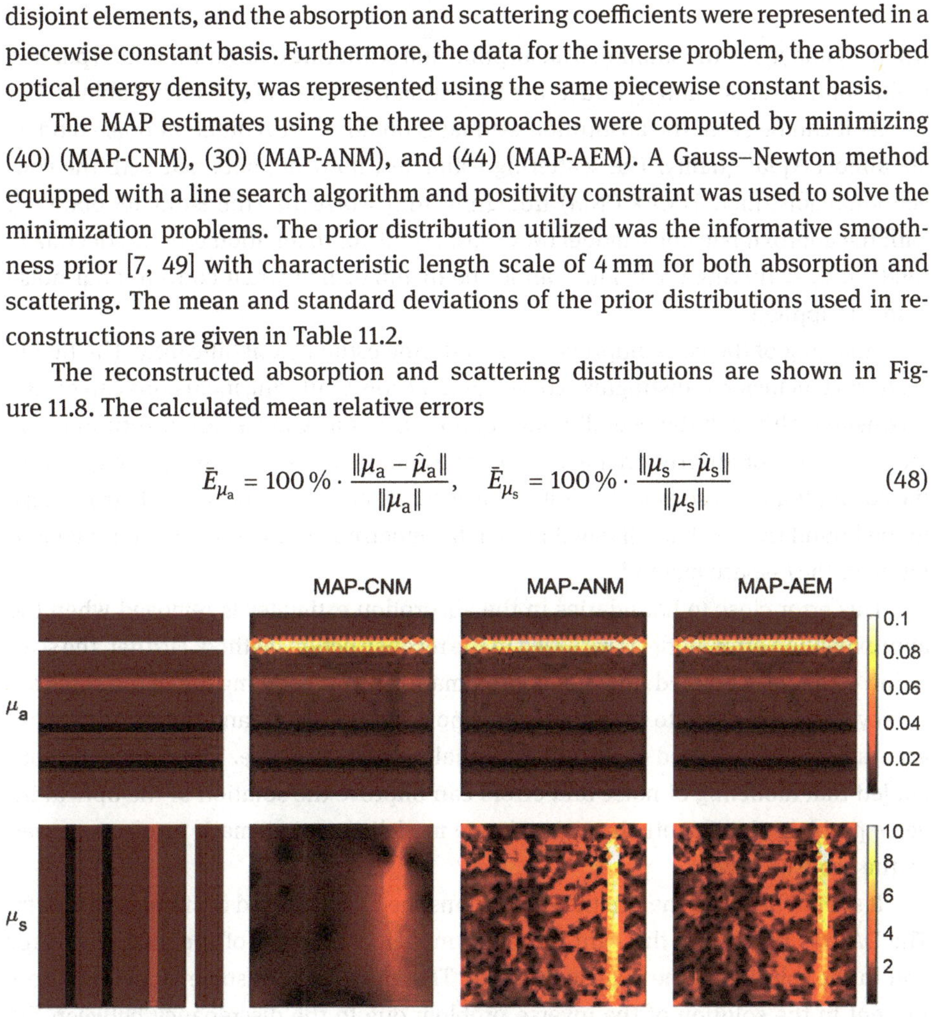

Figure 11.8: Reconstructed absorption (top row) and scattering (bottom row) distributions. Images from left to right: simulated distributions (first column), reconstructions obtained with the conventional noise model MAP-CNM (second column), approximate noise model MAP-ANM (third column), and approximate noise model and error modeling MAP-AEM (fourth column). The units of the color bars are in mm^{-1}.

Table 11.3: Mean relative errors of absorption and scattering MAP estimates $\bar{E}_{\mu_a}$ and $\bar{E}_{\mu_s}$, respectively, obtained with the conventional noise model MAP-CNM, approximate noise model MAP-ANM and approximate noise model and error modeling MAP-AEM.

	MAP-CNM	MAP-ANM	MAP-AEM
$\bar{E}_{\mu_a}$ (%)	17	15	14
$\bar{E}_{\mu_s}$ (%)	72	59	54

where μ_a and μ_s are the simulated absorption and scattering distributions interpolated to the solution space and $\hat{\mu}_a$ and $\hat{\mu}_s$ are the estimated values are given in Table 11.3.

As it can be seen, the absorption estimates obtained with different noise models are almost equal quality. The scattering estimates, however, obtained with the conventional noise model (MAP-CNM) are significantly worse than the estimates obtained with the approximate noise model (MAP-ANM) and approximation error model (MAP-AEM), i. e., scattering inclusions cannot be distinguished when conventional noise model is applied.

Modeling of the noise improves the scattering estimates significantly. The highly scattering inclusion is distinguished well and the low scattering inclusions can be distinguished, although the overall image quality is not as good as for absorption. The relative errors for absorption and scattering estimates are smaller when compared to the conventional noise model results. The absorption images show small errors close to the boundary which are likely due to data smoothing in the numerical implementation of the k-space method.

This error close to boundaries in the absorption estimates is removed when the approximation error model is included in the reconstruction method. Further, the scattering estimates obtained utilizing approximation error modeling are of the same accuracy when compared to the approximate noise model results and significantly more accurate when compared to the conventional noise model case. Thus, it can be concluded that modeling of noise and errors can improve the solution of the optical inverse problem significantly by approximate modeling of systematic artefacts caused by the acoustic solver.

It should be noted that, in the simulations the RTE was used to simulate the data. The DA was utilized as the light transport model in teaching of the modeling error and in the solution of the inverse problem. Therefore, there is some modeling errors present in the solution of the inverse problem due to the discrepancy between the RTE and DA. This error could be examined in the framework of Bayesian approximation error modeling, and although the error may be small in the case of large imaging domain such as used in this work, it may become significant if smaller target size is examined.

11.5 Discussion

In this chapter, quantitative photoacoustic imaging was discussed and approached in the framework of Bayesian inverse problems. The physical models covering photoacoustic effect and their numerical solution methods were shortly reviewed as well as the Bayesian approach to tomographic inverse problems. Two case studies were considered: estimation of the initial pressure and the reliability of the estimates in the acoustic inverse problem and modeling of noise and errors in the optical inverse problem.

In the acoustic inverse problem, solving the posterior distribution was considered in a two-dimensional case which gave the possibility to solve the whole posterior distribution in each pixel of the discretized domain. The results show that Bayesian approach can be used to provide good quality images of the initial pressure. However, if the inverse problem is severely ill-posed, such as in the case of limited view measurement setup, these reconstructions, although they can be computed, they suffer from artefacts. The appearance of the artefacts depend on the modeling errors present in the imaging situation and they are also affected by proper (or lack of proper) prior information. The results also show that evaluating the reliability of photoacoustic images can be performed in Bayesian framework. However, it also seems that these reliability estimates may be very sensitive to the prior information.

In the optical inverse problem, estimation of absorption and scattering was studied. In the optical inverse problem of QPAT, the data is obtained as a solution of the acoustic inverse problem, and thus it may contain artefacts due to the acoustic inverse problem solution method. In this work, the statistics of the noise and the modeling errors in the data were studied. They were approximated as Gaussian and were taken into account in the solution of the optical inverse problem similarly as in Bayesian approximation error modeling. The results show that modeling of noise and errors can improve the quality of the reconstructed images significantly and increase the quantitative accuracy of the estimates.

In this work, Bayesian approach to QPAT was formulated, and the approach was studied using two case studies with numerical simulations. The results give insight into modeling of noise and errors in QPAT and evaluating the reliability of the estimates. The methods could be utilized in other tomographic methodologies as well. However, there are plenty of open questions that need to be studied further. For example, the reliability of the optical parameter estimates was not considered in here, and thus far it has been performed in the case of single stage QPAT [70] but not in other studies. Furthermore, in these studies, the numerical implementations of the forward solvers were assumed to be accurate within measurement accuracy. In practice, however, they may include modeling errors which should be taken into account. Furthermore, evaluating the credibility of the estimates needs to be further studied. In this work, the approaches were studied with two-dimensional simulations, and thus an extension to three dimensions, which sets computational challenges, as well as experiments with real measurement data need to be performed. Thus far, Bayesian image reconstruction in three dimensions and with experimental data has been studied in [94].

Bibliography

[1] M. Agranovsky and P. Kuchment. Uniqueness of reconstruction and an inversion procedure for thermoacoustic and photoacoustic tomography with variable sound speed. *Inverse Probl.*, 23(5):2089–2102, 2007.

[2] G. S. Alberti and H. Ammari. Disjoint sparsity for signal separation and applications to hybrid inverse problems in medical imaging. *Appl. Comput. Harmon. Anal.*, 42(2):319–349, 2017.

[3] H. Ammari, E. Bretin, J. Garnier, and V. Jugnon. Coherent interferometry algorithms for photoacoustic imaging. *SIAM J. Numer. Anal.*, 50(5):2259–2280, 2012.

[4] S. Arridge, P. Beard, M. Betcke, B. Cox, N. Huynh, F. Lucka, O. Ogunlade, and E. Zhang. Accelerated high-resolution photoacoustic tomography via compressed sensing. *Phys. Med. Biol.*, 61:8908–8940, 2016.

[5] S. R. Arridge. Optical tomography in medical imaging. *Inverse Probl.*, 15:R41–R93, 1999.

[6] S. R. Arridge, M. M. Betcke, B. T. Cox, F. Lucka, and B. E. Treeby. On the adjoint operator in photoacoustic tomography. *Inverse Probl.*, 32:115012, 2016.

[7] S. R. Arridge, J. P. Kaipio, V. Kolehmainen, M. Schweiger, E. Somersalo, T. Tarvainen, and M. Vauhkonen. Approximation errors and model reduction with an application in optical diffusion tomography. *Inverse Probl.*, 22:175–195, 2006.

[8] G. Bal. Radiative transfer equation with varying refractive index: a mathematical perspective. *J. Opt. Soc. Am. A*, 23:1639–1644, 2006.

[9] G. Bal and K. Ren. Multi-source quantitative photoacoustic tomography in a diffusive regime. *Inverse Probl.*, 27:075003, 2011.

[10] G. Bal and K. Ren. On multi-spectral quantitative photoacoustic tomography in a diffusive regime. *Inverse Probl.*, 28:025010, 2012.

[11] B. Banerjee, S. Bagchi, R. M. Vasu, and D. Roy. Quantitative photoacoustic tomography from boundary pressure measurements: noniterative recovery of optical absorption coefficient from the reconstructed absorbed energy map. *J. Opt. Soc. Am. A*, 25(9):2347–2356, 2008.

[12] P. Beard. Biomedical photoacoustic imaging. *Interface Focus*, 1:602–631, 2011.

[13] Z. Belhachmi, T. Glatz, and O. Scherzer. A direct method for photoacoustic tomography with inhomogeneous sound speed. *Inverse Probl.*, 32(4):045005, 2016.

[14] M. Bergounioux, X. Bonnefond, T. Haberkorn, and Y. Privat. An optimal control problem in photoacoustic tomography. *Math. Models Methods Appl. Sci.*, 24(12):2525–2548, 2014.

[15] J. Brunker, J. Yao, J. Laufer, and S. E. Bohndiek. Photoacoustic imaging using genetically encoded reporters: a review. *J. Biomed. Opt.*, 22(7):070901, 2017.

[16] P. Burgholzer, G. J. Matt, M. Haltmeier, and G. Paltauf. Exact and approximative imaging methods for photoacoustic tomography using an arbitrary detection surface. *Phys. Rev. E*, 75(4):046706, 2007.

[17] D. Calvetti and E. Somersalo. *Introduction to Bayesian scientific computing*. Springer, New York, 2007.

[18] B. Cox, J. G. Laufer, S. R. Arridge, and P. C. Beard. Quantitative spectroscopic photoacoustic imgaging: a review. *J. Biomed. Opt.*, 17(6):061202, 2012.

[19] B. T. Cox, S. R. Arridge, and P. C. Beard. Estimating chromophore distributions from multiwavelength photoacoustic images. *J. Opt. Soc. Am. A*, 26(2):443–455, 2009.

[20] B. T. Cox, S. R. Arridge, K. P. Köstli, and P. C. Beard. Two-dimensional quantitative photoacoustic image reconstruction of absorption distributions in scattering media by use of a simple iterative method. *Appl. Opt.*, 45:1866–1875, 2006.

[21] B. T. Cox and P. C. Beard. Fast calculation of pulsed photoacoustic fields in fluids using *k*-space methods. *J. Acoust. Soc. Am.*, 117(6):3616–3627, 2005.

[22] B. T. Cox, S. Kara, S. R. Arridge, and P. C. Beard. *k*-space propagation models for acoustically heterogenous media: application to biomedical photoacoustics. *J. Acoust. Soc. Am.*, 121(6):3453–3464, 2007.

[23] B. T. Cox and B. E. Treeby. Artifact trapping during time reversal photoacoustic imaging for acoustically heteregeneous media. *IEEE Trans. Med. Imaging*, 29(2):387–396, 2010.

[24] X. L. Deán-Ben, A. Buehler, V. Ntziachristos, and D. Razansky. Accurate model-based reconstruction algorithm for three-dimensional optoacoustic tomography. *IEEE Trans. Med. Imaging*, 31(10):1922–1928, 2012.

[25] X. L. Deán-Ben, R. Ma, D. Razansky, and V. Ntziachristos. Statistical approach for optoacoustic image reconstruction in the presence of strong acoustic heterogeneities. *IEEE Trans. Med. Imaging*, 30(2):401–408, 2011.

[26] T. Ding, K. Ren, and S. Vallélian. A one-step reconstruction algorithm for quantitative photoacoustic imaging. *Inverse Probl.*, 31:095005, 2015.

[27] D. Finch, S. K. Patch, and Rakesh. Determining a function from its mean values over a family of spheres. *SIAM J. Math. Anal.*, 35(5):1213–1240, 2004.

[28] H. Gao, J. Feng, and L. Song. Limited-view multi-source quantitative photoacoustic tomography. *Inverse Probl.*, 31:065004, 2015.

[29] H. Gao, H. Zhao, and S. Osher. Bregman methods in quantitative photoacoustic tomography. UCLA CAM Report, 10-24, 2010.

[30] M. Haltmeier, R. Kowar, and L. V. Nguyen. Iterative methods for photoacoustic tomography in attenuating acoustic media. *Inverse Probl.*, 33:115009, 2017.

[31] M. Haltmeier, L. Neumann, and S. Rabanser. Single-stage reconstruction algorithm for quantitative photoacoustic tomography. *Inverse Probl.*, 31:065005, 2015.

[32] A. Hannukainen, N. Hyvönen, H. Majander, and T. Tarvainen. Efficient inclusion of total variation type priors in quantitative photoacoustic tomography. *SIAM J. Imaging Sci.*, 9(3):1132–1153, 2016.

[33] A. Hauptmann, F. Lucka, M. Betcke, N. Huynh, J. Adler, B. Cox, P. Beard, S. Ourselin, and S. Arridge. Model based learning for accelerated, limited-view 3-D photoacoustic tomography. *IEEE Trans. Med. Imaging*, 37(6):1382–1393, 2018.

[34] J. Heiskala, V. Kolehmainen, T. Tarvainen, J. P. Kaipio, and S. R. Arridge. Approximation error method can reduce artifacts due to scalp blood flow in optical brain activation imaging. *J. Biomed. Opt.*, 17(9):096012, 2012.

[35] L. G. Henyey and J. L. Greenstein. Diffuse radiation in the galaxy. *Astrophys. J.*, 93:70–83, 1941.

[36] R. Hochuli, S. Powell, S. Arridge, and B. Cox. Quantitative photoacoustic tomography using forward and adjoint Monte Carlo models of radiance. *J. Biomed. Opt.*, 21(12):126004, 2016.

[37] Y. Hristova, P. Kuchment, and L. Nguyen. Reconstruction and time reversal in thermoacoustic tomography in acoustically homogeneous and inhomogeneous media. *Inverse Probl.*, 24:055006, 2008.

[38] C. Huang, L. Nie, R. W. Schoonover, L. V. Wang, and M. A. Anastasio. Photoacoustic computed tomography correcting for heterogeneity and attenuation. *J. Biomed. Opt.*, 17(6):061211, 2012.

[39] A. Ishimaru. *Wave propagation and scattering in random media*, volume 1. Academic Press, New York, 1978.

[40] A. Javaherian and S. Holman. A multi-grid iterative method for photoacoustic tomography. *IEEE Trans. Med. Imaging*, 36(3):696–706, 2017.

[41] T. Jetzfellner, D. Razansky, A. Rosenthal, R. Schulz, and K. H. Englmeier. Performance of iterative optoacoustic tomography with experimental data. *Appl. Phys. Lett.*, 95:013703, 2009.

[42] X. Jin and L. Wang. Thermoacoustic tomography with correction for acoustic speed variations. *Phys. Med. Biol.*, 51:6437–6448, 2006.

[43] J. Kaipio and E. Somersalo. *Statistical and computational inverse problems*. Springer, New York, 2005.

[44] J. Kaipio and E. Somersalo. Statistical inverse problems: discretization, model reduction and inverse crimes. *J. Comput. Appl. Math.*, 198:493–504, 2007.

[45] G. Kanschat. A robust finite element discretization for radiative transfer problems with scattering. *East-West J. Numer. Math.*, 6(4):265–272, 1998.

[46] B. A. Kaplan, J. Buchmann, S. Prohaska, and J. Laufer. Monte-Carlo-based inversion scheme for 3D quantitative photoacoustic tomography. In A. A. Oraevsky and L. V. Wang, editors, *Photons plus ultrasound: imaging and sensing 2017, proc of SPIE*, volume 10064, pages 100645J–1, 2017.

[47] J. J. Kaufman, G. Luo, and R. S. Siffert. Ultrasound simulation in bone. *IEEE Trans. Ultrason. Ferroelectr. Freq. Control*, 55(6):1205–1218, 2008.

[48] L. E Kinsler, A. R. Frey, A. B. Coppens, and J. V. Sanders. *Fundamentals of Acoustics*, 4th edition. John Wiley & Sons, Inc., New York, 2000.

[49] V. Kolehmainen, M. Schweiger, I. Nissilä, T. Tarvainen, S. R. Arridge, and J. P. Kaipio. Approximation errors and model reduction in three-dimensional diffuse optical tomography. *J. Opt. Soc. Am. A*, 26(10):2257–2268, 2009.

[50] J. Koponen, T. Huttunen, T. Tarvainen, and J. P. Kaipio. Bayesian approximation error approach in full-wave ultrasound tomography. *IEEE Trans. Ultrason. Ferroelectr. Freq. Control*, 61(10):1627–1637, 2014.

[51] R. Kowar and O. Scherzer. Photoacoustic imaging taking into account attenuation. In *Mathematics and algorithms in tomography*, volume 18, pp. 54–56, 2010.

[52] P. Kuchment and L. Kunyansky. Mathematics of photoacoustic and thermoacoustic tomography. In O. Scherzer, editor, *Handbook of mathematical methods in imaging*, pp. 1117–1167. Springer, New York, 2015.

[53] L. A. Kunyansky. Explicit inversion formulae for the spherical mean Radon transform. *Inverse Probl.*, 23(1):373–383, 2007.

[54] L. A. Kunyansky. A series solution and a fast algorithm for the inversion of the spherical mean Radon transform. *Inverse Probl.*, 23(6):S11–S20, 2007.

[55] J. Laufer, B. Cox, E. Zhang, and P. Beard. Quantitative determination of chromophore concentrations form 2D photoacoustic images using a nonlinear model-based inversion scheme. *Appl. Opt.*, 49(8):1219–1233, 2010.

[56] K. J. H. Law, A. M. Stuart, and K. C. Zygalakis. *Data assimilation: a mathematical introduction*. Springer, Switzerland, 2015.

[57] O. Lehtikangas, T. Tarvainen, A. D. Kim, and S. R. Arridge. Finite element approximation of the radiative transport equation in a medium with piece-wise constant refractive index. *J. Comput. Phys.*, 282:345–359, 2015.

[58] C. Li and L. V. Wang. Photoacoustic tomography and sensing in biomedicine. *Phys. Med. Biol.*, 54:R59–R97, 2009.

[59] X. Li and H. Jiang. Impact of inhomogeneous optical scattering coefficient distribution on recovery of optical absorption coefficient maps using tomographic photoacoustic data. *Phys. Med. Biol.*, 58:999–1011, 2013.

[60] E. Malone, S. Powell, B. T. Cox, and S. Arridge. Reconstruction-classification method for quantitative photoacoustic tomography. *J. Biomed. Opt.*, 20(12):126004, 2015.

[61] A. V. Mamonov and K. Ren. Quantitative photoacoustic imaging in radiative transport regime. *Commun. Math. Sci.*, 12:201–234, 2014.

[62] K. Mitsuhashi, J. Poudel, T. P. Matthews, A. Garcia-Uribe, L. V. Wang, and M. A. Anastasio. A forward-adjoint operator pair based on the elastic wave equation for use in transcranial photoacoustic computed tomography. *SIAM J. Imaging Sci.*, 10(4):2022–2048, 2017.

[63] P. M. Morse and K. Uno Ingard. *Theoretical acoustics*. Princeton University Press, New Jersey, 1986.

[64] M. Mozumder, T. Tarvainen, S. R. Arridge, J. P. Kaipio, and V. Kolehmainen. Compensation of optode sensitivity and position errors in diffuse optical tomography using the approximation error approach. *J. Biomed. Opt.*, 4(10):2015–2031, 2013.

[65] M. Mozumder, T. Tarvainen, J. P. Kaipio, S. R. Arridge, and V. Kolehmainen. Compensation of modeling errors due to unknown domain boundary in diffuse optical tomography. *J. Opt. Soc. Am. A*, 31(8):1847–1855, 2014.

[66] W. Naetar and O. Scherzer. Quantitative photoacoustic tomography with piecewise constant material parameters. *SIAM J. Imaging Sci.*, 7(3):1755–1774, 2014.

[67] O. Nykänen, A. Pulkkinen, and T. Tarvainen. Quantitative photoacoustic tomography augmented with surface light measurements. *Biomed. Opt. Express*, 8(10):4380–4395, 2017.

[68] S. K. Patch and O. Scherzer. Photo- and thermo-acoustic imaging. *Inverse Probl.*, 23:S1–S10, 2007.

[69] J. Prakash, A. S. Raju, C. B. Shaw, M. Pramanik, and P. K. Yalavarthy. Basis pursuit deconvolution for improving model-based reconstructed images in photoacoustic tomography. *Biomed. Opt. Express*, 5(5):1363–1377, 2014.

[70] A. Pulkkinen, B. T. Cox, S. R. Arridge, H. Goh, J. P. Kaipio, and T. Tarvainen. Direct estimation of optical parameters from photoacoustic time series in quantitative photoacoustic tomography. *IEEE Trans. Med. Imaging*, 35(11):2497–2508, 2016.

[71] A. Pulkkinen, B. T. Cox, S. R. Arridge, J. P. Kaipio, and T. Tarvainen. A Bayesian approach to spectral quantitative photoacoustic tomography. *Inverse Probl.*, 30:065012, 2014.

[72] A. Pulkkinen, B. T. Cox, S. R. Arridge, J. P. Kaipio, and T. Tarvainen. Quantitative photoacoustic tomography using illuminations from a single direction. *J. Biomed. Opt.*, 20(3):036015, 2015.

[73] A. Pulkkinen, V. Kolehmainen, J. P. Kaipio, B. T. Cox, S. R. Arridge, and T. Tarvainen. Approximate marginalization of unknown scattering in quantitative photoacoustic tomography. *Inverse Probl. Imaging*, 8(3):811–829, 2014.

[74] A. Pulkkinen, B. Werner, E. Martin, and K. Hynynen. Numerical simulations of clinical focused ultrasound functional neurosurgery. *Phys. Med. Biol.*, 59(7):1679–1700, 2014.

[75] C. E. Rasmussen and C. K. I. Williams. *Gaussian processes for machine learning*. MIT Press, 2006.

[76] D. Razansky. Multispectral optoacoustic tomography—volumetric color hearing in real time. *IEEE J. Sel. Top. Quantum Electron.*, 18(3):1234–1243, 2012.

[77] D. Razansky, J. Baeten, and V. Ntziachristos. Sensitivity of molecular target detection by multispectral optoacoustic tomography (MSOT). *Med. Phys.*, 36(3):939–945, 2009.

[78] K. Ren, H. Gao, and H. Zhao. A hybrid reconstruction method for quantitative PAT. *SIAM J. Imaging Sci.*, 6(1):32–55, 2013.

[79] J. Ripoll and V. Ntziachristos. Quantitative point source photoacoustic inversion formulas for scattering and absorbing media. *Phys. Rev. E*, 71:031912, 2005.

[80] A. Rosenthal, D. Razansky, and V. Ntziachristos. Quantitative optoacoustic signal extraction using sparse signal representation. *IEEE Trans. Med. Imaging*, 28(12):1997–2006, 2009.

[81] T. Saratoon, T. Tarvainen, S. R. Arridge, and B. T. Cox. 3D quantitative photoacoustic tomography using the δ-Eddington approximation. In A. A. Oraevsky and L. V. Wang, editors, *Proc. of SPIE*, volume 8581, pp. 85810V–1, 2013.

[82] T. Saratoon, T. Tarvainen, B. T. Cox, and S. R. Arridge. A gradient-based method for quantitative photoacoustic tomography using the radiative transfer equation. *Inverse Probl.*, 29:075006, 2013.

[83] O. Scherzer and C. Shi. Reconstruction formulas for photoacoustic imaging in attenuating media. *Inverse Probl.*, 34:015006, 2018.

[84] R. W. Schoonover and M. A. Anastasio. Compensation of shear waves in photoacoustic tomography with layered acoustic media. *J. Opt. Soc. Am. A*, 28(10):2091–2099, 2011.

[85] P. Shao, B. Cox, and R. J. Zemp. Estimating optical absorption, scattering and Grueneisen distributions with multiple-illumination photoacoustic tomography. *Appl. Opt.*, 50(19):3145–3154, 2011.

[86] P. Shao, T. Harrison, and R. J. Zemp. Iterative algorithm for multiple illumination photoacoustic tomography (MIPAT) using ultrasound channel data. *Biomed. Opt. Express*, 3(12):3240–3248, 2012.

[87] N. Song, C. Deumié, and A. Da Silva. Considering sources and detectors distributions for quantitative photoacoustic tomography. *Biomed. Opt. Express*, 5(11):3960–3974, 2014.

[88] A. Tarantola. *Inverse problem theory and methods for model parameter estimation*. SIAM Society for Industrial and Applied Mathematics, Philadelphia, 2005.

[89] T. Tarvainen, B. T. Cox, J. P. Kaipio, and S. R. Arridge. Reconstructing absorption and scattering distributions in quantitative photoacoustic tomography. *Inverse Probl.*, 28:084009, 2012.

[90] T. Tarvainen, V. Kolehmainen, A. Pulkkinen, M. Vauhkonen, M. Schweiger, S. R. Arridge, and J. P. Kaipio. An approximation error approach for compensating for modelling errors between the radiative transfer equation and the diffusion approximation in diffuse optical tomography. *Inverse Probl.*, 26:015005, 2010.

[91] T. Tarvainen, A. Pulkkinen, B. T. Cox, J. P. Kaipio, and S. R. Arridge. Bayesian image reconstruction in quantitative photoacoustic tomography. *IEEE Trans. Med. Imaging*, 32(12):2287–2298, 2013.

[92] T. Tarvainen, M. Vauhkonen, V. Kolehmainen, S. R. Arridge, and J. P. Kaipio. Coupled radiative transfer equation and diffusion approximation model for photon migration in turbid medium with low-scattering and non-scattering regions. *Phys. Med. Biol.*, 50:4913–4930, 2005.

[93] T. Tarvainen, M. Vauhkonen, V. Kolehmainen, and J. P. Kaipio. Hybrid radiative–transfer–diffusion model for optical tomography. *Appl. Opt.*, 44(6):876–886, 2005.

[94] J. Tick, A. Pulkkinen, F. Lucka, R. Ellwood, B. T. Cox, J. P. Kaipio, S. R. Arridge, and T. Tarvainen. Three dimensional photoacoustic tomography in Bayesian framework. *J. Acoust. Soc. Am.*, 144:2061–2071, 2018.

[95] J. Tick, A. Pulkkinen, and T. Tarvainen. Image reconstruction with uncertainty quantification in photoacoustic tomography. *J. Acoust. Soc. Am.*, 139:1951–1961, 2016.

[96] J. Tick, A. Pulkkinen, and T. Tarvainen. Bayesian approach to image reconstruction in photoacoustic tomography. In A. A. Oraevsky and L. V. Wang, editors, *Photons plus ultrasound: imaging and sensing 2017, proc of SPIE*, volume 10064, p. 100643M, 2017.

[97] B. E. Treeby and B. T. Cox. k-Wave: MATLAB toolbox for the simulation and reconstruction of photoacoustic wave fields. *J. Biomed. Opt.*, 15(2):021314, 2010.

[98] B. E. Treeby, J. Jaros, A. P. Rendell, and B. T. Cox. Modeling nonlinear ultrasound propagation in heterogeneous media with power law absorption using a k-space pseudospectral method. *J. Acoust. Soc. Am.*, 131(6):4324–4336, 2012.

[99] B. E. Treeby, E. Z. Zhang, and B. T. Cox. Photoacoustic tomography in absorbing acoustic media using time reversal. *Inverse Probl.*, 26:115003, 2010.

[100] M. Venugopal, P. van Es, S. Manohar, D. Roy, and R. M. Vasu. Quantitative photoacoustic tomography by stochastic search: direct recovery of the optical absorption field. *Opt. Lett.*, 41(18):4202–4205, 2016.

[101] K. Wang and M. A. Anastasio. Photoacoustic and thermoacoustic tomography: image formation principles. In O. Scherzer, editor, *Handbook of mathematical methods in imaging*, pp. 1081–1116. Springer, New York, 2015.

[102] K. Wang, R. Su, A. A. Oraevsky, and M. A. Anastasio. Investigation of iterative image reconstruction in three-dimensional optoacoustic tomography. *Phys. Med. Biol.*, 57:5399–5423, 2012.

[103] L. V. Wang, editor. *Photoacoustic imaging and spectroscopy*. CRC Press, 2009.

[104] L. V. Wang and J. Yao. A practical guide to photoacoustic tomography in the life sciences. *Nat. Methods*, 13(8):627–638, 2016.

[105] J. Weber, P. C. Beard, and S. Bohndiek. Contrast agents for molecular photoacoustic imaging. *Nat. Methods*, 13(8):639–650, 2016.

[106] P. J. White, G. T. Clement, and K. Hynynen. Longitudinal and shear mode ultrasound propagation in human skull bone. *Ultrasound Med. Biol.*, 32(7):1085–1096, 2006.

[107] J. Xia and L. V. Wang. Small-animal whole-body photoacoustic tomography: a review. *Phys. Med. Biol.*, 61(5):1380–1389, 2014.

[108] M. Xu and L. V. Wang. Universal back-projection algorithm for photoacoustic computed tomography. *Phys. Rev. E*, 71(1):016706, 2005.

[109] M. Xu and L. V. Wang. Photoacoustic imaging in biomedicine. *Rev. Sci. Instrum.*, 77:041101, 2006.

[110] Y. Xu and L. V. Wang. Time reversal and its application to tomography with diffracting sources. *Phys. Rev. Lett.*, 92(3):339021–339024, 2004.

[111] Y. Xu, L. V. Wang, G. Ambartsoumian, and P. Kuchment. Reconstructions in limited-view thermoacoustic tomography. *Med. Phys.*, 31(4):724–733, 2004.

[112] L. Yao, Y. Sun, and H. Jiang. Quantitative photoacoustic tomography based on the radiative transfer equation. *Opt. Lett.*, 34(12):1765–1767, 2009.

[113] L. Yao, Y. Sun, and H. Jiang. Transport-based quantitative photoacoustic tomography: simulations and experiments. *Phys. Med. Biol.*, 55:1917–1934, 2010.

[114] R. J. Zemp. Quantitative photoacoustic tomography with multiple optical sources. *Appl. Opt.*, 49(18):3566–3572, 2010.

[115] X. Zhang, W. Zhou, X. Zhang, and H. Gao. Forward-backward splitting method for quantitative photoacoustic tomography. *Inverse Probl.*, 30:125012, 2014.

[116] Y. Zhou, E. Tang, J. Luo, and J. Yao. Deep-tissue temperature mapping by multi-illumination photoacoustic tomography aided by a diffusion optical model: a numerical study. *J. Biomed. Opt.*, 23(1):016014, 2018.

Shari Moskow and John C. Schotland

12 Inverse Born series

Abstract: We present a survey of recent results on the inverse Born series. The convergence and stability of the method are characterized in Banach spaces. Applications to inverse problems in various physical settings are described.

Keywords: Inverse scattering, inverse Born series, optical tomography, radiative transport

MSC 2010: 35R30, 34L25, 78A46

12.1 Introduction

Inverse scattering problems are of fundamental importance in nearly every branch of physics. They also arise in numerous applied fields ranging from biomedical imaging to seismology. Such problems can be formulated in a variety of settings, depending upon the nature of the probing wave field and the length scales of interest. Regardless of such considerations, the fundamental theoretical questions relate to the uniqueness, stability and reconstruction of the solution to the problem. By uniqueness, we mean the injectivity of the forward map from the scattering potential to the scattering data. Stability refers to continuity of the inverse map from scattering data to the potential. We note that inverse scattering problems are typically ill-posed, which means that the inverse map must be suitably regularized to achieve stable inversion.

There are a number of approaches to the problem of recovering the scattering potential. See [6, 8, 9, 12, 16] for a comprehensive overview of inverse scattering theory. Direct reconstruction methods provide an analytic solution to the inverse problems, principally in one-dimension although higher-dimensional methods are also known. Optimization methods iteratively minimize the distance between the scattering data and the solution to the forward problem, viewed as a functional of the scattering potential. Although such techniques are extremely flexible, the presence of local minima and the computational cost of evaluating the forward map limit their practical util-

Acknowledgement: We gratefully acknowledge our collaborations with the following individuals who have helped to shape our understanding of the inverse Born series: Simon Arridge, Scott Carney, Francis Chung, Anna Gilbert, Jeremy Hoskins, Kimberly Kilgore, Manabu Machida, Vadim Markel, George Panasyuk and Hala A. H. Shehadeh.

***Corresponding author: John C. Schotland,** Department of Mathematics and Department of Physics, University of Michigan, Ann Arbor, MI 48109, USA, e-mail: schotland@umich.edu
Shari Moskow, Department of Mathematics, Drexel University, Philadelphia, PA 19104, USA, e-mail: moskow@math.drexel.edu

https://doi.org/10.1515/9783110560855-012

ity. Finally, linear sampling and related qualitative methods can be used to recover the support of the scattering potential for obstacle scattering and the inverse medium problem [4, 5, 17].

The inverse Born series (IBS) is a direct reconstruction method that was initially developed to study the quantum mechanical inverse backscattering problem in one dimension [13, 22] and later extended to higher dimensions [10, 18, 26, 30, 31]. The authors analyzed the convergence, stability and error of the IBS [23]. They have also applied the IBS to various inverse scattering problems, including those of optical tomography, electrical impedance tomography, and acoustic and electromagnetic imaging [1, 14, 15, 19, 20, 24, 25]. The IBS has also been applied to discrete inverse problems on graphs, independent of the continuous setting in which it was initially proposed [7]. Finally, we note that the inverse of the Bremmer series can be investigated using a related approach [29].

It is important to note that the principal computational advantage of the IBS is that it does not make use of a partial differential equation solver. Instead, the IBS obtains the solution to the inverse problem as an explicitly computable functional of the scattering data. This functional can be expressed in terms of the Green's function for the underlying partial differential equation, whose decay governs the convergence and stability of the method.

In this chapter, we present a survey of recent results on the inverse Born series. In Section 12.2, the convergence and stability of the IBS is analyzed in Banach spaces. The results are then applied to a wide range of inverse problems. These include the inverse scattering problem for diffuse waves in Section 12.3, the Calderon problem of electrical impedance tomography in Section 12.4, the inverse radiative transport problem in Section 12.5, and the inverse scattering problem for electromagnetic waves in Section 12.6. Finally, in Section 12.7, we consider the inverse problem for graph diffusion.

We use the following notational conventions throughout this chapter. For $n \geq 2$, Ω denotes a bounded domain in $\mathbb{R}^n$ with a smooth boundary $\partial\Omega$. If X is a Banach space, X^j indicates the j-fold tensor product $X^j = X \otimes \cdots \otimes X$ equipped with the projective norm [27] for $j > 1$. We note that X^j is generally not a Banach space.

12.2 Analysis of the inverse Born series

In this section, we formulate the IBS in a Banach space setting. This formulation will then be applied to various inverse scattering problems later in the chapter. The presentation closely follows [23], where the case of L^p spaces was considered. The extension to Banach spaces was described in [2]. Let X and Y be Banach spaces. We consider the power series

$$\phi = K_1\eta + K_2\eta \otimes \eta + K_3\eta \otimes \eta \otimes \eta + \cdots, \tag{1}$$

where $K_j : X^j \to Y$. The forward problem is to evaluate the map $\mathcal{F} : \eta \mapsto \phi$ defined by (1). We will refer to K_j as forward operators and (1) is called the Born series.

The inverse problem is to determine η assuming that ϕ is known. That is, we seek to construct a map $\mathcal{I} : \phi \mapsto \eta$ which is, in some sense, the inverse of $\mathcal{F}$. Toward this end, we make the ansatz that η may be expressed as a series in tensor powers of ϕ of the form

$$\eta = \mathcal{K}_1\phi + \mathcal{K}_2\phi \otimes \phi + \mathcal{K}_3\phi \otimes \phi \otimes \phi + \cdots . \tag{2}$$

Here, the inverse operators $\mathcal{K}_j : Y^j \to X$ are to be determined. By substituting (1) into (2) and equating terms with the same tensor power of η, we find that the operators $\mathcal{K}_j$ are given by

$$\mathcal{K}_1 K_1 = I, \tag{3}$$

$$\mathcal{K}_2 = -\mathcal{K}_1 K_2 \mathcal{K}_1 \otimes \mathcal{K}_1, \tag{4}$$

$$\mathcal{K}_3 = -(\mathcal{K}_2 K_1 \otimes K_2 + \mathcal{K}_2 K_2 \otimes K_1 + \mathcal{K}_1 K_3)\mathcal{K}_1 \otimes \mathcal{K}_1 \otimes \mathcal{K}_1, \tag{5}$$

$$\mathcal{K}_j = -\left(\sum_{m=1}^{j-1} \mathcal{K}_m \sum_{i_1+\cdots+i_m=j} K_{i_1} \otimes \cdots \otimes K_{i_m} \right) \mathcal{K}_1 \otimes \cdots \otimes \mathcal{K}_1. \tag{6}$$

We will refer to (2) as the inverse Born series (IBS). Here, we note several of its properties. (i) The operator K_1 generally does not have a bounded inverse. Thus $\mathcal{K}_1$ is taken to be the regularized pseudoinverse of K_1, which is defined as follows. Consider the Tikhonov functional T which is of the form

$$T(\eta) = \|K_1\eta - \phi\|_Y^2 + \lambda F(\eta), \tag{7}$$

where F is a convex penalty function and $\lambda > 0$ is a regularization parameter [11, 21, 28]. The minimizer of T is denoted $\eta^\dagger$ and is referred to as the regularized pseudoinverse solution of $K_1\eta = \phi$. The operator $\mathcal{K}_1$ is defined as the map $\mathcal{K}_1 : \phi \mapsto \eta^\dagger$. Here we take $\eta \in \tilde{X}$, where $\tilde{X}$ is a uniformly convex subspace of X. If K_1 is bounded, it follows that $\eta^\dagger$ exists and is unique [28]. (ii) The coefficients in the inverse series have a recursive structure. The operator $\mathcal{K}_j$ is determined by the coefficients of the Born series $K_1, K_2, \ldots, K_j$. (iii) Inversion of only the linear term in the Born series is required to compute the IBS to all orders. Thus a nonlinear inverse problem that is often ill-posed is replaced by an ill-posed linear inverse problem plus a well-posed nonlinear problem, namely the computation of the higher order terms in the series.

We now proceed to analyze the convergence and stability of the IBS. Throughout, we assume that the operator K_j is bounded with

$$\|K_j\| \le \nu\mu^{j-1}, \tag{8}$$

for suitable constants μ and ν. We immediately see that the Born series (1) converges in norm provided that $\|\eta\|_X < 1/\mu$. The following lemma provides an estimate on the norm of the operator $\mathcal{K}_j$.

Lemma 12.2.1. *Let $\|\mathcal{K}_1\| < 1/(\mu + \nu)$. Then the operator $\mathcal{K}_j : X^j \to Y$ defined by (6) is bounded and*

$$\|\mathcal{K}_j\| \le C(\mu+\nu)^j \|\mathcal{K}_1\|, \tag{9}$$

where C is independent of j. Moreover, for all $\phi \in Y$

$$\|\mathcal{K}_j \phi \otimes \cdots \otimes \phi\|_X \le C(\mu+\nu)^j \|\mathcal{K}_1 \phi\|_X^j. \tag{10}$$

Proof. We first prove (9). Using (6), we find that

$$\begin{aligned}\|\mathcal{K}_j\| &\le \sum_{m=1}^{j-1} \sum_{i_1+\cdots+i_m=j} \|\mathcal{K}_m\| \|K_{i_1}\| \cdots \|K_{i_m}\| \|\mathcal{K}_1\|^j \\ &\le \|\mathcal{K}_1\|^j \sum_{m=1}^{j-1} \sum_{i_1+\cdots+i_m=j} \|\mathcal{K}_m\| \nu\mu^{i_1-1} \cdots \nu\mu^{i_m-1},\end{aligned} \tag{11}$$

where we have used (8) to obtain the second inequality. Next, we define $\Pi(j,m)$ to be the number of ordered partitions of the integer j into m parts. It can be seen that

$$\Pi(j,m) = \binom{j-1}{m-1}, \tag{12}$$

$$\sum_{m=1}^{j-1} \Pi(j,m) = 2^{j-1} - 1. \tag{13}$$

It follows that

$$\begin{aligned}\|\mathcal{K}_j\| &\le \|\mathcal{K}_1\|^j \sum_{m=1}^{j-1} \|\mathcal{K}_m\| \Pi(j,m) \nu^m \mu^{j-m} \\ &\le \|\mathcal{K}_1\|^j \left(\sum_{m=1}^{j-1} \|\mathcal{K}_m\|\right)\left(\sum_{m=1}^{j-1} \Pi(j,m)\nu^m\mu^{j-m}\right) \\ &\le \nu\|\mathcal{K}_1\|^j \left(\sum_{m=1}^{j-1} \|\mathcal{K}_m\|\right)\left(\sum_{m=0}^{j-1} \binom{j-1}{m}\nu^m\mu^{j-1-m}\right) \\ &= \nu\|\mathcal{K}_1\|^j (\mu+\nu)^{j-1} \sum_{m=1}^{j-1} \|\mathcal{K}_m\|.\end{aligned} \tag{14}$$

Thus $\|\mathcal{K}_j\|$ is a bounded operator and

$$\|\mathcal{K}_j\| \le (\mu+\nu)^j \|\mathcal{K}_1\|^j \sum_{m=1}^{j-1} \|\mathcal{K}_m\|. \tag{15}$$

The above estimate for $\|\mathcal{K}_j\|$ has a recursive structure. It can be seen that

$$\|\mathcal{K}_j\| \le C_j[(\mu+\nu)\|\mathcal{K}_1\|]^j \|\mathcal{K}_1\|, \tag{16}$$

where, for $j \geq 2$, C_j obeys the recursion relation

$$C_{j+1} = C_j + [(\mu + \nu)\|\mathcal{K}_1\|]^j C_j, \quad C_2 = 1. \tag{17}$$

Evidently,

$$C_j = \prod_{m=2}^{j-1} (1 + [(\mu + \nu)\|\mathcal{K}_1\|]^m). \tag{18}$$

We see that C_j is bounded for all j since

$$\begin{aligned} \ln C_j &\leq \sum_{m=1}^{j-1} \ln(1 + [(\mu + \nu)\|\mathcal{K}_1\|]^m) \\ &\leq \sum_{m=1}^{j-1} [(\mu + \nu)\|\mathcal{K}_1\|]^m \\ &\leq \frac{1}{1 - (\mu + \nu)\|\mathcal{K}_1\|}, \end{aligned} \tag{19}$$

where the final inequality follows if $(\mu + \nu)\|\mathcal{K}_1\| < 1$.

To prove (10), we note that the same reasoning as above leads to the inequality

$$\|\mathcal{K}_j \phi \otimes \cdots \otimes \phi\|_X \leq (\mu + \nu)^j \|\mathcal{K}_1 \phi\|_X^j \sum_{m=1}^{j-1} \|\mathcal{K}_m\|_p. \tag{20}$$

Making use of (20) leads to

$$\|\mathcal{K}_j \phi \otimes \cdots \otimes \phi\|_X \leq C(\mu + \nu)^j \frac{\|\mathcal{K}_1\|}{1 - (\mu + \nu)\|\mathcal{K}_1\|} \|\mathcal{K}_1 \phi\|_X^j, \tag{21}$$

which completes the proof. □

We now establish a basic result that governs the convergence and approximation error of the IBS.

Theorem 12.2.1. *Suppose that* $\|\mathcal{K}_1\| < 1/(\mu + \nu)$, $\|\mathcal{K}_1 \phi\|_X < 1/(\mu + \nu)$. *Let* $\mathcal{M} = \max(\|\eta\|_X, \|\mathcal{K}_1 K_1 \eta\|_X)$ *and assume that* $\mathcal{M} < 1/(\mu + \nu)$. *Then the inverse Born series* (2) *converges in norm and the following error estimate holds:*

$$\left\| \eta - \sum_{j=1}^{N} \mathcal{K}_j \phi \otimes \cdots \otimes \phi \right\|_X \leq C \|(I - \mathcal{K}_1 K_1)\eta\|_X + \tilde{C} \frac{[(\mu + \nu)\|\mathcal{K}_1 \phi\|_X]^N}{1 - (\mu + \nu)\|\mathcal{K}_1 \phi\|_X},$$

where C *and* $\tilde{C}$ *are independent of* N *and* ϕ.

Proof. The hypotheses imply that the series

$$\tilde{\eta} = \sum_j \mathcal{K}_j \phi \otimes \cdots \otimes \phi \tag{22}$$

converges. Here, $\mathcal{K}_1$ is regularized and we denote by $\tilde{\eta}$ the sum of the corresponding IBS. The Born series (1) also converges by hypothesis, so we can substitute it into (22) to obtain

$$\tilde{\eta} = \sum_j \tilde{\mathcal{K}}_j \eta \otimes \cdots \otimes \eta, \tag{23}$$

where

$$\tilde{\mathcal{K}}_1 = \mathcal{K}_1 K_1, \tag{24}$$

and

$$\tilde{\mathcal{K}}_j = \left(\sum_{m=1}^{j-1} \mathcal{K}_m \sum_{i_1+\cdots+i_m=j} K_{i_1} \otimes \cdots \otimes K_{i_m} \right) + \mathcal{K}_j K_1 \otimes \cdots \otimes K_1, \tag{25}$$

for $j \geq 2$. From (6), it follows that

$$\tilde{\mathcal{K}}_j = \sum_{m=1}^{j-1} \mathcal{K}_m \sum_{i_1+\cdots+i_m=j} K_{i_1} \otimes \cdots \otimes K_{i_m} (I - \mathcal{K}_1 K_1 \otimes \cdots \otimes \mathcal{K}_1 K_1). \tag{26}$$

We thus obtain the estimate

$$\|\eta - \tilde{\eta}\|_X \leq \sum_j \sum_{m=1}^{j-1} \sum_{i_1+\cdots+i_m=j} \|\mathcal{K}_m\| \|K_{i_1}\| \cdots \|K_{i_m}\| \|\eta \otimes \cdots \otimes \eta - \mathcal{K}_1 K_1 \eta \otimes \cdots \otimes \mathcal{K}_1 K_1 \eta\|_{X^j}. \tag{27}$$

Next, we put

$$\psi = \eta - \mathcal{K}_1 K_1 \eta \tag{28}$$

and make use of the identity

$$\begin{aligned} &\eta_1 \otimes \cdots \otimes \eta_1 - \eta_2 \otimes \cdots \otimes \eta_2 \\ &\quad = \zeta \otimes \eta_2 \otimes \cdots \otimes \eta_2 + \eta_1 \otimes \zeta \otimes \eta_2 \otimes \cdots \otimes \eta_2 \\ &\quad\quad + \cdots + \eta_1 \otimes \eta_1 \otimes \cdots \otimes \zeta \otimes \eta_2 + \eta_1 \otimes \eta_1 \otimes \cdots \otimes \eta_1 \otimes \zeta, \end{aligned} \tag{29}$$

where $\zeta = \eta_1 - \eta_2$ to obtain

$$\|\eta \otimes \cdots \otimes \eta - \mathcal{K}_1 K_1 \eta \otimes \cdots \otimes \mathcal{K}_1 K_1 \eta\|_{X^j} \leq j \mathcal{M}^{j-1} \|\psi\|_X. \tag{30}$$

We then have

$$\|\eta - \tilde{\eta}\|_X \le \sum_j \sum_{m=1}^{j-1} \sum_{i_1+\cdots+i_m=j} \|\mathcal{K}_m\| \|K_{i_1}\| \cdots \|K_{i_m}\| j\mathcal{M}^{j-1} \|\psi\|_X$$
$$\le \sum_j \sum_{m=1}^{j-1} j\mathcal{M}^{j-1} \|\mathcal{K}_m\| \Pi(j,m) \nu^m \mu^{j-m} \|\psi\|_X, \tag{31}$$

where we have used (8). Making use of (12), we have

$$\|\eta - \tilde{\eta}\|_X \le \nu \sum_j \|\psi\|_X j\mathcal{M}^{j-1} \left(\sum_{m=1}^{j-1} \|\mathcal{K}_m\| \right) \left(\sum_{m=0}^{j-1} \binom{j-1}{m} \nu^m \mu^{j-1-m} \right)$$
$$\le \|\psi\|_X \sum_j \sum_{m=1}^{j-1} j\mathcal{M}^{j-1} (\mu+\nu)^j \|\mathcal{K}_m\|. \tag{32}$$

We now apply Lemma 12.2.1 to obtain

$$\|\eta - \tilde{\eta}\|_X \le C\|\psi\|_X \sum_j \sum_{m=1}^{j-1} j\mathcal{M}^{j-1} (\mu+\nu)^{m+j} \|\mathcal{K}_1\|^m, \tag{33}$$

since the constant C from the lemma is independent of j. Performing the sum over m, we have

$$\|\eta - \tilde{\eta}\|_X \le C\|\psi\|_X \sum_j j\mathcal{M}^{j-1} (\mu+\nu)^j \frac{(\mu+\nu)^j \|\mathcal{K}_1\|^j - 1}{(\mu+\nu)\|\mathcal{K}_1\| - 1}, \tag{34}$$

which is bounded since $\mathcal{M}(\mu+\nu) < 1$ and $(\mu+\nu)\|\mathcal{K}_1\| < 1$. Equation (34) thus becomes

$$\|\eta - \tilde{\eta}\|_X \le C\|(I - \mathcal{K}_1 K_1)\eta\|_X, \tag{35}$$

where C is a new constant which depends on μ, ν, $\mathcal{M}$ and $\|\mathcal{K}_1\|$. Finally, using the triangle inequality and (10), we can account for the error which arises from cutting off the tail of the series. We thus obtain

$$\left\| \eta - \sum_{j=1}^{N} \mathcal{K}_j \phi \otimes \cdots \otimes \phi \right\|_X$$
$$\le \left\| \eta - \sum_j \tilde{\mathcal{K}}_j \eta \otimes \cdots \otimes \eta \right\|_X + \sum_{j=N+1}^{\infty} \|\mathcal{K}_j \phi \otimes \cdots \otimes \phi\|_X$$
$$\le C\|(I - \mathcal{K}_1 K_1)\eta\|_X + \tilde{C} \frac{((\mu+\nu)\|\mathcal{K}_1 \phi\|_X)^{N+1}}{1 - (\mu+\nu)\|\mathcal{K}_1 \phi\|_X}. \tag{36}$$

□

We make two important remarks concerning Theorem 12.2.1. (i) The hypothesis $\|K_1\| \leq \nu$, which corresponds to (8) when $j = 1$, is not generally inconsistent with the condition $\|\mathcal{K}_1\| \leq 1/(\mu + \nu)$. In particular, for the case of Tikhonov regularization, $\|\mathcal{K}_1\|$ can be chosen to be arbitrarily small by proper choice of the regularization parameter. (ii) We note that due to regularization of $\mathcal{K}_1$, the IBS does not converge to η. That is, $\mathcal{K}_1$ is not the true inverse of K_1. However, if it is known a priori that η belongs to a particular finite-dimensional subspace of X, $\mathcal{K}_1$ can be chosen to be a true inverse on this subspace. Then, provided the hypotheses of Theorem 12.2.1 hold, the IBS will recover η exactly.

The next result characterizes the stability of the limit of the IBS under perturbations in the data ϕ.

Theorem 12.2.2. *Let $\|\mathcal{K}_1\| < 1/(\mu + \nu)$ and let ϕ_1 and ϕ_2 be data for which $M\|\mathcal{K}_1\| < 1/(\mu + \nu)$, where $M = \max(\|\phi_1\|_Y, \|\phi_2\|_Y)$. Let η_1 and η_2 denote the corresponding limits of the inverse Born series. Then the following estimate holds:*

$$\|\eta_1 - \eta_2\|_X < C\|\phi_1 - \phi_2\|_Y,$$

where C is a constant which is independent of ϕ_1 and ϕ_2.

Proof. We begin with the estimate

$$\|\eta_1 - \eta_2\|_X \leq \sum_j \|\mathcal{K}_j(\phi_1 \otimes \cdots \otimes \phi_1 - \phi_2 \otimes \cdots \otimes \phi_2)\|_X. \tag{37}$$

Next, we make use of the identity (29) from which it follows that

$$\begin{aligned}\|\eta_1 - \eta_2\|_X &\leq \sum_j \sum_{k=1}^{j} \|\mathcal{K}_j\| \|\phi_1 \otimes \cdots \otimes \phi_1 \otimes \psi \otimes \phi_2 \otimes \cdots \otimes \phi_2\|_{Y^j} \\ &= \sum_j j\|\mathcal{K}_j\| M^{j-1} \|\psi\|_Y,\end{aligned} \tag{38}$$

where $\psi = \phi_1 - \phi_2$ is in the kth position of the tensor product. Using Lemma 12.2.1, we have

$$\begin{aligned}\|\eta_1 - \eta_2\|_X &\leq C\|\mathcal{K}_1\| \|\psi\|_Y \sum_j j[(\mu + \nu)\|\mathcal{K}_1\| M]^j \\ &\leq \|\mathcal{K}_1\| \|\phi_1 - \phi_2\|_Y \frac{C}{[1 - (\mu + \nu)\|\mathcal{K}_1\| M]^2}.\end{aligned} \tag{39}$$

The above series converges when $(\mu + \nu)\|\mathcal{K}_1\| M < 1$, which holds by hypothesis. □

12.3 Diffuse waves

12.3.1 Forward problem

We consider the propagation of a diffuse wave in an absorbing medium. The energy density u of the wave satisfies the time-independent diffusion equation

$$-\nabla^2 u + k^2(1+\eta(x))u = 0 \quad \text{in } \Omega, \tag{40}$$

$$u + \ell\frac{\partial u}{\partial n} = \delta_{x_1} \quad \text{on } \partial\Omega. \tag{41}$$

Here, x_1 is the position of a point source, δ_{x_1} is the Dirac delta at x_1, the diffuse wave number k is a positive constant, n is the outward unit normal to $\partial\Omega$ and ℓ is a positive constant. The function η, which is the spatially varying part of the absorption coefficient, is assumed to be supported in a closed ball B_a of radius a, with $1+\eta(x)$ nonnegative for all $x \in \Omega$. The energy density u obeys the integral equation

$$u(x) = u_i(x) - k^2 \int_\Omega G(x,y)u(y)\eta(y)dy, \tag{42}$$

where u_i is the energy density of the incident diffuse wave which satisfies

$$-\nabla^2 u_i + k^2 u_i = 0 \quad \text{in } \Omega, \tag{43}$$

$$u_i + \ell\frac{\partial u_i}{\partial n} = \delta_{x_1} \quad \text{on } \partial\Omega. \tag{44}$$

Here G is the Green's function for the operator $-\nabla^2 + k^2$, which obey the boundary condition (41). Beginning with the incident wave u_i, we can iterate (42) to obtain the series

$$u(x) = u_i(x) - k^2 \int_\Omega G(x,y)\eta(y)u_i(y)dy + k^4 \int_{\Omega\times\Omega} G(x,y)\eta(y)G(y,y')\eta(y')u_i(y')dydy' + \cdots. \tag{45}$$

Evidently, we can write (45) in the form of the Born series (1), where $\phi = u_i - u$ and the operator K_j is defined by

$$(K_j f)(x_1,x_2) = (-1)^{j+1}k^{2j} \int_{B_a\times\cdots\times B_a} G(x_1,y_1)G(y_1,y_2)\cdots \times G(y_{j-1},y_j)G(y_j,x_2)f(y_1,\ldots,y_j)dy_1\cdots dy_j, \tag{46}$$

where $x_1, x_2 \in \partial\Omega$. The data $\phi(x_1, x_2)$ is proportional to the intensity measured by a point detector at $x_2 \in \partial\Omega$ due to a point source at $x_1 \in \partial\Omega$. In [23] it was shown that K_j is a bounded operator whose norm obeys the estimate (8) with $X = L^2(B_a)$, $Y = L^2(\partial\Omega \times \partial\Omega)$ and

$$\mu = k^2 \sup_{x \in B_a} \|G(x, \cdot)\|_{L^2(B_a)} \leq k^2 e^{-ka/2} \left(\frac{\sinh(ka)}{4\pi k} \right)^{1/2}, \tag{47}$$

$$\nu = k^2 |B_a|^{1/2} \sup_{x \in B_a} \|G(x, \cdot)\|_{L^2(\partial\Omega)} \leq k^2 |\partial\Omega| |B_a|^{1/2} \frac{e^{-2k \operatorname{dist}(\partial\Omega, B_a)}}{(4\pi \operatorname{dist}(\partial\Omega, B_a))^2}. \tag{48}$$

Analogous results for L^p spaces were also obtained for $2 \leq p \leq \infty$.

12.3.2 Inverse problem

The inverse problem is to reconstruct η from boundary measurements of ϕ. We consider a three-dimensional medium which varies only in the radial direction. Here Ω is taken to be a ball of radius R centered at the origin and we consider an absorption coefficient of the form

$$\eta(x) = \begin{cases} \eta_0 & 0 \leq |x| \leq R_1, \\ 0 & R_1 < |x| \leq R, \end{cases} \tag{49}$$

which corresponds to a spherical inclusion of radius R_1 and contrast η_0. The data ϕ is obtained by a series solution to the diffusion equation (40). The solution to the linearized inverse problem is obtained by computing $\mathcal{K}_1$ by regularized singular value decomposition. The details are presented in [24].

The parameters for the reconstructions are chosen as follows: $R = 3\,\text{cm}$, $R_1 = 1.5\,\text{cm}$, $\ell = 0.3\,\text{cm}$ and $k = 1\,\text{cm}^{-1}$. Figure 12.1 shows a series of experiments where the contrast of the inclusion is varied through a series of values of $\eta_0 = 1.1, 1.3, 1.5, 1.7$. In each of the graphs, we show the reconstruction using up to five terms of the IBS. We also display the projection of η, which is given by $\mathcal{K}_1 K_1 \eta$. In some sense, the projection is the best approximation to η that can be expected. Note that at low contrast, the series appears to converge quite rapidly to a reconstruction that is close to the projection. As the contrast is increased, the higher order terms significantly improve the linear reconstruction.

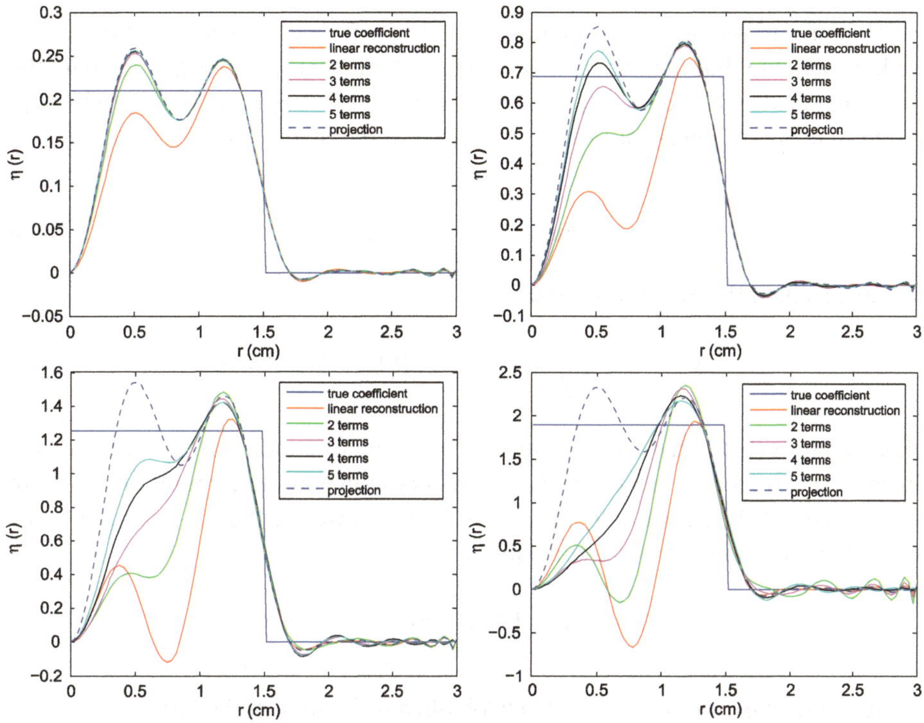

Figure 12.1: Reconstructions of inhomogeneities with $R_1 = 1.5$ cm and $R = 3$ cm. The contrast ranges from: top left $\eta_0 = 1.1$, top right $\eta_0 = 1.3$, bottom left $\eta_0 = 1.5$ and bottom right $\eta_0 = 1.7$.

12.4 Calderon problem

12.4.1 Forward problem

The Calderon problem is the inverse problem of electrical impedance tomography. We consider a scalar field u that obeys the equation

$$\nabla \cdot \sigma(x) \nabla u = 0 \quad \text{in } \Omega, \tag{50}$$

where the coefficient $\sigma(x) > 0$ for all $x \in \Omega$. The field is also taken to satisfy the Robin boundary condition

$$u + z\sigma \frac{\partial u}{\partial n} = g \quad \text{on } \partial\Omega, \tag{51}$$

where σ and z are nonnegative and constant on $\partial\Omega$. In electrical impedance tomography, the field u is identified with the electric potential and the coefficient σ with the conductivity. The coefficient z in (51) is the surface impedance and g is the current density. A typical choice for g is a dipole source of unit strength:

$$g = \delta_{x_1} - \delta_{x_2}, \quad x_1, x_2 \in \partial\Omega. \tag{52}$$

The forward problem is to determine the field u for a given coefficient σ. To proceed, we assume that the conductivity is of the form $\sigma(x) = \sigma_0(1 + \eta(x))$, where the background coefficient $\sigma_0 = \sigma|_{\partial\Omega}$ is constant and $\eta \in L^\infty(B_a)$ is assumed to be supported in a closed ball B_a of radius a centered at the origin. We thus find that (50) becomes

$$-\Delta u = \nabla \cdot \eta(x)\nabla u \quad \text{in } \Omega. \tag{53}$$

The field u obeys the integral equation

$$u(x) = u_0(x) + \int_\Omega G(x,y)\nabla \cdot \eta(y)\nabla u(y)dy, \quad x \in \Omega, \tag{54}$$

where u_0 obeys (53) with $\eta = 0$ and satisfies the boundary condition (86). Here, G is the Green's function for the operator $-\Delta$, which obeys the boundary condition (86) with zero right-hand side. Upon integrating (54) by parts, we see that the solution to the forward problem obeys the integral equation

$$u(x) = u_0(x) - \int_\Omega \nabla_y G(x,y) \cdot \nabla u(y)\eta(y)dy. \tag{55}$$

Beginning with u_0, we can iterate (55) to obtain a series for u of the form

$$u(x) = u_0(x) + u_1(x) + u_2(x) + \cdots, \tag{56}$$

where

$$u_{j+1}(x) = -\int_\Omega \nabla_y G(x,y) \cdot \nabla u_j(y)\eta(y)dy, \quad j = 0, 1, \ldots. \tag{57}$$

We now write (56) in the form of the Born series (1), where $\phi = u_0 - u$ and the operator K_j is defined by

$$\begin{aligned}(K_j f)(x) \\ &= (-1)^j \int_\Omega \nabla_{y_1} G(y_1, x) \cdot \nabla_{y_1} \int_\Omega \nabla_{y_2} G(y_2, y_1) \\ &\cdots \cdot \nabla_{y_{j-1}} \int_\Omega \nabla_{y_j} G(y_j, y_{j-1}) \cdot \nabla_{y_j} u_0(y_j) f(y_1, \ldots, y_j) dy_1 \cdots dy_j.\end{aligned} \tag{58}$$

It was shown in [1] that K_j is a bounded operator whose norm obeys the estimate (8) with $X = L^\infty(B_a)$, $Y = L^2(\partial\Omega)$ and

$$\mu = 1, \tag{59}$$

$$\nu = \sup_{x\in\partial\Omega} \|\nabla G(x,\cdot)\|_{L^2(B_a)} \|\nabla u_0\|_{L^2(B_a)}. \tag{60}$$

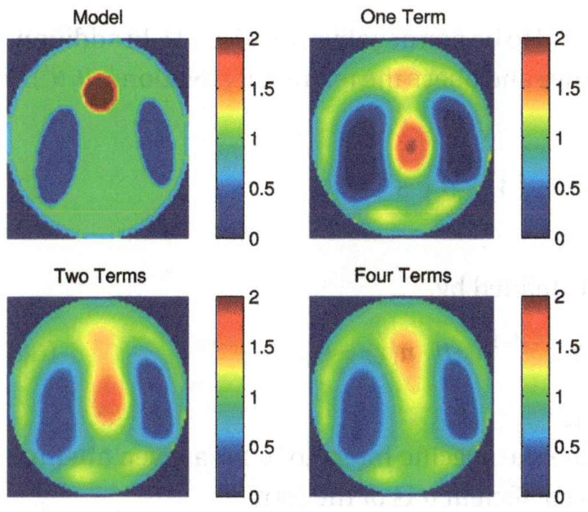

Figure 12.2: Reconstruction of the conductivity with $\sigma_{\text{lungs}} = 0.2$ and $\sigma_{\text{heart}} = 2$.

12.4.2 Inverse problem

The inverse problem is to reconstruct η from measurements of the data ϕ. The function ϕ depends implicitly upon the position of the source. For example, in the case of the dipole source (52), if we fix the point $x_1 \in \partial\Omega$ and vary $x_2 \in \partial\Omega$, then ϕ will depend upon both x_2 and the point x at which we measure the field on $\partial\Omega$. Accordingly, will assume that $\phi \in L^\infty(\partial\Omega \times \partial\Omega)$. We consider a two-dimensional chest phantom, where Ω is taken to be a disk of radius R centered at the origin.

The data ϕ is obtained by solving (21) by the finite-element method. The solution to the linearized inverse problem is obtained by computing $\mathcal{K}_1$ by regularized singular value decomposition. The details are presented in [1]. The parameters for the reconstructions are chosen as follows: $R = 40$, $\sigma_0 = 1$ and $z\sigma_0 = 1$. Figure 12.2 shows reconstructions using up to four terms of the IBS.

12.5 Radiative transport

12.5.1 Forward problem

The physical quantity of interest in radiative transport theory is the specific intensity $u(x,\theta)$ at the point $x \in \Omega$ in the direction $\theta \in S^{d-1}$. The specific intensity obeys the radiative transport equation (RTE)

$$\theta \cdot \nabla u + \sigma(x)u = \int_{S^{d-1}} k(\theta,\theta')u(x,\theta')d\theta' \quad \text{in } \Omega \times S^{d-1}, \tag{61}$$

$$u = g \quad \text{on } \Gamma_-. \tag{62}$$

The attenuation coefficient $\sigma(x)$ is assumed to be nonnegative for all $x \in \Omega$. In addition, the scattering kernel $k(\theta, \theta')$ is nonnegative and obeys the reciprocity relation $k(\theta, \theta') = k(-\theta', \theta)$ and is normalized so that

$$\int_{S^{d-1}} k(\theta, \theta') d\theta' = 1, \quad \theta \in S^{d-1}. \tag{63}$$

We also introduce the sets $\Gamma_\pm$ which are defined by

$$\Gamma_\pm = \{(x, \theta) \in \partial\Omega \times S^{d-1} : \pm\theta \cdot n(x) > 0\}, \tag{64}$$

with n being the outer unit normal to $\partial\Omega$.

The forward problem is to determine the specific intensity u for a given attenuation σ. We assume that the attenuation coefficient σ is of the form

$$\sigma(x) = \sigma_0(1 + \eta(x)), \tag{65}$$

where the background attenuation $\sigma_0 = \sigma|_{\partial\Omega}$ is constant and $\eta(x) > -1$ for all $x \in \Omega$. The function η is the spatially varying part of the attenuation coefficient; it is assumed to be supported in a closed ball B_a of radius a, centered at the origin. The specific intensity u obeys the integral equation

$$u(x, \theta) = u_0(x, \theta) - \sigma_0 \int_{\Omega \times S^{d-1}} G(x, \theta; x', \theta')\eta(x')u(x', \theta')dx'd\theta'. \tag{66}$$

Here, u_0 obeys (61) with $\eta = 0$ and G is the Green's function for the background medium, which satisfies the equation

$$\begin{aligned} &\theta \cdot \nabla_x G(x, \theta; x', \theta') + \sigma_0 G(x, \theta; x', \theta') \\ &= \int_{S^{d-1}} k(\theta, \theta'')G(x, \theta''; x', \theta')d\theta'' + \delta(x - x')\delta(\theta - \theta'), \end{aligned}$$

together with homogeneous boundary conditions on Γ_-. It is easily seen that u_0 is given by the formula

$$u_0(x, \theta) = \int_{\partial\Omega} \int_{\theta' \cdot n < 0} G(x, \theta; x', \theta')|\theta' \cdot n(x')|g(x', \theta')dx'd\theta'. \tag{67}$$

The integral equation (66) has a unique solution. Upon iteration, beginning with $u = u_0$, we obtain an infinite series for u of the form

$$u(x, \theta) = u_0(x, \theta) + u_1(x, \theta) + u_2(x, \theta) + \cdots, \tag{68}$$

where

$$u_{j+1}(x, \theta) = -\sigma_0 \int_{\Omega \times S^{d-1}} G(x, \theta; x', \theta')u_j(x', \theta')\eta(x')dx'd\theta', \quad j = 0, 1, \ldots. \tag{69}$$

The series (68) can be written in the form of the Born series (1), where $\phi = u_0 - u$ and the operator K_j is defined by

$$(K_j f)(x,\theta) = (-1)^{j+1}\sigma_0^j \int\limits_{\Gamma_a\times\cdots\times\Gamma_a} G(x,\theta;x_1',\theta_1')G(x_1',\theta_1';x_2',\theta_2')G(x_2',\theta_2';x_3',\theta_3')\cdots$$
$$\times\, G(x_{j-1}',\theta_{j-1}';x_j',\theta_j')u_0(x_j',\theta_j')f(x_1',\dots,x_j')dx_1'd\theta_1'\cdots dx_j'd\theta_j', \tag{70}$$

where $f \in L^\infty(B_a \times \cdots \times B_a)$. In [19] it was shown that K_j is a bounded operator whose norm obeys the estimate (8) with $X = L^\infty(B_a)$, $Y = L^1(\Gamma_+)$ and

$$\mu = \sigma_0 \sup_{(x',\theta')\in\Gamma_a} \int\limits_{\Gamma_a} G(x,\theta;x',\theta')dxd\theta, \tag{71}$$

$$\nu = \sigma_0 \int\limits_{\Gamma_a} u_0 dxd\theta \sup_{(x',\theta')\in\Gamma_a} \int\limits_{\Gamma_+} G(x,\theta;x',\theta')dxd\theta, \tag{72}$$

where $\Gamma_a = B_a \times S^{d-1}$.

12.5.2 Inverse problem

The inverse problem is to reconstruct the coefficient η everywhere within Ω from measurements of the scattering data Φ on Γ_+. We consider a homogeneous isotropically scattering slab-shaped medium with contrast Δ embedded in a homogeneous infinite medium with attenuation σ_0. We suppose the embedded medium occupies the strip $-a \le z \le a$ inside of a slab of width $2L$ with $-L \le z \le L$. The data ϕ is obtained by solving the RTE by the singular eigenfunction method. The solution to the linearized inverse problem is obtained by computing $\mathcal{K}_1$ by regularized singular value decomposition. The details are presented in [19]. The parameters for the reconstructions are chosen as follows: $L = 10l_s$, $a = 5l_s$, $\sigma_0 = 1$ and $\Delta = 2.3$. Here lengths are measured in units of the scattering length l_s. Figure 12.3 shows reconstructions using up to five terms of the IBS.

12.6 Electromagnetic waves

12.6.1 Forward problem

We consider the scattering of time-harmonic electromagnetic waves in a nonmagnetic medium. The electric field E obeys the wave equation

$$\nabla \times \nabla \times E - k^2\epsilon(x)E = 0, \tag{73}$$

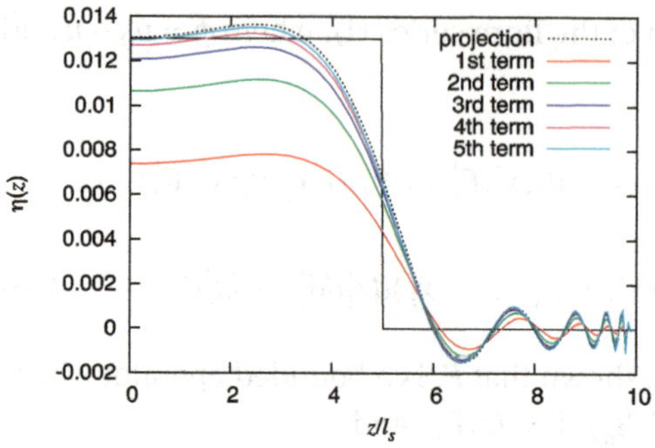

Figure 12.3: Reconstruction of the attenuation coefficient.

where k is the free-space wavenumber and ϵ is the dielectric permittivity. We also impose the radiation condition

$$\lim_{|x|\to\infty} |x| \left[(\nabla \times E^s) \times \frac{x}{|x|} - ikE^s \right] = 0. \tag{74}$$

Here, we have decomposed the total field E into its incident and scattered components E^i and E^s according to

$$E = E^i + E^s, \tag{75}$$

where E^i obeys (73) with $\epsilon = 1$. Evidently, E^s obeys the equation

$$\nabla \times \nabla \times E^s - k^2 E^s = k^2 \eta(x) E, \tag{76}$$

where the susceptibility $\eta = \epsilon - 1$. Eq. (76) is equivalent to the integral equation

$$E^s(x) = (k^2 + \nabla\nabla\cdot) \int_{B_a} G(x,y)\eta(y)E(y)dy, \tag{77}$$

where we have assumed that η is supported in B_a, the ball of radius a centered at the origin. In addition, G is the fundamental solution of the Helmholtz equation, which in three dimensions is given by

$$G(x,y) = \frac{e^{ik|x-y|)}}{4\pi|x-y|}. \tag{78}$$

The field E thus obeys the integral equation

$$E(x) = E^i(x) + (k^2 + \nabla\nabla\cdot) \int_{B_a} G(x,y)\eta(y)E(y)dy. \tag{79}$$

If we iterate (79) beginning with $E = E^i$, we obtain an infinite series for E of the form

$$E = E^i + E^{(1)} + E^{(2)} + \cdots, \tag{80}$$

where

$$E^{(j+1)}(x) = \int_{B_a} G(x,y)\eta(y)E^{(j)}(y)dy, \quad (81)$$

where $E^{(0)} := E^i$. We now write (80) in the form of the Born series (1), where $\phi = E^i - E^*$ and the operator K_j is defined by

$$(K_jf)(x) = (k^2 + \nabla\nabla\cdot) \int_{B_a\times\cdots\times B_a} G(x,y_1)(k^2 + \nabla_{y_1}\nabla_{y_1}\cdot)G(y_1,y_2)\cdots(k^2 + \nabla_{y_{j-1}}\nabla\cdot_{y_{j-1}})$$
$$\times G(y_{j-1},y_j)E_i(y_j)f(y_1,\ldots,y_j)dy_1\cdots dy_j. \quad (82)$$

In the above, the point x will be taken to belong to a compact set which lies outside the support of η. It was shown in [15] that K_j is a bounded operator whose norm obeys the estimate (8) with $X = L^\infty(B_a)$, $Y = [L^2(K)]^3$ and

$$\mu = \frac{17}{2}(ka)^2 + 2\sqrt{74}ka + 105, \quad (83)$$

$$\nu = |B_a|^{1/2}\|E^i\|_{[L^2(B_a)]^3} \sup_{x\in B_a}\|(k^2 + \nabla_{x_1}\nabla_{x_1}\cdot)G(\cdot,x)I\|_{[L^2(K)]^3}. \quad (84)$$

12.6.2 Inverse problem

The inverse problem is to reconstruct η from measurements of ϕ. We consider a scatterer that consists of a sphere of radius $R = 2\lambda = 4\pi/k$ with index of refraction $n = 1.1$ related to the susceptibility by $\eta_0 = (n^2 - 1)/4\pi$. The data ϕ is computed from the Mie solution to the wave equation [3]. The solution to the linearized inverse problem is obtained by computing $\mathcal{K}_1$ by regularized singular value decomposition. The details are presented in [25]. The incident field is polarized in the $\hat{\mathbf{x}}$ direction and the incident plane wave is in the $\hat{\mathbf{y}}$ direction. The plane of detection was located at a distance $\lambda/3$ from the top of the sphere. Figure 12.4 presents the reconstructions obtained using four terms of the IBS. The central column shows the results of reconstructions in the equatorial plane of the sphere. The left and right columns are the results of reconstructions in the planes $0.7R$ above and below the equatorial plane. The first row illustrates the results of linear reconstructions while the second, third and fourth rows show the second, third and fourth order nonlinear reconstructions, respectively. Figure 12.5 shows the one-dimensional profiles of the reconstructed susceptibility along the central line in the equatorial plane.

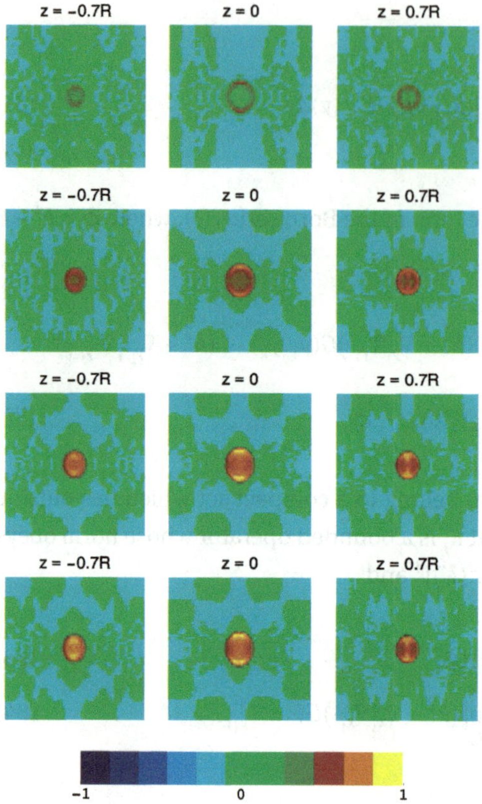

Figure 12.4: Tomographic images of the reconstructed susceptibility.

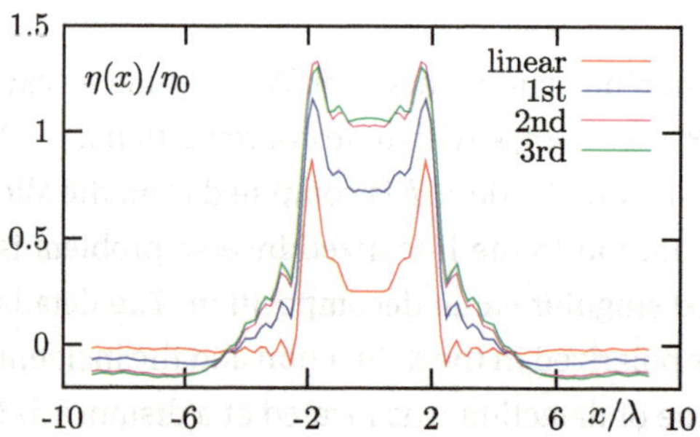

Figure 12.5: One-dimensional profiles of the reconstructed susceptibility.

12.7 Diffusion on graphs

12.7.1 Forward problem

In this section, we consider the discrete analog of the inverse problem of optical tomography with diffuse light. We focus on the problem of recovering vertex properties of a graph from boundary measurements. The results presented here are adapted from [7].

Let $G = (V, E)$ be a finite locally connected loop-free graph with vertex boundary δV. We consider the time-independent diffusion equation

$$(Lu)(x) + \alpha_0[1 + \eta(x)]u(x) = f(x), \quad x \in V, \tag{85}$$

$$t\,u(x) + \partial u(x) = g(x), \quad x \in \delta V. \tag{86}$$

Here, we assume that the absorption of the medium is nearly constant with background absorption α_0 and inhomogeneities represented by the vertex potential η. In place of the Laplace operator, we introduce the combinatorial Laplacian L defined by

$$(Lu)(x) = \sum_{y\sim x} [u(x) - u(y)], \tag{87}$$

where $y \sim x$ if the vertices x and y are adjacent. We make use of the graph analog of Robin boundary conditions, where the normal derivative is defined by

$$\partial u(x) = \sum_{\substack{y\in V \\ y\sim x}} [u(x) - u(y)], \tag{88}$$

and t is an arbitrary nonnegative parameter, which interpolates between Dirichlet and Neumann boundary conditions. If the potential η is nonnegative, then there exists a unique solution to the diffusion equation (85) satisfying the boundary condition (86).

The forward problem is to determine u, given η. The corresponding inverse problem, which we refer to as graph optical tomography, is to recover the potential η from measurements of u on the boundary of the graph. More precisely, let $G = (V, E)$ be a connected subgraph of a finite graph $\Gamma = (\mathcal{V}, \mathcal{E})$ and let δV denote those vertices in $\mathcal{V}$ adjacent to a vertex in V. In addition, let S, R denote fixed subsets of δV. We will refer to elements of S and R as sources and receivers, respectively. For a fixed potential η, source $s \in S$ and receiver $r \in R$, let $u(r, s; \eta)$ be the solution to (85) with vertex potential η and boundary condition (86), where

$$g(x) = \begin{cases} 1 & x = s, \\ 0 & x \neq s. \end{cases} \tag{89}$$

We define the Robin-to-Dirichlet map Λ_η by

$$\Lambda_\eta(s, r) = u(r, s; \eta). \tag{90}$$

The inverse problem is to recover η from the Robin-to-Dirichlet map Λ_η.

The background Green's function for (85) is the matrix G_0 whose i, jth entry is the solution to (85), with $\eta \equiv 0$, at the ith vertex for a unit source at the jth vertex. Under suitable restrictions this matrix can be used to construct the Robin-to-Dirichlet map Λ_η giving the solution of (85) on $R \subset \delta V$ to unit sources located in $S \subset \delta V$. To write a compact expression for Λ_η in terms of G_0, let D_η denote the matrix with entries given by

$$(D_\eta)_{i,j} = \begin{cases} \eta_i & \text{if } i = j, \\ 0 & \text{otherwise.} \end{cases}$$

Additionally, for any two sets $U, W \subset V \cup \delta V$, let $G_0^{U;W}$ denote the submatrix of G_0 formed by taking the rows indexed by U and the columns indexed by W. For η sufficiently small, we may write the Robin-to-Dirichlet map as the series

$$\Lambda_\eta(s,r) = G_0(r,s) - \sum_{j=1}^{\infty} K_j(\eta,\ldots,\eta)\,(r,s), \quad r \in R,\ s \in S, \tag{91}$$

where $K_j : \ell^p(V^n) \to \ell^p(R \times S)$ is defined by

$$K_j(\eta_1,\ldots,\eta_j)\,(r,s) = (-\alpha_0)^j G_0^{r;V} D_{\eta_1} G_0^{V;V} D_{\eta_2} \cdots G_0^{V;V} D_{\eta_j} G_0^{V;s}. \tag{92}$$

Evidently, (91) has the form of the Born series (1).

In order to establish the convergence and stability of (91), we seek appropriate bounds on the operators $K_j : \ell^p(V \times \cdots \times V) \to \ell^p(\delta V \times \delta V)$. Note that if $|V|$ and $|\delta V|$ are finite then all norms are equivalent. However, since we are interested in the rate of convergence of the IBS it will prove useful to establish bounds for arbitrary ℓ_p norms.

Proposition 12.7.1. *Let $p, q \in [1,\infty]$ such that $1/p + 1/q = 1$ and define the constants ν_p and μ_p by*

$$\nu_p = \alpha_0 \|G_0^{R;V}\|_{\ell^q(V)\times\ell^p(R)} \|G_0^{V;S}\|_{\ell^q(V)\times\ell^p(S)}, \quad \mu_p = \alpha_0, C_{G_0^{V;V}}, \tag{93}$$

where

$$C_{G_0^{V;V},q} = \max_{v\in V} \|G_0^{V;v}\|_{\ell^q(V)}. \tag{94}$$

The Born series (91) *converges if*

$$\mu_p \|\eta\|_p < 1. \tag{95}$$

Moreover, the N-term truncation error has the following bound:

$$\left\| \Lambda_\eta - \left(G_0 - \sum_{j=N}^{\infty} K_j(\eta,\ldots,\eta) \right) \right\|_{\ell^p(R\times S)} \le \nu_p \|\eta\|_p^{N+1} \mu_p^N \frac{1}{1 - \mu_p \|\eta\|_p}. \tag{96}$$

12.7.2 Inverse problem

Let $\phi \in \ell^2(R \times S)$ denote the scattering data

$$\phi(r,s) = G_0(r,s) - \Lambda_\eta(r,s), \tag{97}$$

which corresponds to the difference between the measurements in the background medium and those in the medium with the potential present. Note that if the Born series converges, we have

$$\phi(r,s) = \sum_{j=1}^{\infty} K_j(\eta,\dots,\eta). \tag{98}$$

The IBS is now of the form

$$\eta = \mathcal{K}_1(\phi) + \mathcal{K}_2(\phi,\phi) + \mathcal{K}_3(\phi,\phi,\phi) + \cdots, \tag{99}$$

where $\mathcal{K}_j$ is given by (6). Note that ϕ can be thought of as an operator from $\ell^2(R)$ to $\ell^2(S)$, in (99) we treat it as a vector of length $|R| \cdot |S|$. Similarly, though it is often convenient to think of η as a (diagonal) matrix, in (99) it should be thought of as a vector of length $|V|$. With a slight abuse of notation, we also use K_1 to denote the $|R||S| \times |V|$ matrix mapping η (viewed as a vector) to $K_1\eta$, once again thought of as a vector.

The following result provides sufficient conditions for the convergence of the IBS for graphs where $|V| = |R \times S|$, corresponding to the case of a formally determined inverse problem.

Theorem 12.7.1. *Let $|V| = |R \times S|$ and $p \in [1,\infty]$. Suppose that the operator K_1 is invertible. Then the inverse Born series converges to the true potential η if $\|\phi\|_p < r_p$. Here the radius of convergence r_p is defined by*

$$r_p = \frac{C_p}{\mu_p}\left[1 - 2\frac{\nu_p}{C_p}\left(\sqrt{1 + \frac{C_p}{\nu_p}} - 1\right)\right], \tag{100}$$

where

$$C_p = \min_{\|\eta\|_p=1} \|K_1(\eta)\|_p \tag{101}$$

and ν_p, μ_p are defined in (93).

We now consider the stability of the limit of the inverse scattering series under perturbations in the scattering data. The following stability estimate follows immediately from Theorem 12.7.1.

Proposition 12.7.2. *Let E be a compact subset of $\Omega_p = \{\phi \in \mathbb{C}^n \mid \|\phi\|_p < r_p\}$, where r_p is defined in* (100) *and $p \in [1,\infty]$. Let ϕ_1 and ϕ_2 be scattering data belonging to E and*

ψ_1 and ψ_2 denote the corresponding limits of the inverse Born series. Then the following stability estimate holds:

$$\|\psi_1 - \psi_2\|_p \le M\|\phi_1 - \phi_2\|_p,$$

where $M = M(E, p)$ is a constant which is otherwise independent of ϕ_1 and ϕ_2.

Theorem 12.7.1 guarantees convergence of the IBS, but does not provide an estimate of the approximation error. Such an estimate is provided in the next theorem.

Theorem 12.7.2. *Suppose that the hypotheses of Theorem 12.7.1 hold and $\|\phi\|_p < \tau r_p$, where $\tau < 1$. If η is the true potential corresponding to the scattering data ϕ, then*

$$\left\| \eta - \sum_{m=1}^{N} \mathcal{K}_m(\phi, \dots, \phi) \right\| < M\left(\frac{1}{1-\tau}\right)^n \left(\frac{\|\phi\|_p}{\tau r_p}\right)^N \frac{1}{1 - \frac{\|\phi\|_p}{\tau r_p}}.$$

Remark 12.7.1. In the analysis of the IBS in the continuous setting, it was found that certain smallness conditions on both $\|\mathcal{K}_1\|_p$ and $\|\mathcal{K}_1\phi\|_p$ are sufficient to guarantee convergence. Note that such a condition on $\|\mathcal{K}_1\|_p$ is not present in Theorem 12.7.1, Proposition 12.7.2 or Theorem 12.7.2.

Bibliography

[1] S. Arridge, S. Moskow, and J. C. Schotland. *Inverse Probl.*, 28:035003, 2012.

[2] P. Bardsley and F. Guevara Vasquez. *Inverse Probl.*, 30:045014, 2014.

[3] C. F. Bohren and D. R. Huffman. *Absorption and Scattering of Light by Small Particles*. John Wiley & Sons, New York, 1983.

[4] F. Cakoni and D. Colton. *Qualitative Methods in Inverse Scattering Theory: An Introduction*. Springer, New York, 2006.

[5] F. Cakoni and D. Colton. *A Qualitative Approach to Inverse Scattering Theory*. Springer, New York, 2014.

[6] K. Chadan and P. Sabatier. *Inverse Problems in Quantum Scattering Theory*. Springer, New York, 1989.

[7] F. Chung, A. Gilbert, J. Hoskins, and J. C. Schotland. *Inverse Probl.*, 33:055016, 2017.

[8] D. Colton and R. Kress. *Integral Equation Methods in Scattering Theory*. John Wiley and Sons, New York, 1983.

[9] D. Colton and R. Kress. *Inverse Acoustic and Electromagnetic Scattering Theory*. Springer, New York, 2013.

[10] A. Devaney and E. Wolf. *Phys. Lett. A*, 89:269, 1982.

[11] H. W. Engl, M. Hanke, and A. Neubauer. *Regularization of Inverse Problems*. Kluwer Academic Publishers Group, 1996.

[12] V. Isakov. *Inverse Problems for Partial Differential Equations*. Springer, New York, 2006.

[13] R. Jost and W. Kohn. *Phys. Rev.*, 87:977, 1952.

[14] K. Kilgore, S. Moskow, and J. C. Schotland. *J. Comput. Math.*, 30:601, 2012.

[15] K. Kilgore, S. Moskow, and J. C. Schotland. *Appl. Anal.*, 96:1737, 2017.

[16] A. Kirch. *An Introduction to the Mathematical Theory of Inverse Problems*. Springer-Verlag, New York, 2011.

[17] A. Kirsch and N. Grinberg. *The Factorization Method for Inverse Problems*. Oxford University Press, 2008.

[18] A. Louis. *Inverse Probl.*, 12:175, 1996.

[19] M. Machida and J. C. Schotland. *Inverse Probl.*, 31:095009, 2015.

[20] V. Markel, J. A. O'Sullivan, and J. C. Schotland. *J. Opt. Soc. Am. A*, 20:903, 2003.

[21] V. A. Morozov. *Regularization Methods for Ill-Posed Problems*. CRC Press, 1993.

[22] H. Moses. *Phys. Rev.*, 102:550, 1956.

[23] S. Moskow and J. C. Schotland. *Inverse Probl.*, 24:065005, 2008.

[24] S. Moskow and J. C. Schotland. *Inverse Probl.*, 25:095007, 2009.

[25] G. Panasyuk, V. A. Markel, P. S. Carney, and J. C. Schotland. *Appl. Phys. Lett.*, 89:221116, 2006.

[26] R. Prosser. *J. Math. Phys.*, 10:1819, 1969.

[27] R. A. Ryan. *Introduction to Tensor Products of Banach Spaces*. Springer, London, 2002.

[28] T. Schuster, B. Kaltenbacher, B. Hofmann, and K. Kazimierski. *Regularization Methods in Banach Spaces*. De Gruyter, Berlin, 2012.

[29] H. Shehadeh and J. Schotland. *J. Comput. Math.*, 35:586, 2017.

[30] G. Tsihrintzis and A. Devaney. *IEEE Trans. Image Process.*, 9:1560, 2000.

[31] A. B. Weglein, F. V. Arajo, P. M. Carvalho, R. H. Stolt, K. H. Matson, R. T. Coates, D. Corrigan, D. J. Foster, S. A. Shaw, and H. Zhang. *Inverse Probl.*, 19:R27–R83, 2003.

Andreas Alpers and Peter Gritzmann

13 On the reconstruction of static and dynamic discrete structures

Abstract: We study inverse problems of reconstructing static and dynamic discrete structures from tomographic data (with a special focus on the "classical" task of reconstructing finite point sets in $\mathbb{R}^d$). The main emphasis is on recent mathematical developments and new applications, which emerge in scientific areas such as physics and materials science, but also in inner mathematical fields such as number theory, optimization, and imaging. Along with a concise introduction to the field of discrete tomography, we give pointers to related aspects of computerized tomography in order to contrast the worlds of continuous and discrete inverse problems.

Keywords: Discrete inverse problems, discrete tomography, ill-posedness, computational complexity, superresolution imaging, grain mapping, particle tracking

MSC 2010: 49N45, 68Q25, 68R05, 94A08, 90C10, 11Y50, 74A40

13.1 Introduction

We begin with an informal definition of the general field of discrete tomography. As a comprehensive treatise would, however, go far beyond the scope of the present paper, and as we want to limit the overlap with surveys in the literature as much as possible, we will later mainly focus on the most fundamental case of reconstructing finite point sets in $\mathbb{R}^d$ from some of their discrete x-rays. Our special emphasis will be on (a subjective selection of) recent developments and applications. We conclude the introduction with a few comments on the structure of the paper and its bibliography.

13.1.1 What is discrete tomography?

Discrete tomography deals with the problem of retrieving knowledge about an otherwise unknown discrete object from information about its interactions with certain

Acknowledgement: This work was supported in part by the *Deutsche Forschungsgemeinschaft* Grant GR 993/10-2 and the *European COST Network* MP1207. The authors are grateful to Fabian Klemm for his help with producing Figure 13.12 and both, Fabian Klemm and Viviana Ghiglione, for helpful discussions.

***Corresponding author: Peter Gritzmann,** Zentrum Mathematik, Technische Universität München, D-85747 Garching bei München, Germany, e-mail: gritzmann@tum.de
Andreas Alpers, Zentrum Mathematik, Technische Universität München, D-85747 Garching bei München, Germany, e-mail: alpers@ma.tum.de

https://doi.org/10.1515/9783110560855-013

query sets. Of course, this is not a formal definition and, as a matter of fact, the occurring terms leave room for different interpretations. For instance, the *discrete object* can be any set in some $\mathbb{R}^d$ that allows a finite encoding, e. g., a finite point set, a polytope, or even a semialgebraic set. Also functions with finite support are included. *Knowledge* may mean full reconstruction, the detection of certain properties and measures of the object or just a "yes/no" decision whether the object equals (or is close to) a given blueprint. The *query sets* may be windows, affine spaces or certain families of more general manifolds, and *interaction* may simply mean intersection but could also refer to a very different probing procedure.

While results which, in retrospective, belong to this area go back a long time, the name *discrete tomography* and the establishment of the so-named field is of more recent origin. In the past decades, the focus has been on the issues of *uniqueness*, *computational complexity*, and *algorithms*, first under the theoretical assumption that exact x-ray data were available. Later, *stability* and *instability* questions were pursued, and the effect of noise was studied.

Discrete tomography has important applications in physics, materials science, and many other fields. It has, however, also been applied in various other contexts, including scheduling [99], data security [110], image processing [125], data compression [124], combinatorics [52, 79, 90, 93], and graph theory [87, 88]. Closely connected are also several recreational games such as nonograms [73], path puzzles [76], sudokus [85], and color pic-a-pix [68].

13.1.2 Scope of the present paper

In the following, we will concentrate mainly on the "classical" task of reconstructing a finite point set in $\mathbb{R}^d$ from the cardinalities of its intersections with the lines parallel to a (small) finite number of directions. Already in this restricted form, discrete tomography displays important features also known from the continuous world. In particular, discrete tomography is ill-posed in the sense of Hadamard [28]: the data may be inconsistent, the solution need not be unique (Theorem 1), and small changes in the data may result in dramatic changes of the solutions (Theorem 12).

As will become clear, discrete tomography is not simply the discretization of continuous tomography. It derives its special characteristics from the facts that, on the one hand, there are only data in very few directions available, but on the other hand, the classes of objects that have to be reconstructed are rather restricted. Therefore, discrete tomography is based on methods from combinatorics, discrete optimization, algebra, number theory, and other more discrete subfields of mathematics and computer science.

There exist already books and articles, which give detailed accounts of various aspects of discrete tomography and its applications. We single out [20, 24, 25, 27, 35,

36, 51]. The present article differs from these surveys in various ways: the mathematical focus will be on recent developments (which have not been covered in previous surveys). Further, new applications will play a significant role, i. e., applications to other scientific areas like physics and materials science, but also to inner mathematical fields such as number theory, optimization, and imaging. We will, moreover, include pointers to certain related aspects of computerized tomography in order to contrast the worlds of continuous and discrete inverse problems. As general sources for the continuous case and inverse problems, see [32, 41, 43, 44] and [6, 30, 38, 42, 57], respectively.

As a service to the reader and with a view toward a more complete picture we will restate some aspects which are basic for the present article but have been covered before. In order to limit the overlap to other surveys we will, however, neither elude on the tomographic reconstruction of quasicrystals (see [24]) or polyominoes (see [17]), nor on the polyatomic case (see [21, 92]) or point x-rays (see [89]). Also, we will not study general k-dimensional x-rays (see [24, 101]) but concentrate on the case $k = 1$. This means that our exposition will be based on the *x-ray transform* rather than on the *Radon transform* (which is the case $k = d - 1$).

13.1.3 Organization of the present paper and its bibliography

After introducing the basic notation in Section 13.2 we will briefly survey well-known structural results related to the ill-posedness of the problems (Section 13.3) and their computational complexity (Section 13.4). Turning to recent results and applications, Section 13.5 will illustrate some quite unexpected complexity jumps in (a related basic model of) superresolution. Particular emphasis will then be placed on new developments in *dynamic discrete tomography* involving the movement of points over time which are only accessible by very few of their x-ray images. As Section 13.6 will show, aspects of discrete tomography and particle tracking interact deeply. Another more recent issue, which comes up in materials science, is that of multiscale tomographic imaging. Section 13.7 will indicate how different aspects of the reconstruction of polycrystalline materials based on tomographic data lead to very different techniques involving methods from the geometry of numbers, combinatorial optimization and computational geometry. Section 13.8 deals with some inner mathematical connections between discrete tomography and the Prouhet–Tarry–Escott problem from number theory, and Section 13.9 concludes with some final remarks.

Let us point out that (with the exception of some new interpretations and simple observations) the results stated here have all been published in original research papers (which are, of course, cited appropriately). Even more, since we want to use the standard notation and, in particular, a standard framework for expressing the results, some overlap with the above mentioned surveys is unavoidable.

Finally, let us close the Introduction with a comment on the bibliography. Due to the character of the present paper, we included references of different kinds. Of course, we listed all original work quoted in the main body of the paper. However, we felt that for the generally interested reader it would be worthwhile to add sources for general reading. On the other hand, in terms of the included applications we focused mainly on outlining those aspects to which discrete tomography can potentially contribute. While this is in line with the scope of the present paper, readers interested in these fields of applications may appreciate pointers to sources for additional reading. Hence we organized the bibliography in six different parts, namely general reading, papers in tomography, and further reading on particle tracking, tomographic grain mapping, macroscopic grain mapping, and the Prouhet–Tarry–Escott problem, respectively.

13.2 Basic notation

As pointed out before, we will mainly focus on the "classical" inverse problem of reconstructing a finite point set F in $\mathbb{R}^d$ or $\mathbb{Z}^d$ from the cardinalities of its intersections with the lines parallel to a finite number of directions. There are, however, certain aspects which involve weights on the points of F. Hence we will introduce the basic notions for appropriate generalizations of characteristic functions of finite point sets, partly following [24].

As usual, let $\mathbb{N}_0$, $\mathbb{N}$, $\mathbb{Z}$, $\mathbb{Q}$, and $\mathbb{R}$ denote the sets of nonnegative integers, natural numbers, integers, rationals, and reals, respectively. Further, for $n \in \mathbb{N}$, we will often use the notation $[n] = \{1, \ldots, n\}$ and $[n]_0 = [n] \cup \{0\}$.

In the following, let $d, m \in \mathbb{N}$; d denotes the dimension of the space $\mathbb{R}^d$, and m is the number of directions in which images are taken. To exclude trivial cases, we will usually assume that $d, m \geq 2$.

In order to describe the objects of interest, we fix nonempty sets $D \subset \mathbb{R}^d$ and $C \subset \mathbb{R}$ and consider functions $\psi\colon D \to C$ with finite support $\operatorname{supp}(\psi) = \{x \in D : \psi(x) \neq 0\}$. In our context, the most relevant pairs (D, C) of a domain and a codomain are those where $D = \mathbb{R}^d$ or $D = \mathbb{Z}^d$ and $C = \{0, 1\}$. Other standard codomains are $C = \mathbb{N}_0$, $C = \mathbb{Z}$, and also their relaxations $[0, 1]$, $[0, \infty[$, and $\mathbb{R}$.

For any pair (D, C), let $\mathcal{F}(D, C)$ denote the class of all functions $\psi\colon D \to C$ with finite support. Of course, for $C = \{0, 1\}$, such a function ψ can be viewed as the indicator or characteristic function of a finite set F and can therefore be identified with $\operatorname{supp}(\psi)$. We will write $\mathcal{F}(D)$ for $\mathcal{F}(D, \{0, 1\})$ and identify it with the set of all finite subsets of D. In particular, the case $\mathcal{F}(\mathbb{Z}^d)$ encodes the classical *finite lattice sets*. Since this case is particularly important, we will often abbreviate $\mathcal{F}(\mathbb{Z}^d)$ by $\mathcal{F}^d$.

Further, let $\mathcal{S}^d$ denote the set of all 1-dimensional subspaces of $\mathbb{R}^d$, while $\mathcal{L}^d$ is the set of 1-dimensional *lattice lines*, i. e., lines through the origin spanned by an integer vector. For $S \in \mathcal{S}^d$, we use the notation $\mathcal{A}(S)$ for the set of all affine lines in $\mathbb{R}^d$ that are parallel to S. The situation of $(\mathcal{F}^d, \mathcal{L}^d)$ will be referred to as the *lattice case*.

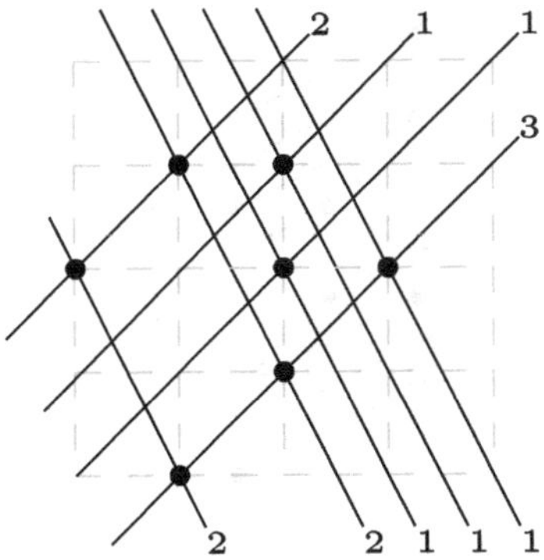

Figure 13.1: A finite lattice set (black dots) and its 1-dimensional x-rays in the two directions $(1,1)^T$ and $(1,-2)^T$.

Now, let $\psi \in \mathcal{F}(D,C)$ and $S \in \mathcal{S}^d$. The *discrete x-ray* of ψ parallel to S (or, in a slight abuse of language, in the direction S) is the function $X_S\psi\colon \mathcal{A}(S) \to \mathbb{R}$ defined by

$$T \longmapsto (X_S\psi)(T) = \sum_{x\in T} \psi(x).$$

Since ψ has finite support all sums are finite. In the case of $C = \{0,1\}$ where ψ can be identified with $F = \mathrm{supp}(\psi)$, we will often write X_SF. See Figure 13.1 for an illustration.

The mapping $\mathcal{X}_\psi$ on $\mathcal{S}^d$ defined by $S \mapsto X_S\psi$ is called the *discrete x-ray transform* of ψ. (In typical applications only very few values of $\mathcal{X}_\psi$ are available.)

Note that it is straightforward to extend this notation to k-dimensional x-rays. Accordingly, for $k = d-1$, we obtain the *discrete Radon transform* of ψ whose measurements come from *hyperplane x-rays*. We will, however, focus on the x-rays defined above, which provide 1-dimensional measurements. The basic task of discrete tomography is then to reconstruct an otherwise unknown function $\psi \in \mathcal{F}(D,C)$ from its x-rays with respect to a finite number m of given lines $S \in \mathcal{S}^d$.

The x-ray information is encoded by means of *data functions*. In fact, the lines $T \in \mathcal{A}(S)$ can be parametrized by vectors $t \in S^\perp$ such that $T = t + S$. Hence, one may regard $X_S\psi$ as a function on $S^\perp$. For algorithmic purposes, it is often preferable to use other representations and encode $X_S\psi$ as a finite set of pairs (x,β) with $x \in D$, $\beta \in C$, and $X_S\psi(x+S) = \beta$.

13.3 Ill-posedness

We begin with some results that deal with the basic issues of uniqueness and stability.

13.3.1 Uniqueness and nonuniqueness

Given a subset $\mathcal{F}$ of $\mathcal{F}(D)$, and $\mathcal{S} \subset \mathcal{S}^d$. We say that two different sets $F_1, F_2 \in \mathcal{F}$ are *tomographically equivalent* with respect to $\mathcal{S}$ if $X_SF_1 = X_SF_2$ for all $S \in \mathcal{S}$. The pair

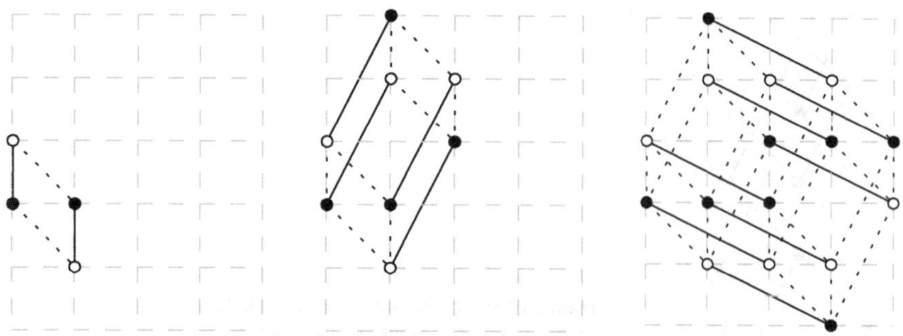

(a) (b) (c)

Figure 13.2: Construction of different lattice sets (black and white points) with equal x-rays in (a) two, (b) three, and (c) four directions.

(F_1, F_2) will then also be referred to as a *switching component*. Further, a set $F \in \mathcal{F}$ is *uniquely determined* within $\mathcal{F}$ by its x-rays parallel to the lines in $\mathcal{S}$ if there does not exist any other set F' in $\mathcal{F}$ that is *tomographically equivalent* to F with respect to $\mathcal{S}$. If the context is clear, we will simply say that $F \in \mathcal{F}$ is *uniquely determined*.

The following classical nonuniqueness result, usually attributed to [118], has been rediscovered several times.

Theorem 1. *For any finite subset $\mathcal{L}$ of $\mathcal{L}^d$ there exist sets in $\mathcal{F}^d$ that cannot be determined by x-rays parallel to the lines in $\mathcal{L}$.*

Figure 13.2 gives an illustration of the typical construction process to obtain different lattice sets with equal x-rays.

Note that Theorem 1 is in accordance with similar results in continuous or geometric tomography. In fact, let $\mathcal{S} \subset \mathcal{S}^d$ be finite, and let $C, K \in \mathbb{R}^d$ be compact and $C \subset \operatorname{int}(K)$. Further, let $f : \mathbb{R}^d \to \mathbb{R}$ be infinitely often differentiable with support K. Then there is a function g with support in K, infinitely often differentiable, but otherwise arbitrary on C such that the continuous x-rays of f and g with respect to all lines in $\mathcal{S}$ coincide; for a proof, see [126]. Also, characteristic functions of compact sets, i. e., functions $f : \mathbb{R}^d \to \{0, 1\}$ with compact support are not determined by their (continuous) x-rays in any finite number of directions; see [20, Theorem 2.3.3].

While nonuniqueness is an undesirable feature for many applications, it will play a positive role for applications in number theory later; see Section 13.8.

In contrast to Theorem 1, uniqueness of the reconstruction can sometimes be guaranteed when certain prior knowledge is available.

Theorem 2. *There exists a line $S \in \mathcal{S}^d$ such that every set $F \in \mathcal{F}^d$ is uniquely determined by its one x-ray $X_S F$.*

At first glance, this result may seem surprising. However, the used a priori information is that the set is contained in $\mathbb{Z}^d$. Then, indeed, the x-ray $X_S F$ for any one line in $\mathcal{S}^d \setminus \mathcal{L}^d$ determines F uniquely, simply because no translate of S can contain more than one lattice point. As this argument shows, Theorem 2 can easily be extended to $\mathcal{F}(\mathbb{Z}^d, C)$.

While Theorem 2 does not seem to be of great practical use, it shows nonetheless that no matter how fine the lattice discretization might be, one x-ray suffices. In the limit, however, finitely many x-rays do in general not suffice. In this sense, discrete tomography does not behave like a discretization of continuous tomography. Note that Theorem 2 is quite different in nature from a result of [126] (see also [41, Theorem 3.148]) that for almost any finite dimensional space of objects its elements can be distinguished by a single x-ray in almost any direction. In fact, $\mathcal{F}^d$ is not finite-dimensional, and moreover, any line $S \in \mathcal{S}^d \setminus \mathcal{L}^d$ works for any set $F \in \mathcal{F}^d$, i. e., S does not depend on the set but is given beforehand.

The next result is due to Rényi [121] for $d = 2$, who attributes an algorithmic proof to Hajós. The generalization to $d \geq 2$ was given by Heppes [106].

Theorem 3 ([121] Rényi). *Let $\mathcal{S}$ be a finite subset of $\mathcal{S}^d$. Then every set $F \in \mathcal{F}(\mathbb{R}^d)$ with $|F| \leq |\mathcal{S}| - 1$ is uniquely determined by its x-rays parallel to the lines in $\mathcal{S}$.*

Since the two color classes of the two-coloring of the vertices of the regular $2m$-gon in the plane are tomographically equivalent with respect to the lines parallel to its edges, Theorem 3 is best possible. A strengthening for (mildly) restricted sets of directions is given in [75]. But even more: generic directions are much better.

Theorem 4 ([120]). *There exist constants $c > 0$ and $m_0 \in \mathbb{N}$ such that for all $m \geq m_0$ the following holds: For almost all sets $\mathcal{S} \subset \mathcal{S}^2$ of m directions any $F \in \mathcal{F}(\mathbb{R}^2)$ with $|F| \leq 2^{cm/\log(m)}$ is uniquely determined by its x-rays parallel to the lines of $\mathcal{S}$.*

Let us point out that in continuous tomography it is well known that a compactly supported, infinitely differentiable function $f : \mathbb{R}^d \to \mathbb{R}$, which does not contain "details of size $2\pi/b$" or smaller can be recovered reliably from its x-rays on special sets of m directions, provided that $m > b$. For a precise statement and proof, see [43, Theorem 2.4]. See also [40]. Such a result can be viewed, to some extent, as an analogue to Rényi's theorem. (Note, however, that the difference to f is measured in an integral norm and hence the difference may get arbitrarily small without ever reaching 0. For a characterization of the null-space, see [40, 119].)

For the special case $m = |\mathcal{S}| = 2$, uniqueness within $\mathcal{F}^2$ is characterized by the work of Ryser [51]. See Figure 13.3 for an example of a set of 12 points in $\mathbb{Z}^2$ that is uniquely determined by its two x-rays in the coordinate directions. For uniqueness results from two directions in geometric tomography, see [116, 118].

In the lattice case, the specialization of Rényi's theorem that any set $F \in \mathcal{F}^d$ is determined by any set of $|F| + 1$ lattice lines is only best possible for the cardinalities $|F| = m \in \{1, 2, 3, 4, 6\}$. For other m, the result can be improved at least by 1.

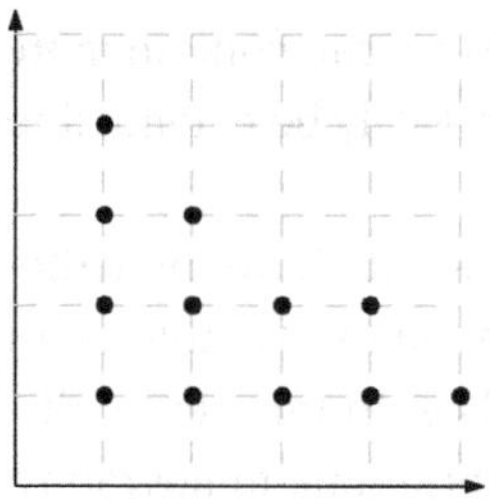

Figure 13.3: A lattice set (black points) uniquely determined by its x-rays in horizontal and vertical directions.

Theorem 5 ([67]). *Let $m \in \mathbb{N} \setminus \{1,2,3,4,6\}$ and let $\mathcal{F}^d(m)$ be the class of sets in $\mathcal{F}^d$ of cardinality less than or equal to m. Let $\mathcal{L} \subset \mathcal{L}^d$ with $|\mathcal{L}| \geq m$. Then the sets in $\mathcal{F}^d(m)$ are determined by their x-rays parallel to the lines in $\mathcal{L}$.*

The question of smallest switching components is widely open in the lattice case.

Problem 1. What is the smallest number $n = n(d,m)$ such that there exist $\mathcal{L} \subset \mathcal{L}^d$ with $|\mathcal{L}| = m$ and two different lattice sets $F_1, F_2 \in \mathcal{F}^d$ of cardinality n that are tomographically equivalent with respect to the lines in $\mathcal{L}$?

Probabilistic arguments of [67] show that, in the lattice case, switching components of a size that is polynomial in m exist for each d. All deterministic constructions so far lead to exponential size switching components. Several small switching components are depicted in Figure 13.4.

We remark that switching components seem to have appeared first in the work of Ryser [51]. Later work on switching components includes [37, 84, 105, 115, 120, 123]. Computational investigations related to the explicit construction of switching components can be found in [80]. Switching components for other projection models are considered in [127–130, 136]. Special types of switching components in the context of

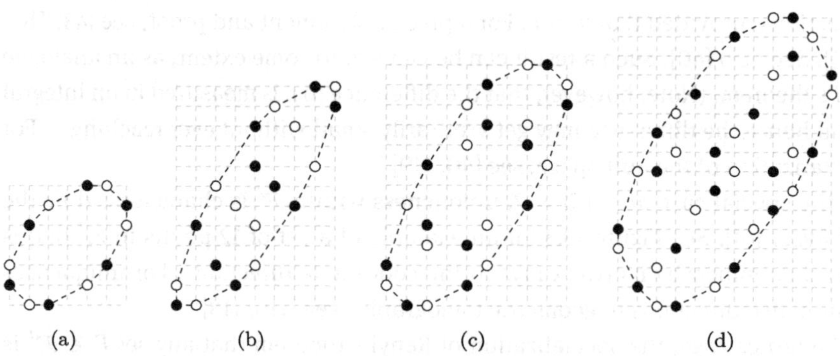

Figure 13.4: Examples of small switching components for (a) $m = 6$, (b) $m = 7$, (c) $m = 8$, (d) $m = 9$, directions (indicated as polygon edges). The switching components are the sets of 6, 10, 12, and 18 black and white points, respectively.

superresolution imaging, *hv*-convex polynomioes, and, in a more algebraic setting, are studied in [65, 71], and [115], respectively.

Quite strong uniqueness results exist for a geometrically motivated more restricted class of lattice sets. A lattice set $F \in \mathcal{F}^d$ is called *convex*, if

$$F = \operatorname{conv}(F) \cap \mathbb{Z}^d.$$

Theorem 6 ([96]). *There are $S_1, S_2, S_3, S_4 \in \mathcal{L}^d$ such that every finite convex lattice set F is uniquely determined by $X_{S_1}F, \ldots, X_{S_4}F$. Further, every set of at least seven coplanar lattice lines always suffices.*

Let us point out that the "good" sets of directions with respect to Theorem 6 need not have coordinates of large absolute value. Examples for $d = 2$ are

$$\{(1,0)^T, (1,1)^T, (1,2)^T, (1,5)^T\} \quad \text{and} \quad \{(1,0)^T, (2,1)^T, (0,1)^T, (-1,2)^T\}.$$

In fact, the sets of four good lattice lines in Thm. 6 are those whose cross-ratio of their slopes does not lie in $\{4/3, 3/2, 2, 3, 4\}$. A converse result for the more general class of *hv-convex lattice sets* is given in [70]. Generalizations of Theorem 6 to so-called *Q-convex lattice sets* and *convex algebraic Delone sets* (in the context of quasicrystals) can be found in [86] and [109], respectively.

As a matter of fact, the directions in Theorem 6 are all coplanar. It is not known how exactly the situation changes if we insist that the lines are in *general position*, i. e., each d of them span $\mathbb{R}^d$.

Problem 2. Let $d \geq 3$. Is there a finite subset $\mathcal{L}$ of lines in $\mathcal{L}^d$ in general position such that each convex set in $\mathcal{F}^d$ is uniquely determined by its x-rays parallel to the lines in $\mathcal{L}$? If so, what is the smallest cardinality?

Is there a smallest number m such that any set $\mathcal{L} \subset \mathcal{L}^d$ of m lines has the property, that each convex set in $\mathcal{F}^d$ is uniquely determined by its x-rays parallel to the lines in $\mathcal{L}$?

The color classes of a 2-coloring of the vertices of the permutahedron in $\mathbb{R}^d$ provide a lower bound on such a universal number, which grows at least quadratically in d [61]. Figure 13.5 depicts the 3-dimensional permutahedron, which is a truncated octahedron.

An analogue to Theorem 6 also holds in the realm of geometric tomography: convex subsets of $\mathbb{R}^d$ are determined by their continuous x-rays from sets of four "good" directions; see [100].

Let us point out that it is the codomain $\{0, 1\}$ which makes the problem difficult. In fact, the case of functions in $\mathcal{F}(\mathbb{Z}^d, \mathbb{Z})$ or, equivalently, lattice sets with integer weights is much simpler, as linear diophantine equations can be solved via the *Hermite normal form*; see, e. g., [53, Sections 4 and 5]). But this also implies that the study of uniqueness for functions $\psi \in \mathcal{F}(\mathbb{Z}^d, \mathbb{N}_0)$ is much easier. In fact, suppose we are given a finite

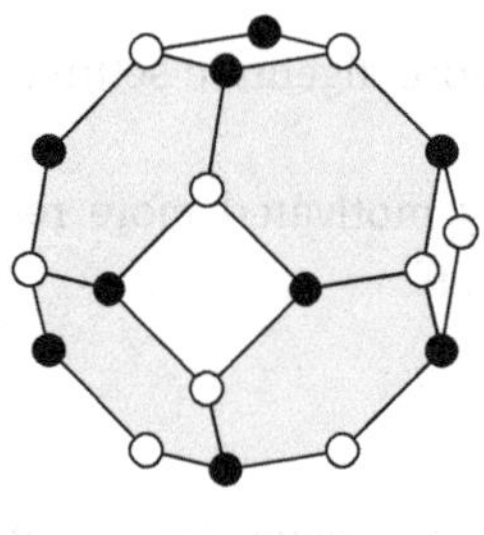

Figure 13.5: Two-coloring of the vertices of the truncated octahedron.

set $\mathcal{L} \subset \mathcal{L}^d$ and a bounded subset B of $\mathbb{Z}^d$ which will act as a superset of all supports we are allowing. Then the corresponding x-ray problem with data functions all identical to 0 can be formulated as a homogenous system of linear diophantine equations and solved efficiently. Let ψ be a nontrivial solution, define $\psi^+ : \mathbb{Z}^d \to \mathbb{R}$ by

$$\psi^+(x) = \begin{cases} \psi(x) & \text{if } \psi(x) > 0; \\ 0 & \text{if } \psi(x) \le 0; \end{cases}$$

for $x \in \mathbb{Z}^d$, and set

$$\psi^- = \psi^+ - \psi.$$

Then, of course, $\psi^+, \psi^- \in \mathcal{F}(\mathbb{Z}^d, \mathbb{N}_0)$, $\psi = \psi^+ - \psi^-$, and $X_S\psi \equiv 0$ for all $S \in \mathcal{L}$. Hence, ψ^+, ψ^- are tomographically equivalent with respect to $\mathcal{L}$.

The uniqueness problem for functions $\psi \in \mathcal{F}(\mathbb{Z}^d, \mathbb{N}_0)$ also permits an algebraic characterization. The subsequently stated result of Hajdu and Tijdeman [29, 105] uses the following notation. A vector $v = (v_1, \dots, v_d)^T \in \mathbb{Z}^d \setminus \{0\}$ is called *reduced* if $\gcd(v_1, \dots, v_d) = 1$. Let v^+ and v^- denote the vectors whose jth component is $v_j^+ = \max\{0, v_j\}$ and $v_j^- = \max\{0, -v_j\}$, respectively. With $\boldsymbol{X}$, we abbreviate the vector $(X_1, \dots, X_d)$ of indeterminants. Accordingly, for $a = (\alpha_1, \dots, \alpha_d)^T \in \mathbb{N}_0^d$, the monomial $X_1^{\alpha_1} \cdot \dots \cdot X_d^{\alpha_d} \in \mathbb{Z}[\boldsymbol{X}]$ is denoted by $\boldsymbol{X}^a$.

Theorem 7 ([29, 105]). *Let $\psi, \varphi \in \mathcal{F}(\mathbb{N}_0^d, \mathbb{N}_0)$, let $v \in \mathbb{Z}^d \setminus \{0\}$ be reduced, and set $S = \text{lin}\{v\}$. Then $X_S\psi = X_S\varphi$ if, and only if, the polynomial*

$$\sum_{a \in \text{supp}(\psi)} \boldsymbol{X}^a - \sum_{b \in \text{supp}(\varphi)} \boldsymbol{X}^b$$

is divisible by $\boldsymbol{X}^{v+} - \boldsymbol{X}^{v^-}$.

Note that the assumption that the functions are defined on $\mathbb{N}_0^d$ rather than on $\mathbb{Z}^d$ is no restriction of generality.

Let $v^{(1)}, \dots, v^{(m)} \in \mathbb{Z}^d \setminus \{0\}$ be reduced, $\mathcal{S} = \{\text{lin}\{v^{(1)}\}, \dots, \text{lin}\{v^{(m)}\}\}$ and

$$q_{\mathcal{S}} = \prod_{v \in \{v^{(1)}, \dots, v^{(m)}\}} (\boldsymbol{X}^{v^+} - \boldsymbol{X}^{v^-}).$$

A consequence of Theorem 7 is that $\psi, \varphi \in \mathcal{F}(\mathbb{N}_0^d, \mathbb{N}_0)$ are tomographically equivalent with respect to $\mathcal{S}$ if, and only if, there is a polynomial p in $\mathbb{Z}[\boldsymbol{X}]$ such that

$$\sum_{a \in \mathrm{supp}(\psi)} \boldsymbol{X}^a - \sum_{b \in \mathrm{supp}(\varphi)} \boldsymbol{X}^b = p \cdot q_{\mathcal{S}}.$$

The algebraic representation by polynomials can be utilized in various ways; examples will be given in Sections 13.3.2 (stability) and 13.8 (number theory). Additional aspects of uniqueness, in particular, concepts of *additivity*, are discussed in [58, 93, 102]. For uniqueness results for functions in $\mathcal{F}(\mathbb{N}_0^d, \{0,1\})$ with several different types of bounded support see [81] and the references cited therein.

13.3.2 Stability and instability

The results of the previous section were based on the assumption that the data functions are given exactly. We will now consider the case that the x-rays may contain errors.

In the following, we will measure the size of a function $\psi \in \mathcal{F}(D, C)$ in terms of its ℓ_1-norm, i. e.,

$$\|\psi\|_1 = \sum_{x \in D} |\psi(x)|.$$

In particular, given a finite set $\mathcal{S} \subset \mathcal{S}^d$ of lines and two sets $F_1, F_2 \in \mathcal{F}(\mathbb{R}^d)$ their *x-ray difference* will be

$$\Delta_{\mathcal{S}}(F_1, F_2) = \sum_{S \in \mathcal{S}} \|X_S F_1 - X_S F_2\|_1.$$

The first result in this section shows that at least some (however marginal) stability is present. In fact, the x-ray difference, if not 0, must jump to at least $2(m-1)$. This means that either two sets are tomographically equivalent or their x-ray difference grows at least linearly in m.

Theorem 8 ([62]). *Let $\mathcal{S} \subset \mathcal{S}^d$, $|\mathcal{S}| = m$, and $F_1, F_2 \in \mathcal{F}(\mathbb{R}^d)$ with $|F_1| = |F_2|$. If $\Delta_{\mathcal{S}}(F_1, F_2) < 2(m-1)$, then F_1 and F_2 are tomographically equivalent with respect to $\mathcal{S}$. The same statement holds in the lattice case.*

As we will see in Theorem 12, this result is, in fact, best possible. First, we use it to give “noisy” variants of some of the uniqueness results of the previous section. We begin with a stable version of Theorem 3.

Theorem 9 ([62]). *Let $\mathcal{S} \subset \mathcal{S}^d$, $F_1, F_2 \in \mathcal{F}(\mathbb{R}^d)$, and $\Delta_{\mathcal{S}}(F_1, F_2) < 2|F_1|$. Further, let $|F_1| = |F_2|$ and $|F_1| + 1 \leq |\mathcal{S}|$, or let $|F_1| \leq |F_2|$ and $2|F_1| \leq |\mathcal{S}|$. Then $F_1 = F_2$. The statement persists in the lattice case.*

The following result is a stable version of Theorem 6.

Theorem 10 ([62]). *There are sets $\mathcal{S} \subset \mathcal{L}^d$ of cardinality 4 for which the following is true: If $F_1, F_2 \in \mathcal{F}^d$ are convex, and $|F_1| = |F_2|$, but $F_1 \neq F_2$, then $\Delta_{\mathcal{S}}(F_1, F_2) \geq 6$. Further, for any set $\mathcal{S} \subset \mathcal{L}^d$ of at least 7 coplanar lattice lines, and sets F_1, F_2 as before, $\Delta_{\mathcal{S}}(F_1, F_2) \geq 2(|\mathcal{S}| - 1)$.*

The following theorem uses the known characterization of the (rather rare) cases of uniqueness in the special case $d = m = 2$ to quantify the deviation of solutions for noisy data; it generalizes a previous result from [59].

Theorem 11 ([133]). *Let $\mathcal{S} \subset \mathcal{L}^2$ with $|\mathcal{S}| = 2$, let $F_1, F_2 \in \mathcal{F}^2$ with $|F_1| = |F_2|$. Further, suppose that F_1 is uniquely determined by $X_S F_1$ for $S \in \mathcal{S}$, and set $\beta = \Delta_{\mathcal{S}}(F_1, F_2)$. Then*

$$4|F_1 \cap F_2| + (\beta + 2)(\beta - 1 + \sqrt{8|F_1 \cap F_2| + (\beta - 1)^2}) \geq 4|F_1|.$$

Stability results in the continuous case with finitely many x-rays typically rely on bounds of the variation of the functions, measured in some weighted Sobolev norms; see [43, Section 4] and [41, Section 5.9] (and references therein). In the realm of geometric tomography, Volčič [134] showed that the problem of reconstructing a convex body from its x-rays in four "good" directions (which guarantee uniqueness) is well-posed. Some further stability estimates are given in [117].

In contrast to Theorem 11, the task of reconstructing finite lattice sets from x-rays taken along $m \geq 3$ directions is highly instable. In particular the following result shows that Theorem 8 is sharp.

Theorem 12 ([2, 66]). *Let $\mathcal{S} \subset \mathcal{S}^d$ with $|\mathcal{S}| \geq 3$, and let $\alpha \in \mathbb{N}$. Then there exist $F_1, F_2 \in \mathcal{F}(\mathbb{R}^d)$ with the following properties:*

(i) *F_1 is uniquely determined by $X_S F_1$ for $S \in \mathcal{S}$;*
(ii) *F_2 is uniquely determined by $X_S F_2$ for $S \in \mathcal{S}$;*
(iii) *$\Delta_{\mathcal{S}}(F_1, F_2) = 2(m - 1)$;*
(iv) *$|F_1| = |F_2| \geq \alpha$;*
(v) *$F_1 \cap F_2 = \emptyset$.*

The statement also holds in the lattice case.

The proof for $d = 2$ is due to [66], while [2] extends the construction to general d. It is actually possible to show that not even affine transformations help much to increase the overlap of the two sets.

13.4 Computational aspects

Next, we deal with algorithmic aspects of actually reconstructing the, one or all sets, that are consistent with the given x-ray data. We will restrict the exposition to functions in $\mathcal{F}(\mathbb{Z}^d, C)$ with $C \subset \mathbb{Q}$ and lines in $\mathcal{L}^d$ since all computational issues can then be studied in the well-known *binary Turing machine model*; see [22, 46] for background information. Again, emphasis will be placed on the lattice case.

13.4.1 Algorithmic problems

Let $\mathcal{S} \subset \mathcal{L}^d$ be finite. From an algorithmic point of view, the following questions are basic: Are the data consistent? If so, reconstruct a solution. Is this solution unique? We will now introduce the corresponding problems more precisely.

CONSISTENCY$_{\mathcal{F}(\mathbb{Z}^d,C)}(\mathcal{S})$.
Instance: *Data functions f_S for $S \in \mathcal{S}$.*
Question: *Does there exist $\psi \in \mathcal{F}(\mathbb{Z}^d, C)$ such that $X_S\psi = f_S$ for all $S \in \mathcal{S}$?*

RECONSTRUCTION$_{\mathcal{F}(\mathbb{Z}^d,C)}(\mathcal{S})$.
Instance: *Data functions f_S for $S \in \mathcal{S}$.*
Task: *Determine a function $\psi \in \mathcal{F}(\mathbb{Z}^d, C)$ such that $X_S\psi = f_S$ for all $S \in \mathcal{S}$, or decide that no such function exists.*

UNIQUENESS$_{\mathcal{F}(\mathbb{Z}^d,C)}(\mathcal{S})$.
Instance: *A function $\psi \in \mathcal{F}(\mathbb{Z}^d, C)$.*
Question: *Does there exist $\varphi \in \mathcal{F}(\mathbb{Z}^d, C) \setminus \{\psi\}$ such that $X_S\psi = X_S\varphi$ for all $S \in \mathcal{S}$?*

Of course, the problem RECONSTRUCTION$_{\mathcal{F}(\mathbb{Z}^d,C)}(\mathcal{S})$ cannot be easier than the problem CONSISTENCY$_{\mathcal{F}(\mathbb{Z}^d,C)}(\mathcal{S})$. Further, note that UNIQUENESS$_{\mathcal{F}(\mathbb{Z}^d,C)}(\mathcal{S})$ actually asks for nonuniqueness in order to place the problem into the class $\mathbb{NP}$; see Theorem 13.

For certain codomains such as $C = \{0, 1\}$, it is reasonable to actually ask for the number of solutions even in the case of nonuniqueness. We will introduce the following problem for general C with the understanding that the (not really interesting) answer "∞" is permitted.

#CONSISTENCY$_{\mathcal{F}(\mathbb{Z}^d,C)}(\mathcal{S})$.
Instance: *Data functions f_S for $S \in \mathcal{S}$.*
Task: *Determine the cardinality of the set of functions $\psi \in \mathcal{F}(\mathbb{Z}^d, C)$ such that $X_S\psi = f_S$ for all $S \in \mathcal{S}$.*

Observe that a given instance $\mathcal{I} = (f_S : S \in \mathcal{S})$ can be consistent only if $\|f_S\|_1$ does not depend on S. Since this condition can be checked efficiently, we will in the following often tacitly assume that this is the case and set

$$n = n(\mathcal{I}) = \|f_S\|_1.$$

Further, for any given instance $\mathcal{I} = (f_S : S \in \mathcal{S})$, the support of all solutions is contained in the *grid*

$$G = G(\mathcal{I}) = \bigcap_{S \in \mathcal{S}} (\mathrm{supp}(f_S) + S)$$

associated with $\mathcal{I}$. Of course, $G(\mathcal{I})$ can be computed from $\mathcal{I}$ by solving polynomially many systems of linear equations. Hence we can associate a variable x_g with every grid point and formulate $\textsc{Consistency}_{\mathcal{F}(\mathbb{Z}^d,C)}(\mathcal{S})$ as a linear (feasibility) problem with the additional constraints that $x_g \in C$ for all $g \in G$. This simple observation shows already that $\textsc{Consistency}_{\mathcal{F}(\mathbb{Z}^d,C)}(\mathcal{S})$ is algorithmically easy for $C \in \{[0,1] \cap \mathbb{Q}, \mathbb{Q}\}$ simply because linear programming can be solved in polynomial time, and also for $C = \mathbb{Z}$ since systems of linear diophantine equations can be solved in polynomial time; see, e. g., [53, Sections 4, 5, 13–15].

Next, we are turning to the other relevant codomains, with a special emphasis on the lattice case.

Theorem 13 ([51, 94, 98]). *$\textsc{Consistency}_{\mathcal{F}(\mathbb{Z}^d,C)}(\mathcal{S})$ and $\textsc{Uniqueness}_{\mathcal{F}(\mathbb{Z}^d,C)}(\mathcal{S})$, $C \in \{\{0,1\}, \mathbb{N}_0\}$, are both in $\mathbb{P}$ if $|\mathcal{S}| \le 2$ whereas they are $\mathbb{NP}$-complete if $|\mathcal{S}| \ge 3$. Also, the problem $\#\textsc{Consistency}_{\mathcal{F}^d}(\mathcal{S})$ is $\#\mathbb{P}$-complete for $|\mathcal{S}| \ge 3$.*

The complexity status of the counting problem for $|\mathcal{S}| = 2$ is still open.

Problem 3. Is $\#\textsc{Consistency}_{\mathcal{F}^d}(\mathcal{S})$, $|\mathcal{S}| = 2$, a $\#\mathbb{P}$-complete problem?

Let us now return to the Rényi setting.

Theorem 14 ([106, 121]). *$\textsc{Reconstruction}_{\mathcal{F}^d}(\mathcal{S})$ is in $\mathbb{P}$ if the input is restricted to those instances $\mathcal{I} = (f_S : S \in \mathcal{S})$ with $n(\mathcal{I}) < |\mathcal{S}|$.*

A similar result holds for convex lattice sets when the lattice lines are chosen according to Theorem 6. So, let $\mathcal{C}^d$ denote the subset of $\mathcal{F}^d$ of convex lattice set, and let $\textsc{Reconstruction}_{\mathcal{C}^d}(\mathcal{S})$ signify the corresponding reconstruction task.

Theorem 15 ([77, 78]). *For any set $\mathcal{S} \subset \mathcal{L}^d$ of at least seven coplanar directions and for suitable such sets of cardinality four $\textsc{Reconstruction}_{\mathcal{C}^d}(\mathcal{S})$ can be solved in polynomial time.*

Let us now turn to the following “noisy” versions of $\textsc{Consistency}_{\mathcal{F}^d}(\mathcal{S})$ and $\textsc{Uniqueness}_{\mathcal{F}^d}(\mathcal{S})$.

X-RAY-CORRECTION$_{\mathcal{F}^d}(\mathcal{S})$.

Instance: *Data functions f_S for $S \in \mathcal{S}$.*

Question: *Does there exist $F \in \mathcal{F}^d$ such that*

$$\sum_{S \in \mathcal{S}} \|X_S F - f_S\|_1 \leq m - 1?$$

SIMILAR-SOLUTION$_{\mathcal{F}^d}(\mathcal{S})$.

Instance: *A set $F_1 \in \mathcal{F}^d$.*

Question: *Does there exist $F_2 \in \mathcal{F}^d$ with $|F_1| = |F_2|$ and $F_1 \neq F_2$ such that $\Delta_{\mathcal{S}}(F_1, F_2) \leq 2m - 3$?*

NEAREST-SOLUTION$_{\mathcal{F}^d}(\mathcal{S})$.

Instance: *Data functions f_S for $S \in \mathcal{S}$.*

Task: *Determine a set $F^* \in \mathcal{F}^d$ such that*

$$\sum_{S \in \mathcal{S}} \|X_S F^* - f_S\|_1 = \min_{F \in \mathcal{F}^d} \sum_{S \in \mathcal{S}} \|X_S F - f_S\|_1.$$

Note that X-RAY-CORRECTION$_{\mathcal{F}^d}(\mathcal{S})$ can also be viewed as the task of deciding, for given data functions f_S, $S \in \mathcal{S}$, whether there exist "corrected" data functions f'_S, $S \in \mathcal{S}$, that are consistent and do not differ from the given functions by more than a total of $m - 1$. NEAREST-SOLUTION$_{\mathcal{F}^d}(\mathcal{S})$ asks for a set $F^* \in \mathcal{F}^d$ that fits the potentially noisy measurements best.

The computational complexity of these tasks is as follows.

Theorem 16 ([62]). *The problems X-RAY-CORRECTION$_{\mathcal{F}^d}(\mathcal{S})$, SIMILAR-SOLUTION$_{\mathcal{F}^d}(\mathcal{S})$, and NEAREST-SOLUTION$_{\mathcal{F}^d}(\mathcal{S})$ are in $\mathbb{P}$ for $|\mathcal{S}| \leq 2$ but are $\mathbb{NP}$-complete for $|\mathcal{S}| \geq 3$.*

13.4.2 Algorithms

Several polynomial-time algorithms for RECONSTRUCTION$_{\mathcal{F}(\mathbb{Z}^d,C)}(\mathcal{S})$, $C \in \{\{0,1\}, \mathbb{N}_0\}$, $|\mathcal{S}| = 2$, can be found in the literature. In addition to Ryser's algorithm [51] for $C = \{0,1\}$, there are approaches based on network-flows [125] or matroid intersections [97]. Moreover, the problem can be modeled as an integer linear program, which involves a totally unimodular coefficient matrix, and which can therefore be solved as a linear program (see, for instance, [53, Sections 16 and 19]). For further comments, see [35, Section 1].

In the presence of $\mathbb{NP}$-hardness, one cannot expect to find generally efficient algorithms. There are, however, various techniques from combinatorial optimization that can and have been applied to solve instances to optimality up to certain sizes; see [103]. Similarly, as for the case $|\mathcal{S}| = 2$, the reconstruction problem

$\text{Reconstruction}_{\mathcal{F}(\mathbb{Z}^d,C)}(\mathcal{S})$, $C \in \{\{0,1\},\mathbb{N}_0\}$, can be formulated as the integer linear program for arbitrary $|\mathcal{S}|$. However, for $|\mathcal{S}| \geq 3$ the coefficient matrix is in general no longer totally unimodular. Of course, we can still solve the corresponding linear programming relaxation (where $\{0,1\}$ is replaced by $[0,1]$ or $\mathbb{N}_0$ by $[0,\infty[$) efficiently. Unless $\mathbb{P} = \mathbb{NP}$, Theorem 13 implies that it will, however, in general not be efficiently possible to convert the obtained fractional solution into a required integer one.

Since, in general, measured data are noisy anyway, research focused on approximate solutions. It is quite natural to try to solve $\text{Reconstruction}_{\mathcal{F}^d}(\mathcal{S})$ even if $|\mathcal{S}| \geq 3$ by using the available polynomial-time algorithms for $|\mathcal{S}| = 2$ in an alternating approach. First, two of the given $|\mathcal{S}|$ data functions are selected and a solution F_0 is computed which is consistent with these. In the jth step, at least one of the two directions is replaced by a different one from $\mathcal{S}$, and a solution is constructed which satisfies the corresponding two constraints and is closest to F_j. While each step of such an *alternating direction approach* can be performed in polynomial time, there are severe limitations on the guaranteed quality of the produced solution. For an analysis of this and other approaches, see [103].

Despite their theoretical limitations there are several approaches that are reported to work very well in practice. Among these are BART [107] and DART [74, 135]. The former, which is a variant of ART as described in [33], is implemented in the open-source software SNARK14 [114] (example code can be found in [60]), the latter is implemented in the open-source ASTRA toolbox [131]. Further algorithms are discussed in [35, Sections 8–14] and [36, Sections 8–11]. For applets illustrating several algorithmic tasks in discrete and geometric tomography, see [91] and [95], respectively.

Let us, finally, point out that for certain applications full reconstructions are not needed. For instance, in quality control for circuit board productions (see Figure 13.6) one may want to certify that the production process actually produced a desired blueprint structure (“verification”). Then one can, of course, compute data functions from the blueprint and compare them with the measured data from the produced board. If the difference is large one would report an error. If, however, the difference is small, the produced board can still be quite different from the blueprint (particularly if the data do not determine the image uniquely). This ambiguity can be reduced by

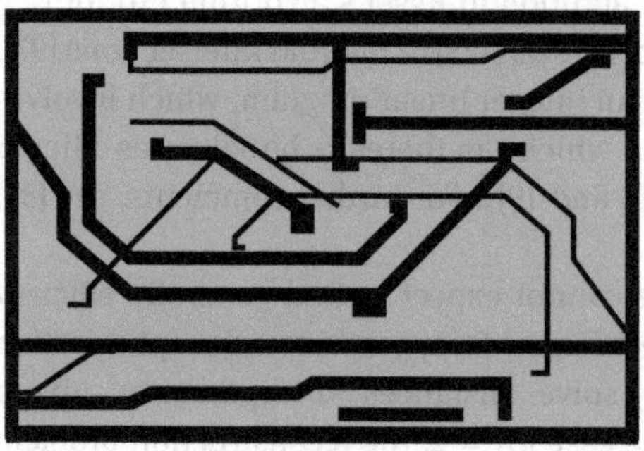

Figure 13.6: (See [26]) An idealized circuit board.

applying a (polynomial-time deterministic) reconstruction heuristic on both sets of data functions and subsequently comparing the reconstructions. In practice, such checks have shown to be able to detect production flaws even on very limited data and quite poor (and very fast) reconstructions algorithms.

13.5 Superresolution and discrete tomography

Electron tomography, pioneered originally in the life sciences (see [19, 34, 45]), is becoming an increasingly important tool in materials science for studying the three-dimensional morphologies and chemical compositions of nanostructures [60, 69, 72, 112]. For various technical reasons, however, atomic resolution tomographic imaging as envisioned in [113, 122] has not become a full reality yet (favorable instances are reported in [111, 132]; see also the surveys [9, 55]). One of the challenges faced by current technology is that tomographic tilt series need to be properly aligned (see, e. g., [56, 108]). Therefore, and also to prevent radiation damage, one might wonder whether it is possible to proceed in a multimodal scheme.

Suppose some reconstruction has been obtained from a (possibly technologically less-demanding) lower-resolution data set. Can one then use limited additional high-resolution data (for instance, acquired from only two directions) to enhance the resolution in a subsequent step? As we will now see the tractability of this approach depends strongly on the reliability of the initial lower-resolution reconstruction. Details of the presented results can be found in [65].

13.5.1 Computational aspects

We have already remarked that a function $\psi \in \mathcal{F}^d$ can be viewed as a characteristic function that encodes a finite lattice set. In a different, yet equivalent, model the function ψ can be viewed as representing a binary image. In this interpretation, the points $x \in \mathbb{Z}^d$ represent the pixel/voxel coordinates while $\psi(x)$ denotes their colors (typically, values 0 and 1 are considered to represent white and black pixels, respectively); see Figure 13.7 for an illustration. Similarly, for $l \in \mathbb{N}$, a function $\rho \in \mathcal{F}(\mathbb{Z}^d, [l]_0)$ can be viewed as representing a gray-scale image with $l + 1$ different gray levels (values 0 and l typically representing the "gray level" white and black, respectively).

For simplicity of the exposition, we restrict our discussion to the case $d = 2$. Now suppose we want to reconstruct a binary image $\psi \in \mathcal{F}^2$ contained in an $n_1 \times n_2$ box

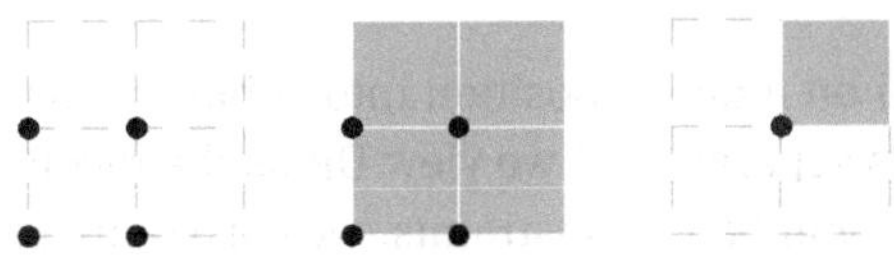

Figure 13.7: Lattice points (left) and pixels (middle); right: pixel associated with its lattice point.

from low-resolution gray scale information and high-resolution x-ray data. The lower resolution is quantified by some $k \in \mathbb{N}\setminus\{1\}$, and we assume that n_1 and n_2 are divisible by k. More precisely, we assume that an $n_1/k \times n_2/k$ low-resolution (gray-scale) image $\rho \in \mathcal{F}(\mathbb{Z}^2, [k^2]_0)$ of ψ is available, and the pixels x in ψ result from a $k \times k$ subdivision of the pixels y of ρ. Hence in any such box B, we have

$$\sum_{x \in B} \psi(x) = \rho(y).$$

For given $\rho(y)$ and unknown $\psi(x)$, $x \in B$, we call the above equation a $k \times k$ *block constraint*. We say that, for some $\varepsilon \in \mathbb{N}_0$, a block constraint is *satisfied within an error of* ε, if

$$\rho(y) - \varepsilon \;\leq\; \sum_{x \in B} \psi(x) \;\leq\; \rho(y) + \varepsilon.$$

We may think of ρ as being the result of some lower-resolution reconstruction of ψ. In order to increase the resolution we want to utilize additional high-resolution x-ray data $X_S\psi$ that are available from the two coordinate directions S_1 and S_2, and we set $\mathcal{S} = \{S_1, S_2\}$. Relatively to ρ the data $X_S\psi$, $S \in \mathcal{S}$, can be considered as k-times finer resolution x-ray data.

For given $k \geq 2$ and $\varepsilon \in \mathbb{N}_0$ the task of *(noisy) superresolution* is as follows:

NSR(k, ε).

Instance: *A gray-level image $\rho \in \mathcal{F}(\mathbb{Z}^2, [k^2]_0)$,*
a subset R of "reliable pixels" of ρ, and
data functions f_{S_1}, f_{S_2} at a k-times finer resolution.

Task: *Determine a function $\psi \in \mathcal{F}^2$ such that*
$X_S\psi = f_S$ for $S \in \{S_1, S_2\}$,
all $k \times k$ block constraints for the pixels in R are satisfied, and
all other $k \times k$ block constraints are satisfied within an error of ε,
or decide that no such function exists.

Since our focus is in the following on double-resolution imaging, i. e., on the case $k = 2$, let us set NDR(ε) = NSR$(2, \varepsilon)$, for $\varepsilon > 0$. In the reliable situation, i. e., for $\varepsilon = 0$ we simply speak of double-resolution and set DR = NSR$(2, 0)$. (Then, of course, the set R can be omitted from the input.) An illustration is given in Figure 13.8.

As it turns out DR is tractable.

Theorem 17 ([65]). *DR and also the corresponding uniqueness problem can be solved in polynomial time.*

The algorithm presented in [65] is based on a decomposition into subproblems, which allows to treat the different gray levels separately. If we view DR as the reconstruction problem for $m = |\mathcal{S}| = 2$ with additional block constraints, we can compare

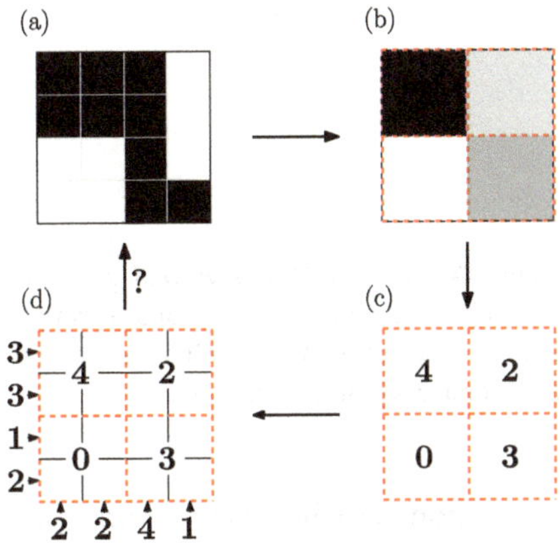

Figure 13.8: (From [65]) The double-resolution imaging task DR. (a) Original (unknown) high-resolution image, (b) the corresponding low-resolution gray-scale image, (c) gray levels converted into block constraints, (d) taken in combination with double-resolution row and column sum data. The task is to reconstruct from (d) the original binary image shown in (a).

Theorem 17 with Theorem 13 and see that bock constraints impose fewer algorithmic difficulties than x-ray data from a third direction.

The next result, which deals with the case that some of the gray levels come with small uncertainties depicts (potentially unexpected) complexity jumps.

Theorem 18 ([65]). *Let* $k \geq 2$ *and* $\varepsilon > 0$.

(i) *NSR*(k, ε) *is* $\mathbb{NP}$*-hard.*

(ii) *The problem of deciding whether a given solution of an instance of* NSR(k, ε) *is not unique is* $\mathbb{NP}$*-complete.*

To put it succinctly: noise in tomographic superresolution imaging does not only affect the quality of a reconstructed image but also the algorithmic tractability of the inverse problem itself.

DR without any block constraints boils down to the reconstruction problem for $m = 2$ and is hence solvable in polynomial time. DR is, however, $\mathbb{NP}$-hard if *several* (but not all) block constraints (which are required to be satisfied with equality) are present (Theorem 18). Possibly less expectedly, if *all* block constraints are included, then the problem becomes polynomial-time solvable again (Theorem 17). If, on the other hand, from *all* block constraints *some* of the data come with *noise* at most 1, then the problem becomes again $\mathbb{NP}$-hard (Theorem 18). And yet again, if from *all* block constraints *all* of the data are *sufficiently noisy*, then the problem is in $\mathbb{P}$ (as this is again the problem of reconstructing binary images from x-ray data taken from two directions). Figure 13.9 gives an overview of these complexity jumps.

It does not seem likely that, but is still open whether the tractability result of Theorem 17 persists for $k \geq 3$.

Problem 4 ([65]). Is the problem NSR$(k, 0)$ $\mathbb{NP}$-hard for $k \geq 3$?

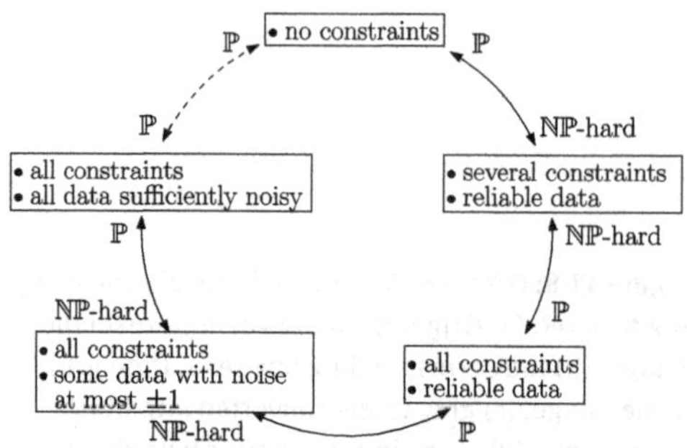

Figure 13.9: (From [65]) Overview of complexity jumps for the problem of reconstructing a binary image from row and column sums and additional 2×2 block constraints.

In the realm of dynamic discrete tomography (see Section 13.6), block constraints play the role of special *window constraints* which can be used to encode velocity information for moving points.

For additional information on discrete tomography problems involving other kinds of constraints, see [14, Section 4].

13.5.2 Stability and instability

Let us now turn to a discussion of the stability of the solutions to DR.

Theorem 19. *Let* $\mathcal{S} = \{\text{lin}(1,0)^T, \text{lin}(0,1)^T\}$, *and* $\alpha \in \mathbb{N}$. *Then there exist instances* $\mathcal{I}_1$ *and* $\mathcal{I}_2$ *of DR with the following properties:*
(i) F_1 *is the unique solution to* $\mathcal{I}_1$;
(ii) F_2 *is the unique solution to* $\mathcal{I}_2$;
(iii) $\Delta_{\mathcal{S}}(F_1, F_2) = 4$;
(iv) $|F_1| = |F_2| \geq \alpha$;
(v) $|F_1 \cap F_2| = \frac{1}{2}|F_1|$.

The proof is based on a construction in [65] of an instance $\mathcal{I}$ of DR that admits precisely two solutions $F_1' \neq F_2'$ with $|F_1'| = |F_2'| \geq \alpha + 2$. From these two solutions, points in one block are deleted to obtain F_1 and F_2; see Figure 13.10 for an illustration.

A small x-ray error of 4 can thus lead to quite different reconstructions (again, see Figure 13.10). It should be noted, however, that the set F_2 has a much larger *total variation (TV)* than F_1 (for some background information see, e. g., [83]). Regularization by total variation minimization, as proposed in [65], would therefore always favor the reconstruction F_1.

It is instructive to compare DR (Theorem 19) with its discrete tomography counterparts for $m = 2$ (Theorem 11) and $m \geq 3$ (Theorem 12), which do not involve any block constraints.

On the one hand, the reconstruction problem for $m = 2$ is much more stable than its double-resolution counterpart. In fact, an easy calculation for $\beta = 4$ shows that

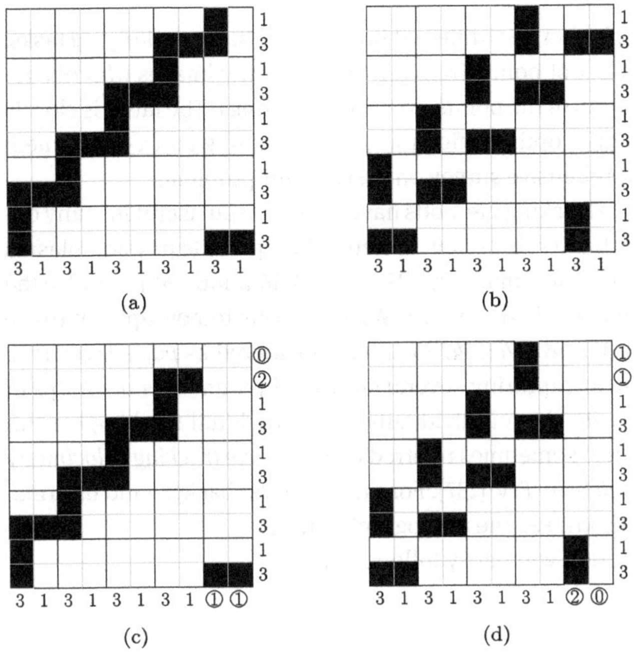

Figure 13.10: An example illustrating instability of DR. (a) and (b): two solutions of the same problem instance; (c) and (d): uniquely determined solutions F_1, F_2 to the two problem instances obtained by deleting points of a block from (a) and (b), respectively. (The x-rays are indicated by the numbers to the bottom and right.) The x-rays differ in the circled numbers yielding an x-ray error $\Delta_{\mathcal{S}}(F_1, F_2) = 4$.

Theorem 11 implies the bound

$$|F_1 \cap F_2| \geq |F_1| - 5\sqrt{|F_1|} - 9.$$

Thus, if the original set F_1 is uniquely determined by its x-rays, then any reconstruction F_2 from x-rays with error 4 needs to coincide with F_1 by an asymptotically much larger fraction than the $|F_1|/2$ provided in Theorem 19 for DR.

On the other hand, the instability result for $m = 3$ (see Theorem 12) is stronger than that of Theorem 19 as for the former an x-ray error of 4 can lead to disjoint reconstructions.

Hence in terms of (in-)stabilities the block constraints seem to play a somewhat weaker role than constraints modeling data from a third direction.

13.6 Dynamics

Let us now turn to dynamic discrete tomography, which, in fact, represents rather recent developments in the field (see [63, 64, 138, 148]). (For dynamic aspects of computerized tomography, see, e. g., [82, 104] and the references cited therein.)

We focus here on the task of *tomographic particle (or point) tracking*, which amounts to determining the paths $\mathcal{P}_1, \ldots, \mathcal{P}_n$ of n points in space over a period of $t \in \mathbb{N}$ moments in time from x-ray images taken from a fixed number m of directions.

This problem comprises, in fact, two different but coupled basic underlying tasks, the reconstruction of a finite set of points from few of their x-ray images (*discrete tomography*) and the identification of the points over time (*tracking*). The latter is closely related to topics in combinatorial optimization including matching and k-assignment problems; see [15] for a comprehensive survey on assignment problems.

Let us remark that particle tracking methods have been proven useful in many different fields such as fluid mechanics, geoscience, elementary particle physics, plasma physics, combustion, and biomedical imaging [137, 141, 143, 144, 146, 148] (see also the monograph [1] and the references cited therein). Most previous tomographic particle tracking methods (such as [139, 140, 142, 147]) can be considered as *particle imaging velocimetry (PIV)* as they aim at capturing several statistical parameters of groups of particles instead of dealing with them individually. The individual tracking considered here is in the literature also sometimes referred to as *particle tracking velocimetry (PTV)* or *low particle number density PIV* [137]. For more general background information on particle tracking methods, see the monographs [1, 54].

The exposition in this section will partly follow [64].

13.6.1 Algorithmic problems

We want to focus here on the interplay between discrete tomography and tracking. Therefore, we will distinguish the cases that for none, some or all of the $\tau \in [t]$ moments in time, a solution $F^{(\tau)} \in \mathcal{F}^d$ of the discrete tomography task at time τ is explicitly available (and is then considered *the* correct solution regardless whether it is uniquely determined by its x-rays). The former case will be referred to as the *(partially)* or *(totally) tomographic* case while we speak of the latter as *positionally determined.* It should be noted that the positionally determined case can be viewed as being the generic case in $\mathbb{R}^d$, $d \geq 3$, because there any two (affine) lines in general position are disjoint, hence x-ray lines meet only in the points of $F^{(\tau)}$.

For simplicity, we assume in the following that there are no particles disappearing or reappearing within the tracked time interval. When $P = \{p_1, \ldots, p_n\}$ denotes the (abstract) set of n particles, we are in the tracking step thus interested in a one-to-one mapping $\pi^{(\tau)} : P \to F^{(\tau)}, \tau \in [t]$, that identifies the points of $F^{(\tau)}$ with the particles. The particle tracks are then given by $\mathcal{P}_i = (\pi^{(1)}(p_i), \ldots, \pi^{(t)}(p_i))$, $i \in [n]$. This identification is referred to as *coupling.*

In typical applications, we would like to incorporate prior knowledge about "physically likely" paths. It seems most natural to input such information in terms of the cost $c(\mathcal{P}_1, \ldots, \mathcal{P}_n)$ of the feasible particle tracks. Note, however, that the number of different particle tracks $(\mathcal{P}_1, \ldots, \mathcal{P}_n)$ is $(n!)^t$, hence *exponential* in n and t. This means that already for moderate problem sizes the costs of all potential tracks cannot be encoded explicitly. There are various ways to deal with this problem. The most general approach is based on the assumption that "an expert knows a good solution if she or he sees it." More technically speaking, it is enough for an algorithm to have access to the cost $c(\mathcal{P}_1,\dots,\mathcal{P}_n)$ only when the particle track $c(\mathcal{P}_1,\dots,\mathcal{P}_n)$ is considered. Accordingly, [64] suggest an oracular model, where such knowledge is available through an algorithm $\mathcal{O}$, called an *objective function oracle*, which computes for any solution $(\mathcal{P}_1,\dots,\mathcal{P}_n)$ its cost $c(\mathcal{P}_1,\dots,\mathcal{P}_n)$ in time that is polynomial in all the other input data. Then the general problem of tomographic particle tracking for $\mathcal{S} \subset \mathcal{S}^d$, can be formulated as follows:

TOMTRAC($\mathcal{O}$; $\mathcal{S}$).

Instance: *$t \in \mathbb{N}$ and data functions $f_S^{(\tau)}$ with $\|f_S^{(\tau)}\|_1 = n$, for $S \in \mathcal{S}$, $\tau \in [t]$.*

Task: *Decide whether, for each $\tau \in [t]$, there exists a set $F^{(\tau)} \in \mathcal{F}^d$ such that $X_S F^{(\tau)} = f_S^{(\tau)}$ for all $S \in \mathcal{S}$. If so, find particle tracks $\mathcal{P}_1,\dots,\mathcal{P}_n$ of minimal cost for $\mathcal{O}$ among all couplings of all tomographic solutions $F^{(1)},\dots,F^{(t)}$.*

In the positionally determined case, the problem TOMTRAC($\mathcal{O}$; $\mathcal{S}$) reduces to the following tracking problem, which can be viewed as a *t-dimensional assignment* problem:

TRAC($\mathcal{O}$; d).

Instance: *$t \in \mathbb{N}$ and sets $F^{(1)},\dots,F^{(t)} \in \mathcal{F}^d$ with $|F^{(1)}| = \dots = |F^{(t)}| = n$.*

Task: *Find particle tracks $\mathcal{P}_1,\dots,\mathcal{P}_n$ of minimal cost for $\mathcal{O}$ among all couplings of the sets $F^{(1)},\dots,F^{(t)}$.*

A priori knowledge may be available in various ways and may then lead to different objective function oracles; see [64]. Here, we focus on information that is actually explicitly available. For instance, we speak of a *path value oracle* if the cost $c(\mathcal{P}_1,\dots,\mathcal{P}_n)$ is just the sum $\sum_{i=1}^n w(\mathcal{P}_i)$ of the weights of the individual paths $\mathcal{P}_i$, $i \in [n]$. Note that the number of different weights is bounded by n^t, and can hence be encoded explicitly for fixed (and small) t; see Theorem 22. If, further, the weights are just the sums of all costs of assigning points between consecutive moments in time the objective function can be described by just $(t-1)n^2$, i. e., polynomially many numbers. In this case, the objective function is of *Markov-type* as it reflects only memoryless dependencies. *Combinatorial models*, which can be viewed as special choices of such parameters, are based on the knowledge that the positions of the particles in the next time step lie in certain *windows*. A particular such situation has be analyzed in Section 13.5. For more results on combinatorial models, see [63, 64].

13.6.2 Algorithms and complexity

We begin with a simple tractability result for the positionally determined case.

Theorem 20 ([64]). *For Markov-type objective function oracles $\mathcal{O}$, the problem* TRAC($\mathcal{O}; d$) *decomposes into uncoupled minimum weight perfect bipartite matching problems and can hence be solved in polynomial time.*

Although the reconstruction problem in discrete tomography for $|\mathcal{S}| = 2$ directions can be solved in polynomial time (see Theorem 13), it turns out that there are severe limitations of extending the previous result already for the following quite restricted partially tomographic case. In fact, the problem becomes hard even if there is only one time step, i. e., $t = 2$, and $F^{(1)}$ is explicitly known while the set $F^{(2)}$ of particle positions for $\tau = 2$ is only accessible through its two x-rays $X_{S_1}F^{(2)}$ and $X_{S_2}F^{(2)}$.

Theorem 21 ([64]). *Even if all instances are restricted to the case $t = 2$, where the solution $F^{(1)}$ is given explicitly,* TOMTRAC($\mathcal{O}; \mathcal{S}$), $|\mathcal{S}| = 2$, *for Markov-type objective function oracles $\mathcal{O}$ is $\mathbb{NP}$-hard. Also the corresponding uniqueness problem is $\mathbb{NP}$-complete and the counting problem is #$\mathbb{P}$-complete.*

Unless $\mathbb{P} = \mathbb{NP}$, there is thus, in general, no efficient algorithm that provides exact solutions to every instance of TOMTRAC($\mathcal{O}; \mathcal{S}$), $|\mathcal{S}| = 2$. A possible remedy is to resort to heuristics, which aim at providing approximate solutions. Before we discuss such a heuristic let us state two additional intractability results, which concern the positionally determined case for non-Markov-type function oracles.

Theorem 22 ([64]). *The problem* TRAC($\mathcal{O}; d$) *is $\mathbb{NP}$-hard, even if all instances are restricted to a fixed $t \geq 3$, and $\mathcal{O}$ is a path value oracle. The $\mathbb{NP}$-hardness persists if the objective function values provided by $\mathcal{O}$ are all encoded explicitly.*

It turns out that even if the particles are expected to move along straight lines, this a priori knowledge cannot be exploited efficiently (unless $\mathbb{P} = \mathbb{NP}$).

Theorem 23 ([64]). *For every fixed $d \geq 2$ and $t \geq 3$, it is an $\mathbb{NP}$-complete problem to decide whether a solution of* TRAC($\mathcal{O}; d$) *exists where all particles move along straight lines.*

The proof of Theorem 23 given in [64] relies on the hardness of the particular variant A3AP of 3D-MATCHING established in [145]. For further results and a discussion of their practical implications, see [64].

The previous complexity results show that even for $t = 3$ and even if there is no tomography involved the coupling becomes hard unless it is of the Markov-type, i. e., it only incorporates information that relate not more than two consecutive moments in time (Theorem 22). But even for $t = 2$, which, of course, is of Markov-type, the problem is hard if tomography is involved at one point in time (Theorem 21). This means that there is not much room for efficient algorithms or "self-suggesting" polynomial-time heuristics.

There are, however, quite involved heuristics for TOMTRAC($\mathcal{O}; S_1, S_2$), which allow to incorporate various different forms of a priori knowledge and different levels of "particle history"; see [64]. Here, we focus only on one basic method, called ROLLING

Horizon Tomography, which was introduced in [138] and applied to the study of the slip velocity of a gliding arc discharge in [148].

The general idea in Rolling Horizon Tomography is to model the time step from τ to $\tau + 1$ as a linear program, based on the assumption that $F^{(1)}$ is known (hence we are dealing with the partially tomographic case). The constraints encode the x-rays provided by the data functions $f_1^{(\tau+1)}, f_2^{(\tau+1)}$. The variables correspond to the points in the grid $G^{(\tau+1)}$ and are collected in a vector $x^{(\tau+1)}$. The x-ray information is encoded by means of a totally unimodular matrix $A^{(\tau+1)}$ and a right-hand side vector $b^{(\tau+1)}$. Further, each point $g_i^{(\tau+1)} \in G^{(\tau+1)}$ carries a weight $\alpha_i^{(\tau+1)}$, which reflects the "distance" to a best point $g_i^{(\tau)} \in F^{(\tau)}$ (which is a likely "predecessor"). These weights are collect in a vector $a^{(\tau+1)}$. Various choices of weights are discussed in [138], which, for instance, model knowledge on the velocity of the particles. The algorithm can then be described as follows.

Beginning with $F^{(1)}$, Rolling Horizon Tomography solves successively for $\tau \in [t-1]$ the linear program

$$\begin{array}{rrcl} & \min\ (a^{(\tau+1)})^T x^{(\tau+1)} & & \\ \text{s.t.} & A^{(\tau+1)} x^{(\tau+1)} & = & b^{(\tau+1)}, \\ & x^{(\tau+1)} & \leq & \mathbb{1}, \\ & x^{(\tau+1)} & \geq & 0, \end{array}$$

in order to determine $F^{(\tau+1)}$ (via its encoding as a 0-1 incidence vector of a basic feasible solution of the linear program). Finally, the paths $\mathcal{P}_1, \ldots, \mathcal{P}_n$ are obtained by a routine that computes a perfect bipartite matching in the graph with vertices $F^{(\tau)} \cup F^{(\tau+1)}$ and edges corresponding to the pairs of vertices that realize the distances $\alpha_i^{(\tau)}$.

Rolling Horizon Tomography runs in polynomial time, is exact in the sense that it is guaranteed to return a solution which matches the data. It also allows to incorporate physical knowledge and it is reported to work quite well in practice (see [138, 148]). However, (and with a view to Theorem 21 not surprisingly), it is only a heuristic, which may fail to reconstruct the correct paths. The reason is that the weights used to measure the quality of the assignment do not incorporate the requirement that no two particles can have originated from the same location at the previous moment in time. Explicit examples are given in [64] which also gives generalizations that combine the general rolling horizon approach with interpolation and backtracking techniques to provide algorithms that incorporate physical knowledge even better while still running in polynomial time.

13.7 Tomographic grain mapping

Tomographic grain mapping deals with the problem of characterizing *polycrystalline materials* from tomographic data. Polycrystalline materials consist of multiple crys-

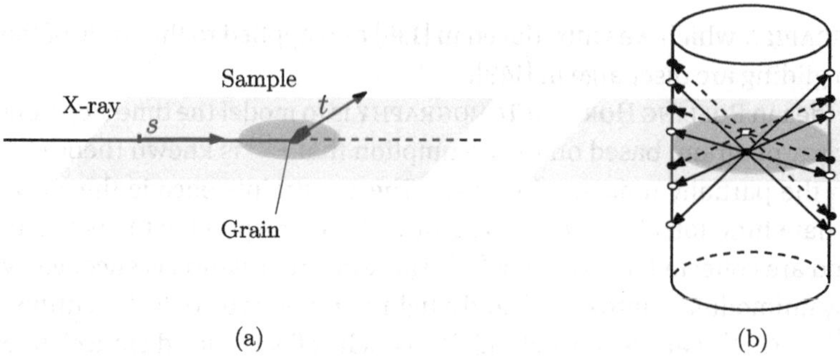

Figure 13.11: [From [149]] (a) Diffraction by a grain (incoming and diffraction directions are s and t, respectively); (b) the general indexing problem (determining the grains from diffraction spots).

tals, called *grains*. These grains, often 10–100 micrometer in diameter, are of central interest in many areas of materials science as most metals, ceramics and alloys are such polycrystalline materials. In fact, the grains determine many of the material's physical, chemical, and mechanical properties (see, e. g., [154, 160, 162, 166] or the monographs [39, 50]).

13.7.1 Diffraction and indexing

There are several nontrivial technological and algorithmical challenges involved in tomographic grain mapping on different scales. Typically, only high-energy x-rays will penetrate the material. In fact, the required x-ray energies are often so large that current experiments need to be conducted at modern synchrotron facilities. For many applications, the data are acquired by diffraction (as, e. g., in the 3-dimensional x-ray diffraction microscopy technique, 3DXRD [3, 10, 47, 48] and in diffraction contrast tomography, DCT [159]). Diffraction occurs, however, only if the grain is in a "favorable" position. This is governed by *Bragg's law* which relates the unit vectors t, s that signify the incoming and the diffraction directions and the wavelength λ of the x-ray with the crystalline structure of the grain encoded by its dual (or reciprocal) lattice L°. More precisely, Bragg's law is as follows:

$$\frac{t-s}{\lambda} = \ell \in L^\circ \setminus \{0\};$$

(see Figure 13.11(a) for an illustration). Consequently, tomographic data are typically only available from a small number of directions (often, 8–10).

The limited number of data, and the fact that multiple grains are simultaneously imaged, poses major algorithmic challenges at quite different scales, from the atomic to the macroscopic level. The problem, commonly referred to as *indexing* [47, 156, 161], is to group the tomographic data according to their grain of origin. This allows often the determination of grain parameters like the lattice (including its orientation), or the center of mass; see Figure 13.11(b). Based on the tomographic data acquired for each single grain, the macroscopic geometric structure of the full collection of different grains is then to be determined. Of course, such tasks can be highly interrelated, and there are also possible cases where it is favorable to reconstruct several of the grains simultaneously. More details can be found in [149].

13.7.2 Macroscopic reconstruction

As in Section 13.5, a single grain g can be considered as a binary image $\psi_g \in \mathcal{F}^3$. The points of $\mathrm{supp}(\psi_g)$ correspond to the pixels that belong to g. The paper [163] describes one of the first attempts of reconstructing multiple grains, the so-called *grain map*. In this paper, the ART algorithm is used, but it is found that often the reconstructions contain unrealistic void spaces between adjacent grains. To overcome this problem, a Monte-Carlo approach based on Gibbs' priors was introduced in [150]. This approach was generalized in [165] (see also [5]) to deal with the task of reconstructing grain maps of moderately deformed grains. More stochastic approaches to grain map reconstruction can be found in [155, 157, 158, 165]. For alternative approaches, see [49, 151–153, 164, 167, 168].

In the following, we describe a linear-programming based method, introduced in [169], which returns approximations of grain maps. It is based on only few input parameters for each grain: (approximations of) its *center-of-mass*, its *volume* and, if available, its *second-order moments*. The centers-of-mass can be determined by the indexing procedure, the grain volume by integration of the respective x-ray data, and the second-order moments by backprojecting the projections acquired from the same grain.

The aim is to reconstruct what we call *generalized balanced power diagrams (GBPDs)*. These diagrams generalize *power diagrams* (which are also known as *Laguerre* or *Dirichlet tessellations*), which in turn generalize *Voronoi diagrams*; see also [7] and [8, Section 6.2].

Any GBPD is specified by a set of distinct *sites* $S = \{s_1, \dots, s_l\} \subset \mathbb{R}^d$, *additive weights* $(\sigma_1, \dots, \sigma_l)^T \in \mathbb{R}^l$, and positive definite matrices $A_1, \dots, A_l \in \mathbb{R}^{d \times d}$. The *jth generalized balanced power cell* P_j is then defined by

$$P_j = \{x \in \mathbb{R}^d \; : \; \|x - s_j\|_{A_j}^2 - \sigma_j \le \|x - s_k\|_{A_k}^2 - \sigma_k, \; \forall k \ne j\},$$

where $\|\cdot\|_{A_j}$, $j \in [l]$, denotes the *ellipsoidal norm*

$$\|x\|_{A_j} = \sqrt{x^T A_j x}.$$

The generalized balanced power diagram P is the l-tuple $P = (P_1, \dots, P_l)$. The proposed method is able to find optimal $\sigma_1, \dots, \sigma_l$ that guarantee that the volumes of each cell are within prescribed ranges.

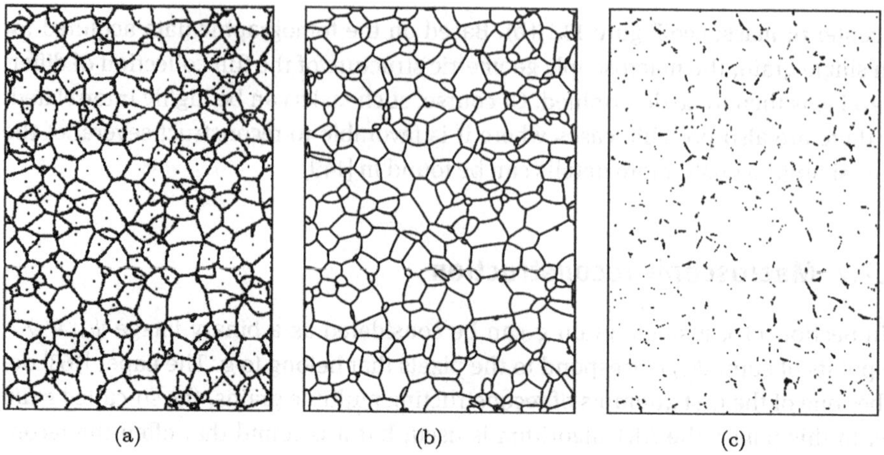

Figure 13.12: (a) Original image from [16, Figure 9.7]. (Black dots represent grain centers.) (b) Reconstructed generalized balanced power diagram. (c) Pixels that are black in (b) but white in (a).

The concept of GBPDs can be viewed as structure-driven weight balanced clusterings; see [171, 172]. For $j \in [l]$, let s_j denote the center of the jth grain, and let κ_j^-, κ_j^+ be lower and upper bounds for its volume, respectively. Further, let $x_1, \dots, x_q$ be the points of the image that has to be partitioned into the grains, and set $\gamma_{i,j} = \|x_i - s_j\|_{A_i}^2$ for all i, j.

Then we can model the assignment problem by the following linear program:

$$\begin{aligned}
\text{(LP)} \qquad & \min \sum_{i=1}^{q} \sum_{j=1}^{l} \gamma_{i,j} \xi_{i,j} & \\
\text{subject to} \quad & \sum_{j=1}^{l} \xi_{i,j} = 1 & (i \in [q]), \\
& \kappa_j^- \leq \sum_{i=1}^{q} \xi_{i,j} \leq \kappa_j^+ & (j \in [l]), \\
& \xi_{i,j} \geq 0 & (i \in [q];\ j \in [l]).
\end{aligned}$$

In general, the variables $\xi_{i,j}$ specify the fraction of the point x_i that is assigned to the center s_j. Since, however, the coefficient matrix is totally unimodular all basic feasible solutions are binary, and we obtain an optimal assignment of pixels to grains in polynomial time.

An example for the quality of reconstruction for planar grain maps (which are easier to visualize) is shown in Figure 13.12. Reports on the favorable performance of the presented approach on various (real-world) data sets can be found in the recent papers [173–177].

We remark that the clustering approach described above was previously applied (in an "isotropic" fashion) in the context of farmland consolidation [11, 170]. For an application in designing electoral districts where municipalities of a state have to be grouped into districts of nearly equal population while obeying certain politically motivated requirements, see [172].

13.8 Switching components and a problem in number theory

Switching components, i. e., pairs of tomographically equivalent sets as introduced in Section 13.3.1, are strongly related to an old problem in Diophantine number theory, called the *Prouhet–Tarry–Escott* or PTE-problem, named after Eugène Prouhet [204], Gaston Tarry [207], and Edward B. Escott [189].

Problem 5 (Prouhet, 1851; Tarry, 1912; Escott, 1910). Given $k, n \in \mathbb{N}$, find two different multisets $X = \{\xi_1, \ldots, \xi_n\} \subset \mathbb{Z}$ and $Y = \{\eta_1, \ldots, \eta_n\} \subset \mathbb{Z}$, such that

$$\xi_1^j + \xi_2^j + \cdots + \xi_n^j = \eta_1^j + \eta_2^j + \cdots + \eta_n^j, \quad \text{for } j \in [k].$$

Pairs (X, Y) satisfying the above equation are called *PTE solutions*. More precisely, we speak of (k, n)-*solutions*, and the numbers k and n, respectively, are referred to as the *degree* and *size* of the PTE solution. Often the notation $X \stackrel{k}{=} Y$ is used to indicate that (X, Y) is a degree k solution. For instance, as an elementary calculation shows

$$\{0, 14, 28, 56, 70, 84\} \stackrel{5}{=} \{4, 6, 40, 44, 78, 80\}.$$

The PTE problem has connections to several other problems in number theory, including the *"easier" Waring problem* [208, 211], [12, Section 12], the *Hilbert–Kamke problem* [194, 195], and a conjecture due to Erdős and Szekeres [187, 188, 197], [12, Section 13]. There are also connections to *Ramanujan identities* [199, 203], other types of multigrade equations [186, 205], problems in algebra [198, 201], geometry [190], combinatorics [178, 179, 182, 185], graph theory [193], and computer science [183, 184, 191]. For background information, see [4, 12, 13, 18, 23].

The PTE problem can be traced back to a correspondence between Goldbach and Euler. In his 1950 letter [192], Goldbach states the identity

$$\begin{aligned}(\alpha + \beta + \delta)^2 + (\alpha + \gamma + \delta)^2 + (\beta + \gamma + \delta)^2 + \delta^2 \\ = (\alpha + \delta)^2 + (\beta + \delta)^2 + (\gamma + \delta)^2 + (\alpha + \beta + \gamma + \delta)^2,\end{aligned}$$

where $\alpha, \beta, \gamma, \delta \in \mathbb{Z}$. In other words,

$$\{\alpha + \beta + \delta, \alpha + \gamma + \delta, \beta + \gamma + \delta, \delta\} \stackrel{2}{=} \{\alpha + \delta, \beta + \delta, \gamma + \delta, \alpha + \beta + \gamma + \delta\}.$$

It was already known to Prouhet, Tarry, and Escott that there exist $(k, 2^k)$-solutions for every k (see [210] and [18, Section 24]). Such solutions can be generated as follows. Express each $p \in [2^{k+1} - 1]_0$ as a binary number. If this binary expression of p contains an even number of 1's, then assign p to the set X, otherwise to Y. Then (X, Y) with $X = \{\xi_1, \ldots, \xi_{2^k}\}$ and $Y = \{\eta_1, \ldots, \eta_{2^k}\}$ is a $(k, 2^k)$-solution. Proofs of this result can be found in [202, 210]. For generalizations, see [196, 206].

On the other hand, there are no (k, n)-solutions whenever $n < k + 1$. This result, commonly attributed to Bastien [181], can be derived from the Newton's identities [31, Section 21.9]. A (k, n)-solution is called *ideal* if $n = k + 1$.

The following is a long-standing open question (see [209] and [12, Section 11]).

Problem 6. Do there exist ideal PTE solutions for every k?

Presently, ideal solutions are only known for $k \in [11] \setminus \{10\}$. Concerning upper bounds on n, the currently best bound (of [200]) guarantees that for any k there exists a (k, n)-solution with $n \leq \frac{1}{2}(k^2 - 3)$ if k is odd and $n \leq \frac{1}{2}(k^2 - 4)$ if k is even. The proofs are nonconstructive. In fact, all currently known constructive proofs yield bounds that are exponential in k.

13.8.1 PTE solutions from switching components

The following explicit connection between the PTE problem and switching components first appeared in [2, Section 6]. Following [180], we will focus on the case $d = 2$ (for general d, see [61]).

For given $M \subset \mathbb{Z}^d$ and $c \in \mathbb{Z}^d$, let $\Pi_c(M)$ denote the multiset

$$\Pi_c(M) = \{c^T x \; : \; x \in X\}.$$

Clearly, $\Pi_c(M) \subset \mathbb{Z}$. Perhaps more surprisingly, the following result holds if we insert the points of a switching component.

Theorem 24 ([180]). *If (X, Y) is an $(m + 1)$-switching component in $\mathbb{Z}^2$ and $c \in \mathbb{Z}^2$ such that $\Pi_c(X) \neq \Pi_c(Y)$, then $(\Pi_c(X), \Pi_c(Y))$ is a degree m solution of the PTE problem.*

This construction of PTE-solutions from switching components can be exemplified, say, for the switching components depicted in Figure 13.4.

For instance, if in Figure 13.4(a) the origin is located in the lower left lattice point (which, of course, is an arbitrary choice) the sets X and Y of white and black points are

$$X = \left\{ \begin{pmatrix} 0 \\ 2 \end{pmatrix}, \begin{pmatrix} 1 \\ 0 \end{pmatrix}, \begin{pmatrix} 2 \\ 5 \end{pmatrix}, \begin{pmatrix} 4 \\ 1 \end{pmatrix}, \begin{pmatrix} 5 \\ 6 \end{pmatrix}, \begin{pmatrix} 6 \\ 4 \end{pmatrix} \right\},$$

$$Y = \left\{ \begin{pmatrix} 0 \\ 1 \end{pmatrix}, \begin{pmatrix} 1 \\ 4 \end{pmatrix}, \begin{pmatrix} 2 \\ 0 \end{pmatrix}, \begin{pmatrix} 4 \\ 6 \end{pmatrix}, \begin{pmatrix} 5 \\ 2 \end{pmatrix}, \begin{pmatrix} 6 \\ 5 \end{pmatrix} \right\}.$$

For $c^T = (1,2)$, we obtain the PTE solution

$$\Pi_c(X) = \{1,4,6,12,14,17\} \stackrel{5}{=} \{2,2,9,9,16,16\} = \Pi_c(Y)$$

of degree 5. As a basic ingredient the standard proof of Theorem 24 uses the encoding of points by polynomials mentioned in connection with Theorem 7.

Theorem 24 allows to derive explicit constructions of families of PTE solutions. As an example, let us consider the result of Prouhet, Tarry, and Escott that $(k, 2^k)$-solutions exist for every k. The proof given in [209] extends over two half-pages. The geometric shortcut via Theorem 24 just uses the construction of switching components (X, Y) with $|X| = |Y| \leq 2^k$ from Figure 13.2.

13.8.2 Generalizations

The geometric point of view also helps in studying other variants of the PTE problem. Naturally, PTE can be considered over arbitrary rings R. For $R = \mathbb{Z}^d$, we obtain PTE_d which can be viewed as a d-dimensional or multinomial version of the original PTE problem.

Problem 7 (PTE_d). Given $d, k, n \in \mathbb{N}$, find two different multisets $X = \{x_1, \ldots, x_n\}$, $Y = \{y_1, \ldots, y_n\} \subset \mathbb{Z}^d$ with $x_l = (\xi_{l1}, \ldots, \xi_{ld})^T$, $y_l = (\eta_{l1}, \ldots, \eta_{ld})^T$ for $l \in [n]$ such that

$$\sum_{l=1}^{n} \xi_{l1}^{j_1} \xi_{l2}^{j_2} \cdot \cdots \cdot \xi_{ld}^{j_d} = \sum_{l=1}^{n} \eta_{l1}^{j_1} \eta_{l2}^{j_2} \cdot \cdots \cdot \eta_{ld}^{j_d}$$

for all nonnegative integers $j_1, \ldots, j_d$ with $j_1 + j_2 + \cdots + j_d \leq k$.

There are trivial ways of generating PTE_d-solution from PTE_1-solutions. For instance, if $\{\alpha_1, \ldots, \alpha_n\} \stackrel{k}{=} \{\beta_1, \ldots, \beta_n\}$ is a PTE_1-solution, then

$$\{(\alpha_1, \ldots, \alpha_1)^T, \ldots, (\alpha_n, \ldots, \alpha_n)^T\} \stackrel{k}{=} \{(\beta_1, \ldots, \beta_1)^T, \ldots, (\beta_n, \ldots, \beta_n)^T\}$$

is a solution to PTE_d. A general method of generating nontrivial solutions for PTE_2 is provided by [180].

Theorem 25 ([180]). *Every $(m+1)$-switching component (X, Y) in $\mathbb{Z}^2$, is a degree m solution of the* PTE_2 *problem.*

For instance, for the sets X and Y corresponding to Figure 13.4(a) it is elementary to verify that

$$\begin{aligned} &0^i 2^j + 1^i 0^j + 2^i 5^j + 4^i 1^j + 5^i 6^j + 6^i 4^j \\ &= 0^i 1^j + 1^i 4^j + 2^i 0^j + 4^i 6^j + 5^i 2^j + 6^i 5^j \end{aligned}$$

for all non-negative i, j with $i + j \leq 5$.

Applying Theorem 25 to the known smallest size switching components (an example for $k+1 = 6$ is depicted in Figure 13.4(a)), one sees that for every degree $k \in \{1, 2, 3, 5\}$ there exist ideal PTE_2 solutions; [180].

The following problem is, however, open already for $k = 4$.

Problem 8. Do there exist ideal PTE_2 for every degree k?

13.9 Concluding remarks

The present paper tried to support the following conviction of the authors: Discrete tomography is a broad and interesting field, both, in terms of its methods and its applications. Discrete tomography has strong links to various areas within mathematics which have the potential to provide new insight in older problems. Discrete tomography has a variety of applications to various other scientific fields and to relevant real-world problems. And, finally, discrete tomography is a rich source of scientific challenges.

Bibliography

General reading: introductions, surveys, books

[1] R. J. Adrian and J. Westerweel. *Particle image velocimetry*. Cambridge University Press, New York, NY, 2010. ISBN: 978-0-521-44008-0.

[2] A. Alpers. Instability and stability in discrete tomography. (Published by Shaker Verlag, ISBN 3-8322-2355-X). PhD thesis. Zentrum Mathematik: Technische Universität München, 2003.

[3] A. Alpers. A short introduction to tomographic grain map reconstruction. *PU.M.A.*, 20(1–2):157–163, 2009. URL: http://puma.dimai.unifi.it/20_1_2/10_Alpers_7.pdf.

[4] A. Alpers. On the tomography of discrete structures: Mathematics, complexity, algorithms, and its applications in materials science and plasma physics. Habilitation thesis. Zentrum Mathematik, Technische Universität München, 2018. URL: http://mediatum.ub.tum.de/1441933.

[5] A. Alpers, L. Rodek, H. F. Poulsen, E. Knudsen, and G. T. Herman. Discrete tomography for generating grain maps of polycrystals. In: G. T. Herman and A. Kuba, editors, *Advances in discrete tomography and its applications*, pages 271–301. Birkhäuser, Boston, 2007. DOI: 10.1007/978-0-8176-4543-4_13.

[6] R. C. Aster, B. Borchers, and C. H. Thurber. *Parameter estimation and inverse problems*, 2nd edition. Academic Press, Boston, MA, 2013. ISBN: 978-0-12-3850485. DOI: 10.1016/B978-0-12-385048-5.00030-6.

[7] F. Aurenhammer. Power diagrams: Properties, algorithms and applications. *SIAM J. Comput.*, 16(1):78–96, 1987. DOI: 10.1137/0216006.

[8] F. Aurenhammer, R. Klein, and D.-T. Lee. *Voronoi diagrams and Delaunay triangulations*. World Scientific, Singapore, 2013. ISBN: 978-981-4447-63-8.

[9] S. Bals, B. Goris, A. De Backer, S. Van Aert, and G. Van Tendeloo. Atomic resolution electron tomography. *Mater. Res. Soc. Bull.*, 41(7):525–530, 2016. DOI: 10.1557/mrs.2016.138.

[10] J. Banhart, ed. *Advanced tomographic methods in materials research and engineering*. Oxford University Press, Oxford, 2008. ISBN: 978-0-19-921324-5.

[11] S. Borgwardt, A. Brieden, and P. Gritzmann. Geometric clustering for the consolidation of farmland and woodland. *Math. Intell.*, 36(2):37–44, 2014. DOI: 10.1007/s00283-014-9448-2.

[12] P. Borwein. *Computational excursions in analysis and number theory*. Springer, New York, NY, 2002. ISBN: 0-387-95444-9. DOI: 10.1007/978-0-387-21652-2.

[13] P. Borwein and C. Ingalls. The Prouhet–Tarry–Escott problem revisited. *Enseign. Math.*, 40(1–2):3–27, 1994. DOI: 10.5169/seals-61102.

[14] R. A. Brualdi. *Combinatorial matrix classes*. Cambridge University Press, Cambridge, 2006. ISBN: 0521865654.

[15] R. Burkard, M. Dell'Amico, and S. Martello. *Assignment problems*. SIAM, Philadelphia, PA, 2009. ISBN: 978-0-898716-63-4. DOI: 10.1137/1.9781611972238.

[16] S. N. Chiu, D. Stoyan, W. S. Kendall, and J. Mecke. *Stochastic geometry and its applications*, 3rd edition. Wiley, Chichester, 2013. ISBN: 978-0-470-66481-0.

[17] A. Del Lungo and M. Nivat. Reconstruction of connected sets from two projections. In: G. T. Herman and A. Kuba, editors, *Discrete tomography*, pages 163–188. Birkhäuser, Boston, 1999. DOI: 10.1007/978-1-4612-1568-4_7.

[18] L. E. Dickson. *History of the theory of numbers*, volume 2. Dover Publications, Mineola, NY, 1920. ISBN: 0-486-44233-0.

[19] L. Gan and G. J. Jensen. Electron tomography of cells. *Q. Rev. Biophys.*, 45(1):27–56, 2012. DOI: 10.1017/S0033583511000102.

[20] R. J. Gardner. *Geometric tomography*. 2nd edition. Cambridge University Press, New York, NY, 2006. ISBN: 978-0-521-68493-4. DOI: 10.1017/CBO9781107341029.

[21] R. J. Gardner and P. Gritzmann. Uniqueness and complexity in discrete tomography. In: G. T. Herman and A. Kuba, editors, *Discrete tomography*, pages 85–113. Birkhäuser, Boston, 1999. DOI: 10.1007/978-1-4612-1568-4_4.

[22] M. R. Garey and D. S. Johnson. *Computers and intractability*. W. H. Freeman and Co., San Francisco, CA, 1979. ISBN: 0-7167-1045-5.

[23] A. Gloden. *Mehrgradige Gleichungen*, 2nd edition. P. Noordhoff, Groningen, 1943.

[24] U. Grimm, P. Gritzmann, and C. Huck. Discrete tomography of model sets: Reconstruction and uniqueness. In: M. Baake and U. Grimm, editors, *Aperiodic order*, volume 2, pages 39–72. Cambridge University Press, Cambridge, 2017. DOI: 10.1017/9781139033862.004.

[25] P. Gritzmann. On the reconstruction of finite lattice sets from their X-rays. In: E. Ahronovitz and C. Fiorio, editors, *Discrete geometry for computer imagery*, pages 19–32. Springer, Berlin, 1997. DOI: 10.1007/BFb0024827.

[26] P. Gritzmann. Discrete tomography: From battleship to nanotechnology. In: M. Aigner and E. Behrends, editors, *Mathematics everywhere*, pages 81–98. American Mathematical Society, 2010. DOI: 10.1136/bmj.c3793.

[27] P. Gritzmann and S. de Vries. Reconstructing crystalline structures from few images under high resolution transmission electron microscopy. In: W. Jäger and H. J. Krebs, editors, *Mathematics: Key technology for the future*, pages 441–459. Springer, Berlin, 2003. DOI: 10.1007/978-3-642-55753-8_36.

[28] J. Hadamard. *Lectures on the Cauchy problem in linear partial differential equations*. Reprint of the 1923 original. Dover Publications Inc., Mineola, NY, 1952. ISBN: 978-0-486-78148-8.

[29] L. Hajdu and R. Tijdeman. Algebraic discrete tomography. In: G. T. Herman and A. Kuba, editors, *Advances in discrete tomography and its applications*, pages 55–81. Birkhäuser, Boston, 2007. DOI: 10.1007/978-0-8176-4543-4_4.

[30] P. C. Hansen. *Discrete inverse problems: Insight and algorithms*. SIAM, Philadelphia, PA, 2010. ISBN: 978-0-89871-696-2. DOI: 10.1137/1.9780898718836.

[31] G. H. Hardy and E. M. Wright. *An introduction to the theory of numbers*, 6th edition. Oxford University Press, Oxford, 2008. ISBN: 0-19-853171-0.

[32] G. T. Herman. *Fundamentals of computerized tomography: Image reconstruction from projections*, 2nd edition. Springer, London, 2009. ISBN: 978-1-85233-617-2. DOI: 10.1007/978-1-84628-723-7.

[33] G. T. Herman. Iterative reconstruction techniques and their superiorization for the inversion of the Radon transform. In: O. Scherzer and R. Ramlau, editors, *The Radon transform: The first 100 years and beyond*. De Gruyter, Berlin, 2019.

[34] G. T. Herman and J. Frank, eds. *Computational methods for three-dimensional microscopy reconstruction*. Springer, New York, NY, 2014. ISBN: 978-1-4614-9520-8. DOI: 10.1007/978-1-4614-9521-5.

[35] G. T. Herman and A. Kuba, eds. *Discrete tomography: Foundations, algorithms, and applications*. Birkhäuser, Boston, MA, 1999. ISBN: 0-8176-4101-7. DOI: 10.1007/978-1-4612-1568-4.

[36] G. T. Herman and A. Kuba, eds. *Advances in discrete tomography and its applications*. Birkhäuser, Boston, MA, 2007. ISBN: 978-0-8176-3614-2. DOI: 10.1007/978-08176-4543-4.

[37] M. B. Katz. *Questions of uniqueness and resolution in reconstruction from projections*. Springer, Berlin, 1978. ISBN: 3-540-09087-8. DOI: 10.1007/978-3-642-45507-0.

[38] A. Kirsch. *An introduction to the mathematical theory of inverse problems*. Springer, New York, NY, 2011. ISBN: 978-1-4419-8473-9. DOI: 10.1007/978-14419-8474-6.

[39] U. F. Kocks, C. N. Tome, and H.-R. Wenk. *Texture and anisotropy: Preferred orientations in polycrystals and their effect on materials properties*. Cambridge University Press, Cambridge, 2000. ISBN: 9780521794206.

[40] A. K. Louis. Uncertainty, ghosts and resolution in Radon problems. In: O. Scherzer and R. Ramlau, editors, *The Radon transform: The first 100 years and beyond*. De Gruyter, Berlin, 2019.

[41] A. Markoe. *Analytic tomography*. Cambridge University Press, New York, NY, 2014. ISBN: 978-1-107-43862-0. DOI: 10.1017/CBO9780511530012.

[42] J. L. Mueller and S. Siltanen. *Linear and nonlinear inverse problems with practical applications*. SIAM, Philadelphia, PA, 2012. ISBN: 978-1-61197-233-7. DOI: 10.1137/1.9781611972344.

[43] F. Natterer. *The mathematics of computerized tomography*. SIAM, Philadelphia, PA, 2001. ISBN: 978-0-89871-493-7. DOI: 10.1137/1.9780898719284.

[44] F. Natterer and F. Wübbeling. *Mathematical methods in image reconstruction*. SIAM, Philadelphia, PA, 2001. ISBN: 978-0-89871-622-1. DOI: 10.1137/1.9780898718324.

[45] O. Öktem. Mathematics of electron tomography. In: O. Scherzer, editor, *Handbook of mathematical methods in imaging*, 2nd edition, pages 937–1031. Springer, New York, NY, 2015. DOI: 10.1007/978-3-642-27795-5_43-2.

[46] C. H. Papadimitriou. *Computational complexity*. Addison-Wesley, Reading, MA, 1995. ISBN: 978-0201530827.

[47] H. F. Poulsen. *Three-dimensional X-ray diffraction microscopy*. Springer, Berlin, 2004. ISBN: 3-540-22330-4.

[48] H. F. Poulsen. An introduction to three-dimensional X-ray diffraction microscopy. *J. Appl. Crystallogr.*, 45(6):1084–1097, 2012. DOI: 10.1107/S0021889812039143.

[49] H. F. Poulsen, S. Schmidt, D. Juul Jensen, H. O. Sørensen, E. M. Lauridsen, U. L. Olsen, W. Ludwig, A. King, J. P. Wright, and G. B. M. Vaughan. 3D X-Ray diffraction microscopy. In: R. Barabash and G. Ice, editors, *Strain and dislocation gradients from diffraction: Spatially-resolved local structure and defects*, pages 205–253. World Scientific, Singapore, 2014.

[50] L. Priester. *Grain boundaries: From theory to engineering*. Springer, Dordrecht, 2013. ISBN: 978-94-007-4968-9. DOI: 10.1007/978-94-007-4969-6.

[51] H. J. Ryser. Combinatorial properties of matrices of zeros and ones. *Can. J. Math.*, 9(1):371–377, 1957. DOI: 10.1007/978-0-8176-4842-8_18.

[52] H. J. Ryser. *Combinatorial mathematics*. MAA, Washington, DC, 1963. ISBN: 978-0883850145.

[53] A. Schrijver. *Theory of linear and integer programming*. Wiley, Chichester, 1986. ISBN: 0-471-90854-1.

[54] A. Schröder and C. E. Willert, eds. *Particle image velocimetry*. Springer, Berlin, 2008. ISBN: 978-3-540-73527-4. DOI: 10.1007/978-3-540-73528-1.

[55] D. J. Smith. Progress and perspectives for atomic-resolution electron microscopy. *Ultramicroscopy*, 108(3):159–166, 2018. DOI: 10.1016/j.ultramic.2007.08.015.

[56] J. C. H. Spence. *High-resolution electron microscopy*, volume 4. Oxford Univ. Press, Oxford, 2013. ISBN: 9780199668632. DOI: 10.1093/acprof:oso/9780199668632.001.0001.

[57] A. Tarantola. *Inverse problem theory and methods for model parameter estimation*. SIAM, Philadelphia, PA, 2005. ISBN: 978-0-898715-72-9.

Citations in tomography

[58] R. Aharoni, G. T. Herman, and A. Kuba. Binary vectors partially determined by linear equation systems. *Discrete Math.*, 171(1–3):1–16, 1997. DOI: 10.1016/S0012-365X(96)00068-4.

[59] A. Alpers and S. Brunetti. On the stability of reconstructing lattice sets from X-rays along two directions. In: E. Andres, G. Damiand, and P. Lienhardt, editors, *Discrete geometry for computer imagery*. LNCS, volume 3429, pages 92–103. Springer, Berlin, 2005. DOI: 10.1007/978-3-540-31965-8_9.

[60] A. Alpers, R. J. Gardner, S. König, R. S. Pennington, C. B. Boothroyd, L. Houben, R. E. Dunin-Borkowski, and K. J. Batenburg. Geometric reconstruction methods for electron tomography. *Ultramicroscopy*, 128(C):42–54, 2013. DOI: 10.1016/j.ultramic.2013.01.002.

[61] A. Alpers, V. Ghiglione, and P. Gritzmann. On switching components. In preparation, 2019.

[62] A. Alpers and P. Gritzmann. On stability, error correction, and noise compensation in discrete tomography. *SIAM J. Discrete Math.*, 20(1):227–239, 2006. DOI: 10.1137/040617443.

[63] A. Alpers and P. Gritzmann. Reconstructing binary matrices under window constraints from their row and column sums. *Fundam. Inform.*, 155(4):321–340, 2017. DOI: 10.3233/FI-2017-1588.

[64] A. Alpers and P. Gritzmann. Dynamic discrete tomography. *Inverse Probl.*, 34(3):034003 (26 pp), 2018. DOI: 10.1088/1361-6420/aaa202.

[65] A. Alpers and P. Gritzmann. On double-resolution imaging and discrete tomography. *SIAM J. Discrete Math.*, 32(2):1369–1399, 2018. DOI: 10.1137/17M1115629.

[66] A. Alpers, P. Gritzmann, and L. Thorens. Stability and instability in discrete tomography. In: G. Bertrand, A. Imiya, and R. Klette, editors, *Digital and image geometry*. LNCS, volume 2243, pages 175–186. Springer, Berlin, 2001. DOI: 10.1007/3-540-45576-0_11.

[67] A. Alpers and D. G. Larman. The smallest sets of points not determined by their X-rays. *Bull. Lond. Math. Soc.*, 47(1):171–176, 2015. DOI: 10.1112/blms/bdu111.

[68] A. Bains and T. Biedl. Reconstructing *hv*-convex multi-coloured polyominoes. *Theor. Comput. Sci.*, 411(34–36):3123–3128, 2010. DOI: 10.1016/j.tcs.2010.04.041.

[69] S. Bals, K. J. Batenburg, J. Verbeeck, J. Sijbers, and G. Van Tendeloo. Quantitative three-dimensional reconstruction of catalyst particles for bamboo-like carbon-nanotubes. *Nano Lett.*, 7(12):3669–3674 2007. DOI: 10.1021/nl071899m.

[70] E. Barcucci, A. Del Lungo, M. Nivat, and R. Pinzani. X-rays characterizing some classes of discrete sets. *Linear Algebra Appl.*, 339(1–3):3–21, 2001. DOI: 10.1016/S0024-3795(01)00431-1.

[71] E. Barcucci, P. Dulio, A. Frosini, and S. Rinaldi. Ambiguity results in the characterization of *hv*-convex polyominoes from projections. In: W. Kropatsch, N. Artner, and I. Janusch, editors, *Discrete geometry for computer imagery*. LNCS, volume 10502, pages 147–158. Springer, Berlin, 2017. DOI: 10.1007/978-3-319-66272-5_13.

[72] K. J. Batenburg, S. Bals, S. Sijbers, C. Kübel, P. A. Midgley, J. C. Hernandez, U. Kaiser, E. R. Encina, E. A. Coronado, and G. Van Tendeloo. 3D imaging of nanomaterials by discrete tomography. *Ultramicroscopy*, 109(6):730–740, 2009. DOI: 10.1016/j.ultramic.2009.01.009.

[73] K. J. Batenburg and W. A. Kosters. A discrete tomography approach to Japanese puzzles. In: *Proceedings of the 16th Belgium–Netherlands conference on artificial intelligence (BNAIC)*, pages 243–250. 2004.

[74] K. J. Batenburg and J. Sijbers. DART: A practical reconstruction algorithm for discrete tomography. *IEEE Trans. Image Process.*, 20(9):2542–2553, 2011. DOI: 10.1109/TIP.2011.2131661.

[75] G. Bianchi and M. Longinetti. Reconstructing plane sets from projections. *Discrete Comput. Geom.*, 5(3):223–242 1990. DOI: 10.1007/BF02187787.

[76] J. Bosboom, E. D. Demaine, M. L. Demaine, A. Hesterberg, R. Kimball, and J. Kopinsky. Path puzzles: Discrete tomography with a path constraint is hard, 2018. URL: http://arxiv.org/abs/1803.01176.

[77] S. Brunetti and A. Daurat. An algorithm reconstructing convex lattice sets. *Theor. Comput. Sci.*, 304(1–3):35–57, 2003. DOI: 10.1016/S0304-3975(03)00050-1.

[78] S. Brunetti and A. Daurat. Reconstruction of convex lattice sets from tomographic projections in quartic time. *Theor. Comput. Sci.*, 406(1–2):55–62, 2008. DOI: 10.1016/j.tcs.2008.06.003.

[79] S. Brunetti, A. Del Lungo, P. Gritzmann, and S. de Vries. On the reconstruction of binary and permutation matrices under (binary) tomographic constraints. *Theor. Comput. Sci.*, 406(1–2):63–71, 2008. DOI: 10.1016/j.tcs.2008.06.014.

[80] S. Brunetti, P. Dulio, L. Hajdu, and C. Peri. Ghosts in discrete tomography. *J. Math. Imaging Vis.*, 53(2):210–224, 2015. DOI: 10.1007/s10851-015-0571-2.

[81] S. Brunetti, P. Dulio, and C. Peri. Discrete tomography determination of bounded sets in $\mathbb{Z}^n$. *Discrete Appl. Math.*, 183:20–30 2015. DOI: 10.1016/j.dam.2014.01.016.

[82] M. Burger, H. Dirks, L. Frerking, T. Hauptmann, A. Helin, and S. Siltanen. A variational reconstruction method for undersampled dynamic X-ray tomography based on physical motion models. *Inverse Probl.*, 33(12):124008, 2017. DOI: 10.1088/1361-6420/aa99cf.

[83] A. Chambolle. An algorithm for total variation minimization and applications. *J. Math. Imaging Vis.*, 20(1):89–97, 2004. DOI: 10.1023/B:JMIV.0000011325.36760.1e.

[84] S.-K. Chang. The reconstruction of binary patterns from their projections. *Commun. ACM*, 14(1):21–25, 1971. DOI: 10.1145/362452.362471.

[85] M. J. Chlond. Classroom exercises in IP modeling: Su Doku and the Log Pile. *INFORMS Trans. Ed.*, 5(2):77–79, 2005. DOI: 10.1287/ited.5.2.77.

[86] A. Daurat. Determination of Q-convex sets by X-rays. *Theor. Comput. Sci.*, 332(1–3):19–45, 2005. DOI: 10.1016/j.tcs.2004.10.001.

[87] D. de Werra, M. C. Costa, C. Picouleau, and B. Ries. On the use of graphs in discrete tomography. *4OR*, 6(2):101–123, 2008. DOI: 10.1007/s10288-008-0077-5.

[88] J. Diemunsch, M. Ferrara, S. Jahanbekam, and J. M. Shook. Extremal theorems for degree sequence packing and the two-color discrete tomography problem. *SIAM J. Discrete Math.*, 29(4):2088–2099, 2015. DOI: 10.1137/140987912.

[89] P. Dulio, R. J. Gardner, and C. Peri. Discrete point X-rays. *SIAM J. Discrete Math.*, 20(1):171–188, 2006. DOI: 10.1137/040621375.

[90] P. Dulio and C. Peri. Discrete Tomography and plane partitions. *Adv. Appl. Math.*, 50(3):390–408, 2013. DOI: 10.1016/j.aam.2012.10.005.

[91] C. Dürr. Discrete tomography applets. Accessed: 2018-09. URL: http://www-desir.lip6.fr/~durrc/Xray.

[92] C. Dürr, F. Guiñez, and M. Matamala. Reconstructing 3-colored grids from horizontal and vertical projections is NP-Hard: A solution to the 2-atom problem in discrete tomography. *SIAM J. Discrete Math.*, 26(1):330–352, 2012. DOI: 10.1137/100799733.

[93] P. C. Fishburn, J. C. Lagarias, J. A. Reeds, and L. A. Shepp. Sets uniquely determined by projections on axes II: Discrete case. *Discrete Math.*, 91(2):149–159, 1991. DOI: 10.1016/0012-365X(91)90106-C.

[94] D. Gale. A theorem on flows in networks. *Pac. J. Math.*, 7(2):1073–1082, 1957.

[95] R. J. Gardner. Geometric tomography website. Accessed: 2018-09. URL: http://www.geometrictomography.com.

[96] R. J. Gardner and P. Gritzmann. Discrete tomography: Determination of finite sets by X-rays. *Trans. Am. Math. Soc.*, 349(6):2271–2295, 1997. DOI: 10.1090/S0002-9947-97-01741-8.

[97] R. J. Gardner, P. Gritzmann, and D. Prangenberg. On the reconstruction of binary images from their discrete Radon transform. In: R. A. Melter, A. Y. Wu, and L. Latecki, editors, *Vision geometry V*. SPIE Proc., volume 2826, pages 121–132. Society of Photo-Optical Instrumentation Engineers, Denver, CO, 1996. DOI: 10.1117/12.251785.

[98] R. J. Gardner, P. Gritzmann, and D. Prangenberg. On the computational complexity of reconstructing lattice sets from their X-rays. *Discrete Math.*, 202(1–3):45–71, 1999. DOI: 10.1016/S0012-365X(98)00347-1.

[99] R. J. Gardner, P. Gritzmann, and D. Prangenberg. On the computational complexity of determining polyatomic structures by X-rays. *Theor. Comput. Sci.*, 233(1–2):91–106, 2000. DOI: 10.1016/S0304-3975(97)00298-3.

[100] R. J. Gardner and P. McMullen. On Hammer's X-ray problem. *J. Lond. Math. Soc. (2)*, 21(1):171–175, 1980. DOI: 10.1112/jlms/s2-21.1.171.

[101] P. Gritzmann and S. de Vries. On the algorithmic inversion of the discrete Radon transform. *Theor. Comput. Sci.*, 281(1–2):455–469, 2002. DOI: 10.1016/S0304-3975(02)00023-3.

[102] P. Gritzmann, B. Langfeld, and M. Wiegelmann. Uniqueness in discrete tomography: Three remarks and a corollary. *SIAM J. Discrete Math.*, 25(4):1589–1599, 2011. DOI: 10.1137/100803262.

[103] P. Gritzmann, D. Prangenberg, S. de Vries, and M. Wiegelmann. Success and failure of certain reconstruction and
uniqueness algorithms in discrete tomography. *Int. J. Imaging Syst. Technol.*, 9(2–3):101–109, 1998. DOI: 10.1002/(SICI)1098-1098(1998)9:2/3<101::AID-IMA6>3.0.CO;2-F.

[104] B. Hahn. Reconstruction of dynamic objects with affine deformations in computerized tomography. *J. Inverse Ill-Posed Probl.*, 22(3):323–339, 2014. DOI: 10.1515/jip-2012-0094.

[105] L. Hajdu and R. Tijdeman. Algebraic aspects of discrete tomography. *J. Reine Angew. Math.*, 534:119–128, 2001. DOI: 10.1515/crll.2001.037.

[106] A. Heppes. On the determination of probability distributions of more dimensions by their projections. *Acta Math. Acad. Sci. Hung.*, 7(3–4):403–410, 1956. DOI: 10.1007/BF02020535.

[107] G. T. Herman. Reconstruction of binary patterns from a few projections. In: A. Günther, B. Levrat, and H. Lipps, editors, *International computing symposium 1973*, pages 371–378. North-Holland, Amsterdam, 1974.

[108] L. Houben and M. Bar Sadan. Refinement procedure for the image alignment in high-resolution electron tomography. *Ultramicroscopy*, 111(9–10):1512–1520, 2011. DOI: 10.1016/j.ultramic.2011.06.001.

[109] C. Huck. Solution of a uniqueness problem in the discrete tomography of algebraic Delone sets. *J. Reine Angew. Math.*, 677:199–224, 2013. DOI: 10.1515/crelle.2012.026.

[110] R. W. Irving and M. R. Jerrum. Three-dimensional statistical data security problems. *SIAM J. Comput.*, 23(1):170–184, 1994. DOI: 10.1137/S0097539790191010.

[111] C. L. Jia, S. B. Mi, J. Barthel, D. W. Wang, R. E. Dunin-Borkowski, K. W. Urban, and A. Thust. Determination of the 3D shape of a nanoscale crystal with atomic resolution from a single image. *Nat. Mater.*, 13:1044–1049 2014. DOI: 10.1038/nmat4087.

[112] J. R. Jinschek, K. J. Batenburg, H. A. Calderon, R. Kilaas, V. Radmilovic, and C. Kisielowski. 3-D reconstruction of the atomic positions in a simulated gold nanocrystal based on discrete tomography: Prospects of atomic resolution electron tomography. *Ultramicroscopy*, 108(6):589–604, 2008. DOI: 10.1016/j.ultramic.2007.10.002.

[113] C. Kisielowski, P. Schwander, F. H. Baumann, M. Seibt, Y. Kim, and A. Ourmazd. An approach to quantitative high-resolution transmission electron microscopy of crystalline materials. *Ultramicroscopy*, 58(2):131–155, 1995. DOI: 10.1016/0304-3991(94)00202-X.

[114] J. Klukowska, R. Davidi, and G. T. Herman. SNARK09—a software package for reconstruction of 2D images from 1D projections. *Comput. Methods Programs Biomed.*, 110(3):424–440, 2013. DOI: 10.1016/j.cmpb.2013.01.003.

[115] T. Y. Kong and G. T. Herman. On which grids can tomographic equivalence of binary pictures be characterized in terms of elementary switching operations? *Int. J. Imaging Syst. Technol.*, 9(2–3):118–125, 1998. DOI: 10.1002/(SICI)1098-1098(1998)9:2/3<118::AID-IMA8>3.0.CO;2-E.

[116] A. Kuba and A. Volčič. Characterisation of measurable plane sets which are reconstructable from their two projections. *Inverse Probl.*, 4(2):513–527, 1988. DOI: 10.1088/0266-5611/4/2/014.

[117] M. Longinetti. Some questions of stability in the reconstruction of plane convex bodies from projections. *Inverse Probl.*, 1(1):87–97, 1985. DOI: 10.1088/0266-5611/1/1/008.

[118] G. G. Lorentz. A problem of plane measure. *Am. J. Math.*, 71(2):417–426, 1949. DOI: 10.2307/2372255.

[119] P. Maass. The X-ray transform: Singular value decomposition and resolution. *Inverse Probl.*, 3(4):729–741, 1987. DOI: 10.1088/0266-5611/3/4/016.

[120] J. Matoušek, A. Přívětivý, and P. Škovroň. How many points can be reconstructed from k projections? *SIAM J. Discrete Math.*, 22(4):1605–1623, 2008. DOI: 10.1016/j.endm.2007.07.069.

[121] A. Rényi. On projections of probability distributions. *Acta Math. Acad. Sci. Hung.*, 3(3):131–142, 1952. DOI: 10.1007/BF02022515.

[122] P. Schwander, C. Kisielowski, F. H. Baumann, Y. Kim, and A. Ourmazd. Mapping projected potential, interfacial roughness, and composition in general crystalline solids by quantitative transmission electron microscopy. *Phys. Rev. Lett.*, 71(25):4150–4153, 1993. DOI: 10.1103/PhysRevLett.71.4150.

[123] A. Shliferstein and Y. T. Chien. Switching components and the ambiguity problem in the reconstruction of pictures from their projections. *Pattern Recognit.*, 10(5–6):327–340, 1978. DOI: 10.1016/0031-3203(78)90004-3.

[124] A. R. Shliferstein and Y. T. Chien. Some properties of image-processing operations on projection sets obtained from digital pictures. *IEEE Trans. Comput.*, C-26(10):958–970, 1977. DOI: 10.1109/TC.1977.1674731.

[125] C. H. Slump and J. J. Gerbrands. A network flow approach to reconstruction of the left ventricle from two projections. *Comput. Vision. Graph*, 18(1):18–36, 1982. DOI: 0.1016/0146-664X(82)90097-1.

[126] K. T. Smith, D. C. Solmon, and S. L. Wagner. Practical and mathematical aspects of the problem of reconstructing objects from radiographs. *Bull. Am. Math. Soc.*, 83(6):1227–1270, 1977. URL: http://projecteuclid.org/euclid.bams/1183539851.

[127] I. Svalbe and M. Ceko. Maximal *N*-Ghosts and minimal information recovery from *N* projected views of an array. In: W. Kropatsch, N. Artner, and I. Janusch, editors, *Discrete geometry for computer imagery*. LNCS, volume 10502, pages 135–146. Springer, Cham, 2017. DOI: 10.1007/978-3-319-66272-5_12.

[128] I. Svalbe and S. Chandra. Growth of discrete projection ghosts created by iteration. In: I. Debled-Rennesson, E. Domenjoud, B. Kerautret, and P. Even, editors, *Discrete geometry for computer imagery*. LNCS, volume 6607, pages 406–416. Springer, Heidelberg, 2011. DOI: 10.1007/978-3-642-19867-0_34.

[129] I. Svalbe, N. Nazareth, N. Normand, and S. Chandra. On constructing minimal ghosts. In: *2010 International conference on digital image computing: Techniques and applications*, pages 276–281. IEEE, Los Alamitos, 2010. DOI: 10.1109/DICTA.2010.56.

[130] I. Svalbe and N. Normand. Properties of minimal ghosts. In: I. Debled-Rennesson, E. Domenjoud, B. Kerautret, and P. Even, editors, *Discrete geometry for computer imagery*. LNCS, volume 6607, pages 417–428. Springer, Heidelberg, 2011. DOI: 10.1007/978-3-642-19867-0_35.

[131] W. Van Aarle, W. J. Palenstijn, J. De Beenhouwer, T. Altantzis, S. Bals, K. J. Batenburg, and J. Sijbers. The ASTRA Toolbox: A platform for advanced algorithm development in electron tomography. *Ultramicroscopy*, 157:35–47, 2015. DOI: 10.1016/j.ultramic.2015.05.002. URL: www.astra-toolbox.com.

[132] S. Van Aert, K. J. Batenburg, M. D. Rossell, R. Erni, and G. Van Tendeloo. Threedimensional atomic imaging of crystalline nanoparticles. *Nature*, 470(7334):374–376, 2011. DOI: 10.1038/nature09741.

[133] B. Van Dalen. Stability results for uniquely determined sets from two directions in discrete tomography. *Discrete Math.*, 309(12):3905–3916, 2009. DOI: 10.1016/j.disc.2008.11.018.

[134] A. Volčič. Well-posedness of the Gardner-McMullen reconstruction problem. In: D. Kölzow and D. Maharam-Stone, editors, *Proceedings of conference on measure theory, Oberwolfach, 1983*, pages 199–210. LNM, volume 1089. 1984. DOI: 10.1007/BFb0072615.

[135] X. Zhuge, W. J. Palenstijn, and K. J. Batenburg. A more robust algorithm for discrete tomography from limited projection data with automated gray value estimation. *IEEE Trans. Image Process.*, 25(1):455–468, 2016. DOI: 10.1109/TIP.2015.2504869.

[136] S. Zopf. Construction of switching components. In: A. Kuba, L. G. Nyúl, and K. Palágyi, editors, *Discrete geometry for computer imagery*. LNCS, volume 4245, pages 157–168. Springer, Berlin, 2006. DOI: 10.1007/11907350_14.

Further reading: particle tracking

[137] R. J. Adrian. Particle-imaging techniques for experimental fluid mechanics. *Annu. Rev. Fluid Mech.*, 23:261–304, 1991. DOI: 10.1146/annurev.fl.23.010191.001401.

[138] A. Alpers, P. Gritzmann, D. Moseev, and M. Salewski. 3D particle tracking velocimetry using dynamic discrete tomography. *Comput. Phys. Commun.*, 187(1):130–136, 2015. DOI: 10.1016/j.cpc.2014.10.022.
[139] R. Dalitz, S. Petra, and C. Schnörr. Compressed motion sensing. In: F. Lauze, Y. Dong, and A. Dahl, editors, *Proc. SSVM*. LNCS, volume 10302, pages 602–613. Springer, 2017. DOI: 10.1007/978-3-319-58771-4_48.
[140] G. E. Elsinga, F. Scarano, B. Wieneke, and B. W. Oudheusden. Tomographic particle image velocimetry. *Exp. Fluids*, 41(6):933–947, 2006. DOI: 10.1007/s00348-006-0212-z.
[141] R. A. Jamison, A. Fouras, and R. J. Bryson-Richardson. Cardiac-phase filtering in intracardiac particle image velocimetry. *J. Biomed. Opt.*, 17(3):036007, 2012. DOI: 10.1117/1.JBO.17.3.036007.
[142] M. Novara, K. J. Batenburg, and F. Scarano. Motion tracking-enhanced MART for tomographic PIV. *Meas. Sci. Technol.*, 21(3):035401, 2010. DOI: 10.1088/0957-0233/21/3/035401.
[143] J. F. Pusztaszeri, P. E. Rensing, and T. M. Liebling. Tracking elementary particles near their primary vertex: A combinatorial approach. *J. Glob. Optim.*, 9(1):41–64, 1996. DOI: 10.1007/BF00121750.
[144] D. Reuss, R. Adrian, and C. Landreth. Two-dimensional velocity measurements in a laminar flame using particle image velocimetry. *Combust. Sci. Technol.*, 67(4–6):73–83, 1986. DOI: 10.1080/00102208908924062.
[145] F. C. R. Spieksma and G. J. Woeginger. Geometric three-dimensional assignment problems. *Eur. J. Oper. Res.*, 91(3):611–618, 1996. DOI: 10.1016/0377-2217(95)00003-8.
[146] M. Umeyama and S. Matsuki. Measurements of velocity and trajectory of water particle for internal waves in two density layers. *Geophys. Res. Lett.*, 38(3):L03612, 2011. DOI: 10.1029/2010GL046419.
[147] J. Williams. Application of tomographic particle image velocimetry to studies of transport in complex (dusty) plasma. *Phys. Plasmas*, 18(5):050702, 2011. DOI: 10.1063/1.3587090.
[148] J. Zhu, J. Gao, A. Ehn, M. Aldén, Z. Li, D. Moseev, Y. Kusano, M. Salewski, A. Alpers, P. Gritzmann, and M. Schwenk. Measurements of 3D slip velocities and plasma column lengths of a gliding arc discharge. *Appl. Phys. Lett.*, 106(4):044101, 2015. DOI: 10.1063/1.4906928.

Further reading: tomographic grain mapping

[149] A. Alpers, P. Gritzmann, C. G. Heise, and A. Taraz. On the mathematics of grain reconstruction I: Modeling and computational complexity. Manuscript, 2019.
[150] A. Alpers, H. F. Poulsen, E. Knudsen, and G. T. Herman. A discrete tomography algorithm for improving the quality of 3DXRD grain maps. *J. Appl. Crystallogr.*, 39(4):582–588, 2006. DOI: 10.1107/S002188980601939X.
[151] N. R. Barton and J. V. Bernier. A method for intragranular orientation and lattice strain distribution determination. *J. Appl. Crystallogr.*, 45(6):1145–1155, 2012. DOI: 10.1107/S0021889812040782.
[152] K. J. Batenburg, J. Sijbers, H. F. Poulsen, and E. Knudsen. DART: A robust algorithm for fast reconstruction of three-dimensional grain maps. *J. Appl. Crystallogr.*, 43(6):1464–1473 2010. DOI: 10.1107/S0021889810034114.
[153] Y. Hayashi, Y. Hirose, and Y. Seno. Polycrystal orientation mapping using scanning three-dimensional X-ray diffraction microscopy. *J. Appl. Crystallogr.*, 48(4):1094–1101, 2015. DOI: 10.1107/S1600576715009899.
[154] B. Jakobsen, H. F. Poulsen, U. Lienert, J. Almer, S. D. Shastri, H. O. Sørensen, C. Gundlach, and W. Pantleon. Formation and subdivision of deformation structures during plastic deformation. *Science*, 312(5775):889–892, 2006. DOI: 10.1126/science.1124141.

[155] A. K. Kulshreshth, A. Alpers, G. T. Herman, E. Knudsen, L. Rodek, and H. F. Poulsen. A greedy method for reconstructing polycrystals from three-dimensional X-ray diffraction data. *Inverse Probl. Imaging*, 3(1):69–85, 2009. DOI: 10.3934/ipi.2009.3.69.

[156] E. M. Lauridsen, S. Schmidt, R. M. Suter, and H. F. Poulsen. Tracking: A method for structural characterization of grains in powders or polycrystals. *J. Appl. Crystallogr.*, 34:744–750, 2001. DOI: 10.1107/S0021889801014170.

[157] H. Li, N. Chawla, and Y. Jiao. Reconstruction of heterogeneous materials via stochastic optimization of limited-angle X-ray tomographic projections. *Scr. Mater.*, 86(1),48–51, 2014. DOI: 10.1016/j.scriptamat.2014.05.002.

[158] H. Li, S. Kaira, N. Chawla, and Y. Jiao. Accurate stochastic reconstruction of heterogeneous microstructures by limited X-ray tomographic projections. *J. Microsc.*, 264(3):339–350, 2016. DOI: 10.1111/jmi.12449.

[159] W. Ludwig, S. Schmidt, E. M. Lauridsen, and H. F. Poulsen. X-ray diffraction contrast tomography: A novel technique for three-dimensional grain mapping of polycrystals. I. Direct beam case. *J. Appl. Crystallogr.*, 41(2):302–309, 2008. DOI: 10.1107/S002188980800168.

[160] L. Margulies, G. Winther, and H. F. Poulsen. In situ measurement of grain rotation during deformation of polycrystals. *Science*, 291(5512):2392–2394, 2001. DOI: 10.1126/science.1057956.

[161] M. Moscicki, P. Kenesei, J. Wright, H. Pinto, T. Lippmann, A. Borbely, and A. R. Pyzalla. Friedel-pair based indexing method for characterization of single grains with hard X-rays. *Mater. Sci. Eng. A*, 524(1–2):64–68, 2009. DOI: 10.1016/j.msea.2009.05.002.

[162] S. E. Offerman, N. H. van Dijk, J. Sietsma, S. Grigull, E. M. Lauridsen, L. Margulies, H. F. Poulsen, M. T. Rekveldt, and S. van der Zwaag. Grain nucleation and growth during phase transformations. *Science*, 298(5595):1003–1005, 2002. DOI: 10.1126/science.1076681.

[163] H. F. Poulsen and X. Fu. Generation of grain maps by an algebraic reconstruction technique. *J. Appl. Crystallogr.*, 36(4):1062–1068, 2003. DOI: 10.1107/S0021889803011063.

[164] P. Reischig, A. King, L. Nervo, N. Viganò, Y. Guilhem, W. J. Palenstijn, K. J. Batenburg, M. Preuss, and W. Ludwig. Advances in X-ray diffraction contrast tomography: Flexibility in the setup geometry and application to multiphase materials. *J. Appl. Crystallogr.*, 46(2):297–311, 2013. DOI: 10.1107/S002188981300260.

[165] L. Rodek, H. F. Poulsen, E. Knudsen, and G. T. Herman. A stochastic algorithm for reconstruction of grain maps of moderately deformed specimens based on X-ray diffraction. *J. Appl. Crystallogr.*, 40(2):313–321, 2007. DOI: 10.1107/S0021889807001288.

[166] S. Schmidt, S. F. Nielsen, C. Gundlach, L. Margulies, X. Huang, and D. Juul Jensen. Watching the growth of bulk grains during recrystallization of deformed metals. *Science*, 305(5681):229–232, 2004. DOI: 10.1126/science.1098627.

[167] R. M. Suter, D. Hennessy, C. Xiao, and U. Lienert. Forward modeling method for microstructure reconstruction using X-ray diffraction microscopy: Single-crystal verification. *Rev. Sci. Instrum.*, 77(12):123905, 2006. DOI: 10.1063/1.2400017.

[168] N. Viganò, W. Ludwig, and K. J. Batenburg. Reconstruction of local orientation in grains using a discrete representation of orientation space. *J. Appl. Crystallogr.*, 47(6):1826–1840, 2014. DOI: 10.1107/S1600576714020147.

Further reading: macroscopic grain mapping

[169] A. Alpers, A. Brieden, P. Gritzmann, A. Lyckegaard, and H. F. Poulsen. Generalized balanced power diagrams for 3D representations of polycrystals. *Philos. Mag.*, 95(9):1016–1028, 2015. DOI: 10.1080/14786435.2015.1015469.

[170] A. Brieden and P. Gritzmann. A quadratic optimization model for the consolidation of farmland by means of lend-lease agreements. In: D. Ahr, R. Fahrion, M. Oswald, and G. Reinelt, editors, *Operations research proceedings 2003: Selected papers of the international conference on operations research (OR 2003)*, pages 324–331. Springer, Heidelberg, 2004. DOI: 10.1007/978-3-642-17022-5_42.

[171] A. Brieden and P. Gritzmann. On optimal weighted balanced clusterings: Gravity bodies and power diagrams. *SIAM J. Discrete Math.*, 26(2):415–434, 2012. DOI: 10.1137/110832707.

[172] A. Brieden, P. Gritzmann, and F. Klemm. Constrained clustering via diagrams: A unified theory and its applications to electoral district design. *Eur. J. Oper. Res.*, 263(1):18–34, 2017. DOI: 10.1016/j.ejor.2017.04.018.

[173] S. N. Chiu, D. Stoyan, W. Kendall, and J. Mecke. Accompanying web page for the book: Stochastic geometry and its applications, 3rd edition. [Accessed: 2018-09]. URL: http://www.math.hkbu.edu.hk/~snchiu/cskm/cskm2013.html.

[174] O. Šedivý, T. Brereton, D. Westhoff, L. Polívka, V. Beneš, V. Schmidt, and A. Jäger. 3D reconstruction of grains in polycrystalline materials using a tessellation model with curved grain boundaries. *Philos. Mag.*, 96(18):1926–1949, 2016. DOI: 10.1080/14786435.2016.1183829.

[175] O. Šedivý, J. Dake, C. E. Krill III, V. Schmidt, and A. Jäger. Description of the 3D morphology of grain boundaries in aluminum alloys using tessellation models generated by ellipsoids. *Image Anal. Stereol.*, 36(1):5–13, 2017. DOI: 10.5566/ias.1656.

[176] A. Spettl, T. Brereton, Q. Duan, T. Werz, C. E. Krill III, D. P. Kroese, and V. Schmidt. Fitting Laguerre tessellation approximations to tomographic image data. *Philos. Mag.*, 96(2):166–189, 2016. DOI: 10.1080/14786435.2015.1125540.

[177] K. Teferra and D. J. Rowenhorst. Direct parameter estimation for generalized balanced power diagrams. *Philos. Mag. Lett.*, 98(2):79–87, 2018. DOI: 10.1080/09500839.2018.1472399.

Further reading: the Prouhet–Tarry–Escott problem

[178] A. Adler and S.-Y. R. Li. Magic cubes and Prouhet sequences. *Am. Math. Mon.*, 84(8):618–627, 1977. DOI: 10.2307/2321011.

[179] J. Aliste-Prieto, A. de Mier, and J. Zamora. On trees with the same restricted U-polynomial and the Prouhet–Tarry–Escott problem. *Discrete Math.*, 340(6):1435–1441, 2017. DOI: 10.1016/j.disc.2016.09.019.

[180] A. Alpers and R. Tijdeman. The two-dimensional Prouhet–Tarry–Escott problem. *J. Number Theory*, 123(2):403–412, 2007. DOI: 10.1016/j.jnt.2006.07.001.

[181] L. Bastien. Impossibilité de $u + v = x + y + z$. *Sphinx-Oedipe*, 8(1):171–172, 1913.

[182] E. D. Bolker, C. Offner, R. Richman, and C. Zara. The Prouhet–Tarry–Escott problem and generalized Thue-Morse sequences. *J. Comb.*, 7(1):117–133, 2016. DOI: 10.4310/JOC.2016.v7.n1.a5.

[183] B. Borchert, P. McKenzie, and K. Reinhardt. Few product gates but many zeroes. *Chic. J. Theor. Comput. Sci.*, 2013(2):1–22, 2013. DOI: 10.1007/978-3-642-03816-7_15.

[184] A. Černý. On Prouhet's solution to the equal powers problem. *Theor. Comput. Sci.*, 491(17):33–46, 2013. DOI: 10.1016/j.tcs.2013.04.001.

[185] A. Černý. Solutions to the multi-dimensional Prouhet–Tarry–Escott problem resulting from composition of balanced morphisms. *Inf. Comput.*, 253(3):424–435, 2017. DOI: 10.1016/j.ic.2016.06.008.

[186] A. Choudhry. A new approach to the Tarry–Escott problem. *Int. J. Number Theory*, 13(2):393–417, 2017. DOI: 10.1142/S1793042117500233.

[187] M. Cipu. Upper bounds for norms of products of binomials. *LMS J. Comput. Math.*, 7(1):37–49, 2004. DOI: 10.1112/S1461157000001030.

[188] P. Erdős and G. Szekeres. On the product $\prod_{k=1}^{n}(1 - z^{a_k})$. *Acad. Serbe Sci. Publ. Inst. Math.*, 13(1):29–34, 1959. URL: http://eudml.org/doc/271936.

[189] E. B. Escott. The calculation of logarithms. *Quart. J. Math.*, 41(2):147–167, 1910.

[190] M. Filaseta and M. Markovich. Newton polygons and the Prouhet–Tarry–Escott problem. *J. Number Theory*, 174(1):384–400, 2017. DOI: 10.1016/j.jnt.2016.10.009.

[191] V. Gandikota, B. Ghazi, and E. Grigorescu. NP-hardness of Reed–Solomon decoding, and the Prouhet–Tarry–Escott problem. In: *IEEE 57th annual symposium on foundations of computer science (FOCS)*, volume 1, pages 760–769. IEEE, 2016. DOI: 10.1109/FOCS.2016.86.

[192] C. Goldbach. Letter to Euler, July 18, 1750. In: *Corresp. math. phys. (ed. Fuss)*, volume. 1. St. Petersburg, 1843, pages 525–526. URL: http://eulerarchive.maa.org/correspondence/.

[193] S. Hernández and F. Luca. Integer roots chromatic polynomials of nonchordal graphs and the Prouhet–Tarry–Escott problem. *Graphs Comb.*, 21(3):319–323, 2005. DOI: 10.1007/s00373-005-0617-0.

[194] E. Kamke. Verallgemeinerungen des Waring–Hilbertschen Satzes. *Math. Ann.*, 83(1–2):85–112, 1921.

[195] H. Kleiman. A note on the Tarry–Escott problem. *J. Reine Angew. Math.*, 278–279:48–51, 1975. DOI: 10.1515/crll.1975.278-279.48.

[196] D. H. Lehmer. The Tarry–Escott problem. *Scr. Math.*, 13(1):37–41, 1947.

[197] R. Maltby. Pure product polynomials and the Prouhet–Tarry–Escott problem. *Math. Compet.*, 66(219):1323–1340, 1997. DOI: 10.1090/S0025-5718-97-00865-X.

[198] R. Maltby. A combinatorial identity of subset-sum powers in rings. *Rocky Mt. J. Math.*, 30(1):325–329, 2000. DOI: 10.1216/rmjm/1022008994.

[199] J. McLaughlin. An identity motivated by an amazing identity of Ramanujan. *Fibonacci Q.*, 48(1):34–38, 2010.

[200] Z. A. Melzak. A note on the Tarry–Escott problem. *Can. Math. Bull.*, 4(3):233–237, 1961. DOI: 10.4153/CMB-1961-025-1.

[201] G. Myerson. How small can a sum of roots of unity be? *Am. Math. Mon.*, 93(6):457–459, 1986. DOI: 10.2307/2323469.

[202] H. D. Nguyen. A new proof of the Prouhet–Tarry–Escott problem. *Integers*, 16(A1):1–9, 2016.

[203] P. A. Panzone. On a formula of S. Ramanujan. *Am. Math. Mon.*, 122(1):65–69, 2015. DOI: 10.4169/amer.math.monthly.122.01.65.

[204] M. E. Prouhet. Mémoire sur quelques relations entre les puissances des nombres. *C. R. Math. Acad. Sci. Paris*, 33:225, 1851.

[205] T. N. Sinha. A relation between the coefficients and roots of two equations and its application to diophantine problems. *J. Res. Natl. Bur. Stand. B*, 74B(1):31–36, 1970. DOI: 10.6028/jres.074B.002.

[206] T. N. Sinha. A note on a theorem of Lehmer. *J. Lond. Math. Soc.*, s2-4(3):541–544, 1972. DOI: 10.1112/jlms/s2-4.3.541.

[207] G. Tarry. Question 4100. *Interméd. Math.*, 19(1):200, 1912.

[208] E. M. Wright. An easier Waring's problem. *J. Lond. Math. Soc.*, 9(4):267–272, 1934. DOI: 10.1112/jlms/s1-9.4.267.

[209] E. M. Wright. On Tarry's problem (I). *Quart. J. Math.*, 6(1):261–267, 1935. DOI: 10.1093/qmath/os-6.1.261.

[210] E. M. Wright. Prouhet's 1851 solution of the Tarry–Escott problem of 1910. *Am. Math. Mon.*, 66(3):199–201, 1959. DOI: 10.2307/2309513.

[211] E. M. Wright. The Tarry–Escott and the "easier" Waring problem. *J. Reine Angew. Math.*, 309:170–173, 1979. DOI: 10.1515/crll.1979.311-312.170.

Radon Series on Computational and Applied Mathematics

Volume 21
Hamilton-Jacobi-Bellman Equations. Numerical Methods and Applications in Optimal Control
Dante Kalise, Karl Kunisch, Zhiping Rao (Eds.), 2018
ISBN: 978-3-11-054263-9, e-ISBN: 978-3-11-054359-9

Volume 20
Fluid-Structure Interaction. Modeling, Adaptive Discretisations and Solvers
Stefan Frei, Bärbel Holm, Thomas Richter, Thomas Wick, Huidong Yang (Eds.), 2017
ISBN: 978-3-11-049527-0, e-ISBN: 978-3-11-049425-9

Volume 19
Tensor Numerical Methods in Scientific Computing
Boris N. Khoromskij, 2018
ISBN: 978-3-11-037013-3, e-ISBN: 978-3-11-036591-7

Volume 18
Variational Methods. In Imaging and Geometric Control
Maitine Bergounioux, Gabriel Peyré, Christoph Schnörr, Jean-Baptiste Caillau, Thomas Haberkorn (Eds.), 2016
ISBN: 978-3-11-043923-6, e-ISBN: 978-3-11-043039-4

Volume 17
Topological Optimization. Optimal Transport in the Applied Sciences
Maitine Bergounioux, Édouard Oudet, Martin Rumpf, Filippo Santambrogio, Guilaume Carlier, Thierry Champion (Eds.), 2016
ISBN: 978-3-11-043926-7, e-ISBN: 978-3-11-043041-7

Volume 16
Algebraic Curves and Finite Fields. Cryptography and Other Applications
Harald Niederreiter, Alina Ostafe, Daniel Panario, Arne Winterhof (Eds.), 2014
ISBN: 978-3-11-031788-6, e-ISBN: 978-3-11-031791-6

Volume 15
Uniform Distribution and Quasi-Monte Carlo Methods. Discrepancy, Integration and Applications
Peter Kritzer, Harald Niederreiter, Friedrich Pillichshammer, Arne Winterhof (Eds.), 2014
ISBN: 978-3-11-031789-3, e-ISBN: 978-3-11-031793-0

www.degruyter.com